Lecture Notes in Computer Science

Lecture Notes in Artificial Intelligence 15392

Founding Editor

Jörg Siekmann

Series Editors

Randy Goebel, *University of Alberta, Edmonton, Canada*
Wolfgang Wahlster, *DFKI, Berlin, Germany*
Zhi-Hua Zhou, *Nanjing University, Nanjing, China*

The series Lecture Notes in Artificial Intelligence (LNAI) was established in 1988 as a topical subseries of LNCS devoted to artificial intelligence.

The series publishes state-of-the-art research results at a high level. As with the LNCS mother series, the mission of the series is to serve the international R & D community by providing an invaluable service, mainly focused on the publication of conference and workshop proceedings and postproceedings.

Quan Z. Sheng · Gill Dobbie · Jing Jiang ·
Xuyun Zhang · Wei Emma Zhang ·
Yannis Manolopoulos · Jia Wu ·
Wathiq Mansoor · Congbo Ma
Editors

Advanced Data Mining and Applications

20th International Conference, ADMA 2024
Sydney, NSW, Australia, December 3–5, 2024
Proceedings, Part VI

Editors
Quan Z. Sheng
Macquarie University
Sydney, NSW, Australia

Jing Jiang
Australian National University
Canberra, ACT, Australia

Wei Emma Zhang
The University of Adelaide
Adelaide, SA, Australia

Jia Wu
Macquarie University
Sydney, NSW, Australia

Congbo Ma
Macquarie University
Sydney, NSW, Australia

Gill Dobbie
University of Auckland
Auckland, New Zealand

Xuyun Zhang
Macquarie University
Sydney, NSW, Australia

Yannis Manolopoulos
Open University of Cyprus
Nicosia, Cyprus

Wathiq Mansoor
University of Dubai
Dubai, United Arab Emirates

ISSN 0302-9743   ISSN 1611-3349 (electronic)
Lecture Notes in Artificial Intelligence
ISBN 978-981-96-0849-2   ISBN 978-981-96-0850-8 (eBook)
https://doi.org/10.1007/978-981-96-0850-8

LNCS Sublibrary: SL7 – Artificial Intelligence

© The Editor(s) (if applicable) and The Author(s), under exclusive license
to Springer Nature Singapore Pte Ltd. 2025

This work is subject to copyright. All rights are solely and exclusively licensed by the Publisher, whether the whole or part of the material is concerned, specifically the rights of translation, reprinting, reuse of illustrations, recitation, broadcasting, reproduction on microfilms or in any other physical way, and transmission or information storage and retrieval, electronic adaptation, computer software, or by similar or dissimilar methodology now known or hereafter developed.
The use of general descriptive names, registered names, trademarks, service marks, etc. in this publication does not imply, even in the absence of a specific statement, that such names are exempt from the relevant protective laws and regulations and therefore free for general use.
The publisher, the authors and the editors are safe to assume that the advice and information in this book are believed to be true and accurate at the date of publication. Neither the publisher nor the authors or the editors give a warranty, expressed or implied, with respect to the material contained herein or for any errors or omissions that may have been made. The publisher remains neutral with regard to jurisdictional claims in published maps and institutional affiliations.

This Springer imprint is published by the registered company Springer Nature Singapore Pte Ltd.
The registered company address is: 152 Beach Road, #21-01/04 Gateway East, Singapore 189721, Singapore

If disposing of this product, please recycle the paper.

Preface

The 20th International Conference on Advanced Data Mining Applications (ADMA 2024) was held in Sydney, Australia, during December 3–5, 2024. Researchers and practitioners from around the world came together at this leading international forum to share innovative ideas, original research findings, case study results, and experienced insights into advanced data mining and its applications. With the ever-growing importance of appropriate methods in these data-rich times, ADMA has become a flagship conference in this field.

ADMA 2024 received a total of 422 valid submissions to the following tracks: Early Research Paper, Main Research Paper, Special Sessions, Industry Paper, and Posters/Encore Presentations. After a rigorous review process by 269 reviewers, 159 regular papers were accepted to be published in the proceedings, corresponding to a full research paper acceptance rate of 37.6%. 14 papers were accepted as extended abstract and poster.

The Program Committee (PC), composed of international experts in relevant fields, did a thorough and professional job of reviewing the papers submitted to ADMA 2024, and each paper was reviewed by at least three PC members. With the growing importance of data in this digital age, papers accepted by ADMA 2024 covered a wide range of research topics in the field of data mining, including machine learning, text mining, graph mining, predictive data analytics, recommender systems, natural language processing, analytics-based applications, and privacy and security analytics. It is worth mentioning that ADMA 2024 was the 20th anniversary of the ADMA conference and special events such as a female scientist forum, young scientist forum, and 20th anniversary celebration were arranged.

We thank the PC and Senior PC members for completing the review process and providing valuable comments within tight schedules. The high-quality program would not have been possible without the expertise and dedication of our PC and Senior members. Moreover, we would like to take this valuable opportunity to thank all authors who submitted technical papers and contributed to the tradition of excellence at ADMA. We are confident that the papers presented in these proceedings will be both exciting and beneficial for colleagues seeking to advance their research. We extend our gratitude to Microsoft for providing the complimentary CMT system for conference organization and to Springer for their enduring support. Special thanks are also due to the Australian Academy of Science for generously funding the first travel grant for the ADMA conference, and to Macquarie University and The University of Adelaide for sponsoring ADMA 2024. Additionally, we appreciate the generous contributions from Data Horizons Research Center, The Centre of Applied AI, Macquarie University and Lightwheel AI.

We are grateful for the guidance of the steering committee members, Xue Li, Osmar R. Zaiane, Chengqi Zhang, Jianxin Li, Guodong Long, and Weitong Chen, whose leadership and support helped the conference to run smoothly. We also would like to acknowledge the support of the other members of the organising committee. All of them helped to make ADMA 2024 a success. Finally, we would like to thank all the session chairs and volunteers who contributed to the conference organisation.

December 2024

Quan Z. Sheng
Gill Dobbie
Jing Jiang
Xuyun Zhang
Wei Emma Zhang
Yannis Manolopoulos
Jia Wu
Wathiq Mansoor
Congbo Ma

Organisation

General Chairs

Quan Z. Sheng   Macquarie University, Australia
Gill Dobbie   University of Auckland, New Zealand
Jing Jiang   Australian National University, Australia

Program Chairs

Xuyun Zhang   Macquarie University, Australia
Wei Emma Zhang   University of Adelaide, Australia
Yannis Manolopoulos   Open University of Cyprus, Cyprus

Proceedings Chairs

Jia Wu   Macquarie University, Australia
Wathiq Mansoor   University of Dubai, UAE
Congbo Ma   Macquarie University, Australia

Local Chairs

Adnan Mahmood   Macquarie University, Australia
Xianzhi Wang   University of Technology Sydney, Australia
Yimeng Feng   Macquarie University, Australia

Publicity Chairs

Jun Shen   University of Wollongong, Australia
Yanjun Shu   Harbin Institute of Technology, China
Ling Jian   China University of Petroleum, China
Chen Li   Nagoya University, Japan

20th Anniversary Celebration Chairs

Wenjie Zhang	UNSW Sydney, Australia
Weitong Chen	University of Adelaide, Australia
Chuan Shi	Beijing University of Posts and Telecommunications, China

Finance Chair

Erin Chen	Macquarie University, Australia

Sponsorship Chairs

Amin Beheshti	Macquarie University, Australia
Hao Zhao	Tsinghua University, China
Yun Li	Data61, CSIRO, Australia

Panel Discussion Chairs

Guanfeng Liu	Macquarie University, Australia
Wen Hua	Hong Kong Polytechnic, China
Victor Sheng	Texas Tech University, USA

Award and Grant Chairs

Hongzhi Yin	University of Queensland, Australia
Xiaolong Xu	Nanjing University of Information Sciences & Technology, China
Henry Nguyen	Griffith University, Australia

Journal Special Issue Chairs

Lianyong Qi	China University of Petroleum, Australia
Muhammad Bilal	Lancaster University, UK
Shoujin Wang	University of Technology Sydney, Australia

Special Session Chairs

Hongsheng Hu	University of Newcastle, Australia
Maia Angelova	Aston University, UK
Yoshihiro Yamanishi	Nagoya University, Japan
Chen Li	Nagoya University, Japan
Yuncheng Jiang	South China Normal University, China
Gang Li	Deakin University, Australia
Aswani Kumar Cherukuri	Vellore Institute of Technology, India
Qingyi Zhu	Chongqing University of Posts and Telecommunications, China
Yipeng Zhou	Macquarie University, Australia
Zhiwang Zhang	Ningbo Technology University, China
Shui Yu	University of Technology Sydney, Australia

Industry Track Chairs

Lingfei Wu	Anytime AI, USA
Siwei (Sherry) Xu	Data61, CSIRO, Australia
Zhao Li	Zhejiang University, China

Tutorial Chairs

Yipeng Zhou	Macquarie University, Australia
Guansong Pang	Singapore Management University, Singapore
Ye Zhu	Deakin University, Australia

Poster Track Chairs

Huaming Chen	University of Sydney, Australia
Mengdi Huai	Iowa State University, USA
Xiaoyong Li	National University of Defense Technology, China

Encore Track Chairs

Shan (Emma) Xue	Macquarie University, Australia
Chenliang Li	Wuhan University, China
Abdulwahab Aljubairy	Umm Al Qura University, Saudi Arabia

Early Career Researcher Forum Chairs

Amanda Li	Macquarie University, Australia
Xiaoling Wang	East China Normal University, China
Haojie Zhuang	University of Adelaide, Australia

Registration Management Chairs

Taotao Xai	University of Southern Queensland, Australia
Xiaocong Chen	Data61, CSIRO, Australia

Web Chairs

Adeem Ali Anwar	Macquarie University, Australia
Rafiullah Khan	University of Agriculture, Peshawar, Pakistan
Yang Zhang	Macquarie University, Australia

Senior Program Committee

Taotao Cai	University of Southern Queensland, Australia
Weitong Chen	University of Adelaide, Australia
Lianhua Chi	La Trobe University, Australia
Hongsheng Hu	University of Newcastle, Australia
Cheqing Jin	East China Normal University, China
Bohan Li	Nanjing University of Aeronautics and Astronautics, China
Jianxin Li	Deakin University, Australia
Xiang Lian	Kent State University, USA
Guodong Long	University of Technology Sydney, Australia
Xueping Peng	University of Technology Sydney, Australia
Yanjun Shu	Harbin Institute of Technology, China
Chaokun Wang	Tsinghua University, China
Hongzhi Wang	Harbin Institute of Technology, China
Shoujin Wang	University of Technology Sydney, Australia
Xianzhi Wang	University of Technology Sydney, Australia
Xiaoling Wang	East China Normal University, China
Lin Yue	University of Adelaide, Australia
Yang Zhang	Macquarie University, Australia
Zheng Zhang	Harbin Institute of Technology, China

Huaijie Zhu Sun Yat-sen University, China
Zhigang Lu Western Sydney University, Australia

Program Committee

Adita Kulkarni SUNY Brockport, USA
Alex Delis University of Athens, Greece
Ali Shakeri University of Adelaide, Australia
An Liu Soochow University, China
Anan Du Nanjing Vocational University of Industry Technology, China
Baoling Ning Heilongjiang University, China
Bin Zhao Nanjing Normal University, China
Bing Wang University of New South Wales, Australia
Bowen Liu Nanjing University, China
Buqing Cao Hunan University of Science and Technology, China
Carson K. Leung University of Manitoba, Canada
Chang Dong University of Adelaide, Australia
Chao Zhang Renmin University of China, China
Chaokun Wang Tsinghua University, China
Chaoran Huang University of New South Wales, Australia
Chen Wang Shaoxing University, China
Cheng Wu Tsinghua University, China
Chengcheng Yang East China Normal University, China
Chenhao Zhang University of Queensland, Australia
Cheqing Jin East China Normal University, China
Chuan Ma Chongqing University, China
Chuan Xiao Osaka University, Nagoya University, Japan
Congbo Ma Macquarie University, Australia
Dechang Pi Nanjing University of Aeronautics and Astronautics, China
Derong Shen Northeastern University, China
Dong Li Liaoning University, China
Dong Wen University of New South Wales, Australia
Eiji Uchino Yamaguchi University, Japan
Elaf A. Alhazmi Macquarie University, Australia
Faming Li Northeastern University, China
Gong Cheng Nanjing University, China
Guanfeng Liu Macquarie University, Australia
Guan-Nan Dong Nanjing Tech University, China

Guohao Sun	Donghua University, China
Haïfa Nakouri	ISG Tunis, Tunisia
Hao Feng	Tsinghua University, China
Hao Tian	Nanjing University, China
Haojie Zhuang	University of Adelaide, Australia
Haolong Xiang	Nanjing University of Information Science and Technology, China
Hongsheng Hu	University of Newcastle, Australia
Hongzhi Wang	Harbin Institute of Technology, China
Hongzhi Yin	University of Queensland, Australia
Hu Wang	Mohamed bin Zayed University of Artificial Intelligence, UAE
Huaibing Peng	Nanjing University of Science and Technology, China
Huaijie Zhu	Sun Yat-sen University, China
Huaming Chen	University of Sydney, Australia
Hui Yin	Swinburne University of Technology, Australia
Indika Priyantha Kumara Dewage	Tilburg University, Netherlands
Jagat Challa	BITS Pilani, India
Jia Wu	Macquarie University, Australia
Jiajie Xu	Soochow University, China
Jiali Mao	East China Normal University, China
Jian Li	Northeastern University, China
Jianxing Yu	Sun Yat-sen University, China
Jiazun Chen	Peking University, China
Jie Shao	University of Electronic Science and Technology of China, China
Jun Gao	Peking University, China
Junchang Xin	Northeastern University, China
Junhu Wang	Griffith University, Australia
Junhua Fang	Soochow University, China
Kai Peng	Huaqiao University, China
Kai Wang	Shanghai Jiao Tong University, China
Kun Yue	Yunnan University, China
Lei Duan	Sichuan University, China
Lei Li	Hong Kong University of Science and Technology (Guangzhou), China
Li Li	Southwest University, China
Liangwei Zheng	University of Adelaide, Australia
Linlin Ding	Liaoning University, China
Lipin Guo	University of Adelaide, Australia
Lishan Yang	University of Adelaide, Australia

Long Yuan	Nanjing University of Science and Technology, China
Lu Chen	Swinburne University of Technology, Australia
Lyuyang Tong	Wuhan University, China
Maneet Singh	IIT Ropar, India
Mariusz Bajger	Flinders University, Australia
Mehmet Ali Kaygusuz	Middle East Technical University, Turkey
Meiyan Teng	Macquarie University, Australia
Ming Zhong	Wuhan University, China
Minghe Yu	Northeastern University, China
Mingzhe Zhang	University of Queensland, Australia
Mirco Nanni	CNR-ISTI Pisa, Italy
Mong Yuan Sim	University of Adelaide, Australia
Mourad Ellouze	Lille Catholic University, France
Nicolas Travers	Léonard de Vinci Pôle Universitaire, France
Peng Cheng	East China Normal University, China
Peng Peng	Hunan University, China
Pengpeng Zhao	Soochow University, China
Ping Lu	Beihang University, China
Qing Liao	Harbin Institute of Technology (Shenzhen), China
Qing Liu	Data61, CSIRO, Australia
Qing Xie	Wuhan University of Technology, China
Quoc Viet Hung Nguyen	Griffith University, Australia
Renhe Jiang	University of Tokyo, Japan
Riccardo Cantini	University of Calabria, Italy
Rong Gao	Hubei University of Technology, China
Rui Zhou	Swinburne University of Technology, Australia
Shaofei Shen	University of Queensland, Australia
Shi Feng	Northeastern University, China
Shi-ting Wen	Ningbo Tech University, China
Shuchao Pang	Nanjing University of Science and Technology, China
Shucun Fu	Southeast University, China
Shuhao Zhang	Nanyang Technological University, Singapore
Shunmei Meng	Nanjing University of Science and Technology, China
Shuo Wang	Shanghai Jiao Tong University, China
Silvestro Roberto Poccia	University of Turin, Italy
Simi Job	University of Southern Queensland, Australia
Siyuan Wu	Nanjing University, China
Tarique Anwar	University of York, England
Thanh Tam Nguyen	Griffith University, Australia

Tianrui Li	Southwest Jiaotong University, China
Tieke He	Nanjing University, China
Tiezheng Nie	Northeastern University, China
Wei Hu	Nanjing University, China
Wei Li	Harbin Engineering University, China
Wei Shen	Nankai University, China
Wei Song	Wuhan University, China
Wencheng Yang	University of Southern Queensland, Australia
Wenda Tang	China Telecom eSurfing Cloud, China
Wenhao Liang	University of Adelaide, Australia
Wenmin Lin	Hangzhou Normal University, China
Wenpeng Lu	Qilu University of Technology (Shandong Academy of Sciences), China
Wenwen Gong	Tsinghua University, China
Wenzhou Chen	Hangzhou Dianzi University, China
Xi Guo	University of Science and Technology Beijing, China
Xiangyu Song	Pengcheng Laboratory, China
Xianzhi Wang	University of Technology Sydney, Australia
Xianzhi Zhang	Sun Yat-sen University, China
Xiao Pan	Shijiazhuang Tiedao University, China
Xiaofeng Gao	Shanghai Jiao Tong University, China
Xiaojun Xie	Nanjing Agricultural University, China
Xiaolei Zhang	Zhengzhou University of Aeronautics, China
Xiaoxiao Chi	Macquarie University, Australia
Xin Yuan	CSIRO, Australia
Xiu Fang	Donghua University, China
Xiujuan Xu	Dalian University of Technology, China
Xunxiang Yao	Shandong University of Finance and Economics, China
YajunYang	Tianjin University, China
Yanan Cai	James Cook University, Australia
Yanda Wang	Nanjing University of Aeronautics and Astronautics, China
Yanfeng Zhang	Northeastern University, China
Yang-Sae Moon	Kangwon National University, South Korea
Ye Zhu	Deakin University, Australia
Yibo Sun	University of Adelaide, Australia
Yihong Yang	China University of Geosciences (Beijing), China
Ying Liu	Northwest University, China
Yingbo Li	Blue

Yingxia Shao	Beijing University of Posts and Telecommunications, China
Yiping Wen	Hunan University of Science and Technology, China
Yixiang Fang	Chinese University of Hong Kong, Shenzhen, China
Yong Zhang	Tsinghua University, China
Yongpan Sheng	Southwest University, China
Yongzhe Jia	Nanjing University, China
Yu Gu	Northeastern University, China
Yu Liu	Huazhong University of Science and Technology, China
Yu Yang	Education University of Hong Kong, China
Yunjun Gao	Zhejiang University, China
Yurong Cheng	Beijing Institute of Technology, China
Yutong Han	Dalian Minzu University, China
Yutong Qu	University of Adelaide, Australia
Yuwei Peng	Wuhan University, China
Yuxiang Zeng	Hong Kong University of Science and Technology, China
Zekun Xu	Northeast Forestry University, China
Zesheng Ye	University of Melbourne, Australia
Zhaojing Luo	Beijing Institute of Technology, China
Zheng Li	Nanjing University, China
Zheng Liu	Nanjing University of Posts and Telecommunications, China
Zhengyang Li	University of Adelaide, Australia
Zhengyi Yang	University of New South Wales, Australia
Zhenying He	Fudan University, China
Zhongnan Zhang	Xiamen University, China
Zhongrui Zhao	James Cook University, Australia

Special Session Program Committee

Muhammad Naveed Aman	University of Nebraska Lincoln, USA
Ali Ismail Awad	United Arab Emirates University, UAE
Aladdin Ayesh	University of Aberdeen, UK
Laura E. Barnes	University of Virginia, USA
Jichao Bi	Zhejiang Institute of Industry and Information Technology, China
Zhenghao Chen	University of Sydney, Australia

Long Chen	Guangdong University, China
Raghavendra Kumar Chunduri	Concentrix, USA
Longchao Da	Arizona State University, USA
Arghya Kusum Das	University of Alaska, USA
Praveen Kumar Donta	Vienna University of Technology, Austria
Chenquan Gan	Chongqing University of Posts and Telecommunications, China
Debasis Ganguly	University of Glasgow, UK
Jinyang Guo	Beihang University, China
Miao Hu	Sun Yat-sen University, China
Xiaowei Jia	University of Pittsburgh, USA
Yuncheng Jiang	South China Normal University, China
Firuz Kamalov	Canadian University of Dubai, UAE
Naresh Kshetri	Emporia State University, USA
Chen Li	Nagoya University, Japan
Ming Li	Henan Normal University, China
Xianyong Li	Xihua University, China
Weiru Liu	University of Bristol, UK
Mengchi Liu	Carleton University, Canada
Yassine Maleh	Sultan Moulay Slimane University, Morocco
Sindhuja Rao	CISCO Bengaluru, India
Yinuo Ren	Stanford University, USA
Amit Kumar Singh	Siksha 'O' Anusandhan University, India
Rui Su	Shanghai AI Lab, China
Weifeng Sun	Dalian University of Technology, China
Man Wu	Keio University, Japan
Lu-Xing Yang	Deakin University, Australia
Zhiwen Yu	South China University of Technology, China
Jinli Zhang	Beijing University of Technology, China
Tianwei Zhang	Nanyang Technological University, Singapore
Yushu Zhang	Nanjing University of Aeronautics and Astronautics, China
Leo Yu Zhang	Griffith University, Australia
Jun Zhou	RIKEN, Japan

Contents – Part VI

Recommendation Systems

New Contrastive Learning Method Using Embedding Space Data
Augmented for Sequence Recommendation 3
   Zhenhai Wang, Yunlong Guo, Weimin Li, and Hongyu Tian

Context-Augmented Contrastive Learning Method for Session-based
Recommendation 19
   *Xianlan Sun, Xiangyun Gao, Subin Huang, Haibei Zhu, Chen Xu,
Pingfu Chao, and Chao Kong*

Graph Contrastive Learning for Multi-behavior Recommendation 34
   Haiying Li, Huihui Wang, Shunmei Meng, and Xingguo Chen

UID-Net: Enhancing Click-Through Rate Prediction in Trigger-Induced
Recommendation Through User Interest Decomposition 49
   *Jiazhen Lou, Zhao Li, Hong Wen, Jingsong Lv, Jing Zhang, Fuyu Lv,
Zulong Chen, and Jia Wu*

PDC-FRS: Privacy-Preserving Data Contribution for Federated
Recommender System 65
   Chaoqun Yang, Wei Yuan, Liang Qu, and Thanh Tam Nguyen

Explicit and Implicit Counterfactual Data Augmentation for Sequential
Recommendation 80
   *Zhouying Xu, Xuejun Liu, Zhuoya Xing, Jiasheng Cao, Tao He,
and Xiaoyang Huang*

Attention-Based Causal Graph Convolutional Collaborative Filtering 95
   Youhan Qi, Xinglin Liu, Chenyu Li, and Ying Wang

Sequential Recommendation with Diverse Supervised Contrastive Views 110
   Zitong Zhu, Meixiu Long, Junfa Lin, and Jiahai Wang

Disentangled Causal Embedding with Unbiased Knowledge Distillation
for Recommendation 128
   Nan Liu, Shunmei Meng, Xiao Liu, Qianmu Li, and Yu Jiang

Multi-Attribute Sequential Recommendation 143
  Shuhan Qiu, Shanming Wei, and Qianmu Li

Layer Transformer-Powered Graph Convolutional Networks for Enhanced
Recommendation 158
  Shicong Lin, Zhilong Shan, and Su Mu

Adaptive Disentangled Contrastive Collaborative Filtering 174
  Sujie Yu, Junnan Zhuo, Lvying Chen, Hailian Yin, and Bohan Li

Security and Privacy Issues

Enhancing IoT Security: Hybrid Machine Learning Approach for IoT
Attack Detection 193
  *Alavikunhu Panthakkan, S. M Anzar, Dina J. M. Shehada,
  Wathiq Mansoor, and Hussain Al Ahmad*

Boosting Adversarial Transferability by Uniform Scale and Mix Mask
Method 208
  *Tao Wang, Qianmu Li, Zhichao Lian, Zijian Ying, Fan Liu,
  and Shunmei Meng*

A Differential Privacy Decision Forest Algorithm for Reducing the Effect
of Noise 224
  *Runfei Liu, Mingze Chu, Yuming Jiang, Xuefeng Ding, Yuncheng Shen,
  and Dasha Hu*

CDGM: Controllable Dataset Generation Method for Cybersecurity 238
  Yushun Xie, Haiyan Wang, Runnan Tan, Xiangyu Song, and Zhaoquan Gu

Enhancing Privacy in Big Data Publishing: η-Inference Model 254
  Zhenyu Chen, Lin yao, Guowei Wu, and Shisong Geng

Secure Why-Not Spatial Keyword Top-k Queries in Cloud Environments 269
  *Yiping Teng, Miao Li, Shiqing Wang, Huan Wang, Chuanyu Zong,
  and Chunlong Fan*

FreqAT: An Adversarial Training Based on Adaptive Frequency-Domain
Transform 287
  *Denghui Zhang, Yanming Liang, Qiangbo Huang, Xvxin Huang,
  Peixin Liao, Ming Yang, and Liyi Zeng*

Enhancing Network Intrusion Detection with VAE-GNN 302
  Junyu Li and Haoxi Wang

Semantic-Integrated Online Audit Log Reduction for Efficient Forensic
Analysis 318
  Wenhao Liao, Jia Sun, Haiyan Wang, Zhaoquan Gu, and Jianye Yang

Extended Abstracts

Truth Discovery for Spatio-Temporal Data from Multiple Sources 337
  He Zhang, Shuang Wang, Yufei Wang, Tianxing Wu, Lu Jixiang, and Long Chen

An Efficient Wind Power Prediction Based on Improved Feature Crossover
Mechanism, N-BEATSx and LightGBM 340
  Kaibo Zhang, Feng Ye, Lina Wang, Nadia Nedjah, Xuejie Zhang, and Shulei Yu

A Novel Hash Hypercube-Based Attribute Reduction Approach
For Neighborhood Rough Sets 343
  Qin Si, Haoyan Qiu, Yuanyuan Dong, Xiang Ding, Haibo Li, and Xiaojun Xie

MGMFF: Efficient Spatio-Temporal Forecasting via Multi-graph
and Multi-feature Fusion 346
  Jianqiao Hu, Qian Tao, Songbo Wang, Chenghao Liu, Lusi Li, Hao Yang, and Xiuhang Shi

A Few-Shot Relation Extraction Approach for Threat Intelligence Field 349
  Haiyan Wang, Junchi Bao, Liyi Zeng, Wenying Feng, Weihong Han, and Zhaoquan Gu

A Subgroup Framework of Interpatient Arrhythmia Classification Using
Deep Learning Network 352
  Xia Yu, Zi Yang, Ning Shen, Youhe Huang, and Hongru Li

Text and Image Multimodal Dataset for Fine-Grained E-Commerce
Product Classification 356
  Ajibola Obayemi and Khuong Nguyen

Enhanced Network Traffic Prediction with Transformer-Based Models
and Clustering 359
  Peiqi Jin, Kun Zheng, Yixin Che, Chenming Qiu, Ling Jin, Yunkai Wang, Tian Hu, Shouguo Du, and Yiming Tang

BSCL: A Model for Solving Math Word Problems Based on Contrastive Learning 363
Tiancheng Zhang, Yuyang Wang, Yijia Zhang, Minghe Yu, Fangling Leng, and Ge Yu

Predicting Student Success in Learning Management Systems: A Case Study of the Madrasati Platform 367
Abdullrahman Alabdali, Mohammad A. Alshehri, Matthew Stephenson, and Paulo Santos

IEP: An Intelligent Event Prediction Model Based on Momentum 370
Lize Zheng and Yanxi Li

Identifying Sources in Complex Dynamic Networks Through Label Propagation and ODE Integration 373
Fuyuan Ma, Yuhan Wang, Junhe Zhang, and Ying Wang

Unsupervised Domain Adaptation for Entity Blocking 376
Yaoshu Wang and Mengyi Yan

Author Index 381

Recommendation Systems

New Contrastive Learning Method Using Embedding Space Data Augmented for Sequence Recommendation

Zhenhai Wang[1], Yunlong Guo[1(✉)], Weimin Li[2], and Hongyu Tian[3]

[1] School of Computer Science and Engineering, Linyi University, Linyi 276000, China
wangzhenhai@lyu.edu.cn, gyl0805123@gmail.com
[2] School of Computer Engineering and Science, Shanghai University, Shanghai 200444, China
wmli@shu.edu.cn
[3] School of Physics and Electronic Engineering, Linyi University, Linyi 276000, China
tianhongyu@lyu.edu.cn

Abstract. Sequence recommendation has achieved excellent results in recent years, with an ability to capture changes in user interests over time. With the introduction of the self-attention mechanism, the sequence recommendation model can better capture changes in user interests and obtain better user embedding representations. Many studies have combined sequence recommendation models with contrastive learning, enabling the sequence model to capture more information and improve the expression quality of user embeddings. However, problems associated with existing contrastive learning techniques for sequence recommendation still remain. Multiple forward calculations bring an additional training burden. Meanwhile, performing data enhancement of the input sequence destroys information in the dataset. To solve these problems, we propose a contrastive learning model, ESRec, based on enhancing embedding space data. Data enhancement in the embedding space can avoid multiple forwardta calculations and reduce the burden of model training. Ultimately, two types of random data enhancement are used, simulating the impact of noise in the dataset and improving the robustness of the model. Extensive experimental results on four public datasets illustrate that our model outperforms all baseline models with significantly reduced training time.

Keywords: Recommender Systems · Sequential Recommendation · Contrastive Learning · Data Augmentation

1 Introduction

In recent years, recommendation systems have been widely used in various fields, including e-commerce platforms such as Alibaba and Amazon and video platforms such as Douyin and YouTube [5,7,24]. Recommendation systems make personalized recommendations by predicting user interests. User interests are inseparable from time attributes, with interests often changing with time. To better match user interests, existing recommendation models often consider a sequence of user historical interactions to predict changes in user interests [14,17,25].

In sequence recommendation, interactive items are analyzed with time attributes to infer user interests at the next time step, whereby a representation is obtained that matches user behavioral interests and personalized recommendations to the user are formulated [15,41,42]. With the rise of deep learning, many deep learning models have been applied to recommendation systems. Earlier deep learning models used Recurrent Neural Network(RNN) [21] and Transformer [6,26,35] for training. The popularity of Graph Neural Network(GNN) [10,12,30] in recent years has made user-item interaction no longer limited to simple linear data structures. Currently, more complex graphical data structures can be used to predict user interests.

Although these methods effectively improve the expression quality of user embeddings, problems still remain. Recently, self-supervised learning technology has achieved great success in fields such as computer vision and natural language processing. Its advantage is that it can be trained on unlabeled data, with the contrast learning approach being one of the most widely used [16,31,33].

However, contrastive learning models contain some deficiencies. For example, the process of building an enhanced view requires multiple forward calculations, which increase the burden of model training and increase resource consumption. Furthermore, for enhanced views, existing data enhancement methods lead to a loss of original information, which limits the expressive ability of user embedding to a certain extent [38].

To address these problems, we propose a new contrastive learning model based on embedded noise. Embedded noise is noise information is added to the user's embedded spatial representation rather than to the sequence of user behaviors. Different from previous contrastive learning models used for sequence recommendation, we reduce training expense and attempt to avoid the training burden caused by contrastive learning. Meanwhile, we use the embedding space for data augmentation, which mitigates the information corruption in the augmented view to a certain extent and retains the best performance of the original model, while greatly reducing the training time of the model.

Specifically, our work is mainly reflected in the following aspects:

We propose two data enhancement methods in embedding space to replace the destruction of original data and retain the information contained in the data to the maximum extent.

We change the structure of the contrastive model, using only one forward calculation during the data enhancement process, which reduces the training burden of the contrastive learning model. Compared with other contrastive learning models, the training time required for our model is greatly reduced.

After extensive experimental verification, the results demonstrate that our model is effective. Compared with the best baselines, our model achieves significant improvements for multiple public datasets.

2 Related Work

2.1 Sequence Recommendation

Early recommendation systems used matrix decomposition to model latent vectors as representations of user interests. With the rise of deep learning, NeuMF [13] began to use neural networks to model user interests. However, these early recommendation system models did not consider the correlation between time attributes and user interests.

With the gradual development of deep learning, some neural networks that are strongly related to the input time step have been derived, which has facilitated the introduction of time attributes for modeling. Among them, RNNs, such as Gated Recurrent Unit (GRU) [4] and Long Short-Term Memory [9] have been used in sequence recommendations that focus on time series. For example, GRU4Rec [14] applies GRU modifications to the recommendation system, capturing user interests that change over time and improving the expressiveness of user embeddings.

The emergence of Transformer has introduced the self-attention mechanism to the field of deep learning. The self-attention mechanism allows the model to pay more attention to the impact of input items with heavy weights. SASRec [17] uses a transformer to automatically assign weights to the input sequence; meanwhile, BERT4SRec [25] uses a bidirectional Transformer structure, which presents a significant improvement in sequence recommendation, providing more information for fitting user interests.

In addition, GNN have also become popular in recent years. This makes the input sequences no longer simply belong to linear relationships, but have more flexible and complex connections. SRGNN [32] combines GNN with sequence recommendation, models the sequence as a directed graph, and then uses a gated unit GNN and attention network to model user interests. GC-SAN [36] further adopts the self-attention mechanism on this basis, achieving satisfactory results.

2.2 Self-supervised Learning

In recent years, self-supervised learning has achieved great success in the field of deep learning. Self-supervised learning aims to design auxiliary tasks such that the model can be trained on unlabeled data to obtain excellent embedding expressions [16,29]. Contrastive learning is one of the most popular self-supervised learning methods and has achieved remarkable results in the fields of computer vision and natural language processing. The main idea in contrastive learning is to generate enhanced images through data enhancement methods, such as rotating or cropping images and using grayscale images, to maximize shared information between views [1,2,11,39].

Recent research shows that the effectiveness of this method may be related to the distribution of embeddings. Contrastive learning improves the alignment and uniformity of the embedding distribution by reducing the distance between positive samples and identifying negative samples [27,28].

In the field of recommendation systems, there are also some studies on contrastive learning [20,31,37,38]. In general recommendation, SGL [31] combines contrast learning with graph neural networks to randomly discard the input graph data structure to obtain enhanced graphs, which improves the expression quality of user embeddings in general recommendation and obtains performance improvement, but this operation performs a huge damage to the original graph information. SimGCL [38] interferes with embeddings of different convolutional phases in graph convolutional models to reduce the damage to the information destruction and improves the alignment and uniformity of the model embedding.

In sequence recommendation, CL4SRec [34] employs cropping and masking operations on the input sequences for data enhancement, and constructs the enhancement map for information sharing, which improves the model performance, while DuoRec [23] employs dropout operations for data enhancement, only needs to carry out two forward computations to obtain two different enhancement maps. ICLRec [3] mines the user's latent intent and maximizes the consistency of the user's intent and the sequence view with contrast learning. IOCRec [19], on the other hand, integrates global and local intents and proposes contrast learning for multiple intents.

3 Model

In this section, we introduce in detail the sequence recommendation contrastive learning model based on the embedding space. This model converts the data enhancement module from the input data into the embedding space, using a method to increase embedding information and a method to reduce embedding information, while ensuring that the input information is constant. Different user embedding representations are obtained, and information is shared through

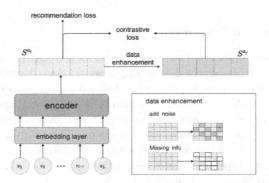

Fig. 1. ESRec model diagram and data enhancement method based on embedding space

comparative learning to improve the embedding expression quality and enhance the robustness of the model. The model structure diagram is shown in Fig. 1.

3.1 Contrastive Learning Framework in Sequence Recommendation

Our model is based on the architecture of the SASRec [5] model, which uses a transformer as the user embedding encoder to introduce the self-attention mechanism into sequence recommendation.

First, different items are encoded into embedding vectors $v_1, v_2, \ldots, v_L$. Then, to add position information in the sequence to the input information, the model generates position encodings to combine the position information for training. Therefore, the final embedding of an item is expressed as the sum of the item embedding and position encoding,

$$h_i^0 = v_i + p_i. \tag{1}$$

where v_i is the item embedding and p_i represents the embedding code of the corresponding position of the item. After obtaining the final embedding of the items, the transformer uses a self-attention mechanism to assign different attention weights to the sequence to represent the importance of the items. Transformers use a multi-head attention mechanism, which provides more information for model training. The actual formula for the self-attention operation is

$$ATT(Q, K, V) = softmax(QK^T/\sqrt{(d/n)})V, \tag{2}$$

where Q, K, and V represent the query, key, and value in the attention mechanism. $\sqrt{(d/n)}$ is the weight that controls the size of the output value, where d represents the length of the quey and key in the scaled dot product attention and n represents the number of heads in the multi-head attention. The inner product value caused by the dot product operation may be too large; thus, this method is used to control the size of the attention weight.

In deep learning, deep network structures often improve performance; thus, we improve the fitting ability of the model by stacking multiple self-attention modules. To prevent problems in training caused by overly deep networks, the transformer uses residual connection, dropout, and normalization techniques, as represented in the following formula where H^l represents each self-attention module output,

$$H^l = LN(H^l + Dropout(S(H^l))). \tag{3}$$

Finally, we get the input of each time step for T time steps, and we need to predict the output at the next time step at each time step t. Therefore, we set the final representation of the user to be then the vector representation of the user item interaction sequence at the moment of time step $t+1$:

$$H_u = Trm^l(s_t), \tag{4}$$

where Trm^l represents the output of the lth self-attention module and s_t represents the input sequence of items in time step t. The loss of the recommendation task is calculated as:

$$L_{main}(H_u) = -\log \frac{exp(H_u, v_{t+1}^+)}{exp(H_u, v_{t+1}^+) + exp(H_u, v_{t+1}^-)}, \tag{5}$$

where H_u, v_{t+1}^+ and v_{t+1}^- represent the user embedding representation of the model output, the item embedding that the user actually interacted with at the $t+1$ time step and the randomly sampled negative items at the $t+1$ time step, respectively.

After we obtain the output of the self-attention module, we use data augmentation to generate the enhanced embedding,

$$\widehat{H_u} = DA(H_u), \tag{6}$$

where DA represents the data augmentation operation used in this study. Finally, we compare the loss function to achieve information sharing between the original embedding sequence and enhanced embedding sequence,

$$L_{cl}(H_u, \widehat{H_u}) = -\log \frac{exp(sim(H_u, \widehat{H_u}))}{exp(sim(H_u, \widehat{H_u})) + exp(sim(H_u, H_u^-))}, \tag{7}$$

where sim represents the dot product operation that measures the embedding similarity and H^- represents other enhanced embeddings in the same batch. We use joint training as a multi-task training strategy,

$$L = L_{main} + \lambda L_{cl}. \tag{8}$$

where λ represents the weighting parameter that controls the loss of contrast to balance the impact of the recommendation task with the auxiliary task.

3.2 Data Enhancement Based on Embedding Space

Using the above deep learning model, we propose two embedding space-based data enhancement methods to replace the data enhancement of the input sequence, namely, applying masks to reduce the embedding information and applying noise to expand the interference information.

Missing Information. During the model training process, a lack of interactive items can seriously affect the expression quality of the user embedding. To adapt the model to handle this situation, we mask the embedded information to a certain extent to simulate the training situation in which there is less interactive information. Specifically, we first generate a mask sequence according to the mask ratio θ and then set the embedding vector at the mask sequence in the embedding space as the mask vector [mask]. The formulation is as follows:

$$\Gamma_u = (t_1, t_2, \ldots, t_n), \tag{9}$$

$$\widehat{H_u} = MASK(H_u) = [\widehat{v_1}, \widehat{v_2}, \ldots, \widehat{v_u}], \tag{10}$$

$$\widehat{v_x} = \begin{cases} v_x, x \in \Gamma_u \\ [mask], x \notin \Gamma_u \end{cases}. \tag{11}$$

where Γ_u represents the generated mask sequence and its length is the product of the mask ratio θ and sequence length h. $\widehat{H_u}$ mask represents the masked user embedding vector. Here masked items will be replaced by masked items [mask], thus reducing the information in the user embedding.

Addition of Noise. As the recommendation dataset mainly uses interaction as training information, it is difficult to clearly determine user preferences, introducing much noise into the system. To improve the ability of the model to resist noise disturbance, we simulate the impact of the noise present in the dataset by adding a small amount of meaningless noise to the embedding vector.

$$\Delta_u = (t_1, t_2, \ldots, t_l,), \tag{12}$$

$$noise_u = (n_1, n_1, \ldots, n_l), \tag{13}$$

$$\widehat{H_u} = ADD(H_u) = [\widehat{v_1}, \widehat{v_2}, \ldots, \widehat{v_u}], \tag{14}$$

$$\widehat{v_x} = \begin{cases} v_x, x \in \Delta_u \\ v_x + n_x, x \notin \Delta_u \end{cases}. \tag{15}$$

where Δ_u represents the generated noi se sequence whose length is the product of the generation ratio μ and the sequence length h. The uniform noise sequence $noise_u$ is then generated and added to the original sequence in random positions. The effect of noisy data on the embedding is modeled by the noise addition.

With the above data enhancement technique, the original user-embedded expression will be introduced with some minor noise disturbance as shown in the following equation:

$$\widehat{v_x} = v_x + \varepsilon_x, \tag{16}$$

where v_x represents the original embedding without added noise and ε_x represents the added noise.

4 Experiments

In this section, we explain the following questions through extensive experiments:

RQ1. How does our model perform compared to the latest baselines for sequential recommendations?

RQ2. What role does the contrastive learning module play? Is it effective to randomly apply two data enhancement methods?

RQ3. What impact do the parameters in contrastive learning have on model performance?

RQ4. Does our model perform better in training time compared to other contrastive learning models?

4.1 Experimental Setup

Dataset. We selected four common datasets to conduct comparative experiments on the models. MovieLens-1M, a dataset containing ratings of movies from multiple users and commonly used in recommendation systems, was selected. Another dataset is yelp2018, which contains data such as user ratings and evaluations of various merchants on the Yelp website [22]. The last two datasets are product reviews and product ratings from Amazon. The datasets are divided into multiple categories through product category classification. In this study, two datasets consisting of clothes and toys are used for experiments.

Table 1. Dataset information

	ml-1m	Yelp2018	Clothing	Toys
User	5,986	26,752	22,553	15,529
Item	2,975	19,857	13,886	9,697
Action	832,045	1,001,901	154,902	133,837
Avg. length	139.0	37.5	6.9	8.6
Density	4.68%	0.19%	0.05%	0.09%

Baseline. To verify the effectiveness of our model, we applied models commonly used in sequence recommendation as baselines for experimental comparison as listed below:

BPR [24]: This is a well-known non-sequential recommendation model that serves as the basis of many other recommendation models.

GRU4Rec [14]: This model uses variant GRUs in RNNs to improve the network and utilizes a two-layer GRU to model user interaction sequences as user representation embeddings.

SASRec [17]: This model adopts a self-attention neural network and uses a transformer to model users, improving the expression quality of user embeddings.

GC-SAN [36]: This model uses a combination of GNN and self-attention to improve the sequence recommendation model.

CL4SRec [34]: This is a model that uses contrastive learning for sequence recommendation, performing data augmentation on input user historical interaction sequences to generate the enhanced views required for contrastive learning.
DuoRec [23]: This model uses dropout, commonly considered in models, as a data enhancement method in contrastive learning to obtain better semantic retention.
ICLRec [19]: This model maximizes the consistency between sequential views and their corresponding intents by mining user intents and leveraging latent intents for comparative learning.

Other Settings. For all models in comparative experiments, we uniformly use recbole [40] for model implementation. We set all user embedding sizes to 64, the maximum user interaction history to 50, and the batch size to 1,024. We sort all user interactions by interaction time, select the 50 most recent interaction items to form an interaction sequence, and populate the user interaction sequence to 50 when it is less than 50. We use Xavier [8] for initialization and apply the Adam [18] optimizer for optimization, with the learning rate set to 0.001. For all models, Hit Ratio (HR) and Normalized Discounted Cumulative Gain (NDCG) serve as evaluation indicators, early stopping strategy is applied in the training of the models, and select top-k 10 and 20.

4.2 Model Performance Comparison (RQ1)

To answer question RQ1, we experimentally compared ESRec with the above baseline, and the results are shown in Table 2.

Table 2. Dataset information

Dataset	Metric	BPR-MF	GRU4Rec	SASRec	GC-SAN	CL4SRec	DuoRec	ICLRec	ESRec	Imp
ml-1m	HR10	0.0510	0.2304	0.2473	0.2331	0.2520	0.2501	0.2558	**0.2608 ± 0.0050**	1.95%
	HR20	0.0959	0.3387	0.3505	0.3310	0.3551	0.3551	0.3589	**0.3612 ± 0.0109**	0.64%
	NDCG10	0.0244	0.1256	0.1305	0.1275	0.1327	0.1368	0.1367	**0.1415 ± 0.0044**	3.32%
	NDCG20	0.0356	0.1529	0.1566	0.1522	0.1588	0.1633	0.1625	**0.1668 ± 0.0054**	2.10%
Yelp2018	HR10	0.0231	0.0958	0.0902	0.0849	0.1002	0.1020	0.0855	**0.1071 ± 0.0051**	4.76%
	HR20	0.0406	0.1517	0.1458	0.1367	0.1624	0.1589	0.1373	**0.1720 ± 0.0064**	5.58%
	NDCG10	0.0113	0.0485	0.0441	0.0431	0.0509	0.0510	0.0434	**0.0531 ± 0.0023**	3.95%
	NDCG20	0.0157	0.0626	0.0581	0.0561	0.0665	0.0654	0.0564	**0.0694 ± 0.0026**	4.18%
Clothing	HR10	0.0057	0.0167	0.0292	0.0276	0.0294	0.0298	0.0291	**0.0316 ± 0.0004**	5.70%
	HR20	0.0087	0.0265	0.0423	0.0399	0.0412	0.0436	0.0426	**0.0449 ± 0.0010**	2.90%
	NDCG10	0.0028	0.0111	0.0138	0.0138	0.0140	0.0140	0.0136	**0.0146 ± 0.0001**	4.11%
	NDCG20	0.0035	0.0151	0.0172	0.0168	0.0169	0.0175	0.0170	**0.0180 ± 0.0003**	2.78%
Toys	HR10	0.0147	0.0575	0.0757	0.0811	0.0773	0.0878	0.0744	**0.0934 ± 0.0006**	6.00%
	HR20	0.0220	0.0841	0.1043	0.1087	0.1052	0.1231	0.1014	**0.1277 ± 0.0005**	3.60%
	NDCG10	0.0081	0.0335	0.0417	0.0435	0.0424	0.0445	0.0423	**0.0462 ± 0.0001**	3.68%
	NDCG20	0.0099	0.0402	0.0489	0.0504	0.0495	0.0534	0.0491	**0.0549 ± 0.0001**	2.73%

An analysis of the experimental results reveal the following:

(1) Compared with the traditional non-sequential recommendation method BPR, time-sequential sequence recommendation can achieve better results. This may be because user interests change over time, necessitating the inclusion of temporal attributes in the user embedded expressions. SASRec uses a self-attention mechanism and can obtain more information.
(2) Compared with SASRec, GC-SAN uses a GNN combined with an attention mechanism to capture high-order hidden information between items through the GNN. However, in experiments, GC-SAN did not demonstrate better performance. One possible reason is that there is no clear high-order connectivity in a single user sequence in sequence recommendation.
(3) The sequence recommendation model that applies contrastive learning achieves better results. DuoRec uses the randomness of the original dropout operation in the model to perform comparative learning data enhancement. Compared with CL4SRec, which performs data enhancement on the input sequence, DuoRec retains more original information and thus achieves better results. And ICLRec is able to capture the underlying intent of the user, which provides more information for the model's prediction.
(4) Compared with all models, our model can achieve excellent results for all datasets. This is because ESRec performs data enhancement in the embedding space, which improves model resistance to noise while retaining the original information. By using contrastive learning, model robustness is improved without affecting the recommendation task.

4.3 Impact of Data Enhancement Module (RQ2)

In this section, we demonstrate the effectiveness of our proposed contrastive learning module through experiments on the contrastive learning module. Specifically, we demonstrate the effectiveness of the contrastive learning module by canceling the contrastive loss and using a single data augmentation method to answer the two questions in RQ2.

First, we retained the random data enhancement operation in the embedding space. When all experimental settings were identical, we set up SASRec, ESRec, and removed the contrastive loss, leaving only the variant $SASRec_{aug}$ of the data enhancement operation. The experimental results are shown in Table 3.

Table 3. Performance comparison of SASRec, SASRecaug, and ESRec on the four datasets

	ml-1 m		Yelp2018		Clothing		Toys	
	HR10	NDCG10	HR10	NDCG10	HR10	NDCG10	HR10	NDCG10
SASRec	0.2473	0.1305	0.0902	0.0441	0.0292	0.0138	0.0757	0.0417
$SASRec_{aug}$	0.2576	0.1365	0.0976	0.0494	0.0306	0.0144	0.0875	0.0426
ESRec	0.2608	0.1415	0.1071	0.0531	0.0316	0.0146	0.0934	0.0462

Experimental results demonstrate that SASRecaug exhibits improved performance compared to SASRec, which shows that the data enhancement can improve model resistance to noise disturbance and improve the robustness of the model. ESRec demonstrates the best performance on all datasets, showing that the comparative learning task can share information without interfering with the recommendation task, improving the expressive ability of user embedding.

Meanwhile, to verify the effectiveness of randomly adopting the two data enhancements of noise addition and missing information, we use ESRec with noise addition alone, missing information alone, and random adoption of the two enhancements. A comparison of the experimental results is shown in Table 4.

Table 4. Comparison of model performance between SASRec, ESRec, and two types of data enhancement alone

	ml-1m		Yelp2018		Clothing		Toys	
	HR10	NDCG10	HR10	NDCG10	HR10	NDCG10	HR10	NDCG10
SASRec	0.2473	0.1305	0.0902	0.0441	0.0292	0.0138	0.0757	0.0417
Add noise	0.2475	0.1356	0.1003	0.0502	0.0312	0.0143	0.0936	0.0462
Miss info	0.2506	0.1377	0.0992	0.0499	0.031	0.0145	0.0905	0.0444
ESRec	0.2608	0.1415	0.1071	0.0531	0.0316	0.0146	0.0934	0.0462

As shown in the table, in the ml-1 m dataset, the data enhancement method with missing information is slightly better. Using missing information can make the model more resistant to disturbance. Adding noise in yelp2018 achieves better results, showing that the existence of noise in yelp2018 causes greater disturbance to the model. In the Toys dataset, the effect of adding noise is slightly better than the result of ESRec, and in all other cases, randomly adopting two data augmentation methods achieves better results. This shows that the two types of data enhancement can improve model robustness, enabling the model to better adapt to different disturbance situations and achieve better performance.

4.4 Impact of Contrastive Learning Parameters (RQ3)

To answer RQ3, we analyzed the impact of contrastive loss control parameters λ on model performance in contrastive learning. The experimental results are shown in Fig. 2. As shown in the figure, ESRec exhibits better model performance in most parameter ranges, which also reflects the better robustness of the model. However, there is a small degradation in model performance when the contrast loss is small, which may be due to the fact that the contrast loss is too small to make a large impact.

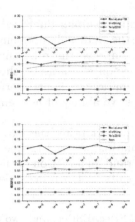

Fig. 2. Impact of contrastive loss control parameters on the model

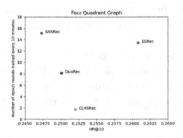

Fig. 3. Model performance versus training time (epoch rounds per 10 minutes)

4.5 Analysis of Model Runtime (RQ4)

To answer RQ4, we performed a time complexity analysis of the model.

As shown in Table 5, compared to CL4SRec, ESRec possesses less computational cost in both data augmentation as well as forward computation process, and thus it possesses several times less training time and training cost than CL4SRec in the training process.Although, DuoRec does not need data augmentation operation, it has one more forward computation process than ESRec, and since ESRec uses data augmentation in the embedding space, the training cost spent in the data augmentation phase is greatly reduced, which is much less than the training cost required for one forward computation, and thus the training cost of ESRec is less than that of DuoRec.We measured the training time of SASRec, CL4SRec, DuoRec, and ESRec for one epoch in each dataset.

Table 5. Model time complexity analysis

	CL4SRec	DuoRec	ESRec
Data Enhancement	O(BNL)	-	O(BNL)
Forward Calculation	3O(BNL)	2O(BNL)	O(BNL)
Loss Calculation	O(BL)	O(BL)	O(BL)

Table 6. Model training time per epoch

	ml-1m	Yelp2018	Clothing	Toys
SASRec	39.6s	45.7s	4.4s	4.3s
CL4SRec	354.5s(8.95x)	396.0s(8.67x)	30.8s(7.00x)	33.4s(7.77x)
DuoRec	73.8s(1.86x)	85.8s(1.88x)	8.2s(1.86x)	8.1s(1.88x)
ESRec	44.6s(1.13x)	50.3s(1.10x)	4.7s(1.07x)	4.7s(1.09x)

As shown in Table 6, the training time of CL4SRec is on average 7 to 8 times slower than that of SASRec. Although DuoRec uses the original dropout operation of the model, it still requires two forward operations, resulting in a training time that is 1.8 times slower than SASRec. ESRec uses only one forward operation, and the embedded spatial data augmentation used greatly reduces the time required for data augmentation, resulting in a training time that is only about 1.1 times slower than SASRec.

To visualize the training time and model performance, we consider the evaluation metric HR@10 as the horizontal coordinate and number of epoch rounds trained every 10 min as the vertical coordinate to evaluate the model effect. The results are shown in Fig. 3. Compared to the other models, ESRec consumes less training time while achieving the best results; therefore, we argue that ESRec is better than the other baselines.

4.6 Analysis of Model Generalizability

In order to analyze the generalizability of the model, we partitioned the dataset in terms of the number of item interactions into three datasets: popular, average, and unpopular, and then we conducted experiments in the three datasets to compare the HR@10 and NDCG@10 metrics of the model. The results of the experiment are shown in Fig. 4.

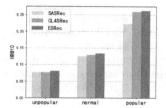
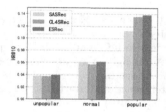

Fig. 4. Performance comparison over different item groups

The experimental results show that ESRec guarantees good prediction performance in both unpopular datasets with too few item interactions and popular datasets with dense interactions, which demonstrates that ESRec has excellent generalization and robustness.

5 Conclusion

In this study, we propose a contrastive learning model based on embedded spatial data augmentation, which simulates the effect on the model when the dataset is affected by noise as well as when data are missing and improves the ability of the model to resist disturbance. Meanwhile, we also directly compared the original view with the contrasted view for information sharing in different views. We find that data augmentation in the embedding space reduces the complexity of the model and the burden of model training. After extensive experiments, our results are shown to be valid and interpretable. How to generalize and summarize the way noise affects user-embedded expressions in more detail and improve the robustness of the model by means of targeted data augmentation is the direction we will further investigate.

References

1. Caron, M., Misra, I., Mairal, J., Goyal, P., Bojanowski, P., Joulin, A.: Unsupervised learning of visual features by contrasting cluster assignments. Adv. Neural. Inf. Process. Syst. **33**, 9912–9924 (2020)
2. Chen, T., Kornblith, S., Norouzi, M., Hinton, G.: A simple framework for contrastive learning of visual representations. In: International conference on machine learning. pp. 1597–1607. PMLR (2020)
3. Chen, Y., Liu, Z., Li, J., McAuley, J., Xiong, C.: Intent contrastive learning for sequential recommendation. In: Proceedings of the ACM Web Conference 2022. pp. 2172–2182 (2022)
4. Cho, K., Van Merriënboer, B., Gulcehre, C., Bahdanau, D., Bougares, F., Schwenk, H., Bengio, Y.: Learning phrase representations using rnn encoder-decoder for statistical machine translation. arXiv preprint arXiv:1406.1078 (2014)
5. Covington, P., Adams, J., Sargin, E.: Deep neural networks for youtube recommendations. In: Proceedings of the 10th ACM conference on recommender systems. pp. 191–198 (2016)
6. Devlin, J., Chang, M.W., Lee, K., Toutanova, K.: Bert: Pre-training of deep bidirectional transformers for language understanding. arXiv preprint arXiv:1810.04805 (2018)
7. Ebesu, T., Shen, B., Fang, Y.: Collaborative memory network for recommendation systems. In: The 41st international ACM SIGIR conference on research & development in information retrieval. pp. 515–524 (2018)
8. Glorot, X., Bengio, Y.: Understanding the difficulty of training deep feedforward neural networks. In: Proceedings of the thirteenth international conference on artificial intelligence and statistics. pp. 249–256. JMLR Workshop and Conference Proceedings (2010)
9. Greff, K., Srivastava, R.K., Koutník, J., Steunebrink, B.R., Schmidhuber, J.: Lstm: A search space odyssey. IEEE transactions on neural networks and learning systems **28**(10), 2222–2232 (2016)
10. Hamilton, W., Ying, Z., Leskovec, J.: Inductive representation learning on large graphs. Advances in neural information processing systems **30** (2017)
11. He, K., Fan, H., Wu, Y., Xie, S., Girshick, R.: Momentum contrast for unsupervised visual representation learning. In: Proceedings of the IEEE/CVF conference on computer vision and pattern recognition. pp. 9729–9738 (2020)
12. He, X., Deng, K., Wang, X., Li, Y., Zhang, Y., Wang, M.: Lightgcn: Simplifying and powering graph convolution network for recommendation. In: Proceedings of the 43rd International ACM SIGIR conference on research and development in Information Retrieval. pp. 639–648 (2020)
13. He, X., Liao, L., Zhang, H., Nie, L., Hu, X., Chua, T.S.: Neural collaborative filtering. In: Proceedings of the 26th international conference on world wide web. pp. 173–182 (2017)
14. Hidasi, B., Karatzoglou, A., Baltrunas, L., Tikk, D.: Session-based recommendations with recurrent neural networks. arXiv preprint arXiv:1511.06939 (2015)
15. Hou, Y., Hu, B., Zhang, Z., Zhao, W.X.: Core: simple and effective session-based recommendation within consistent representation space. In: Proceedings of the 45th international ACM SIGIR conference on research and development in information retrieval. pp. 1796–1801 (2022)
16. Jing, L., Tian, Y.: Self-supervised visual feature learning with deep neural networks: A survey. IEEE Trans. Pattern Anal. Mach. Intell. **43**(11), 4037–4058 (2020)

17. Kang, W.C., McAuley, J.: Self-attentive sequential recommendation. In: 2018 IEEE international conference on data mining (ICDM). pp. 197–206. IEEE (2018)
18. Kingma, D.P., Ba, J.: Adam: A method for stochastic optimization. arXiv preprint arXiv:1412.6980 (2014)
19. Li, X., Sun, A., Zhao, M., Yu, J., Zhu, K., Jin, D., Yu, M., Yu, R.: Multi-intention oriented contrastive learning for sequential recommendation. In: Proceedings of the sixteenth ACM international conference on web search and data mining. pp. 411–419 (2023)
20. Lin, Z., Tian, C., Hou, Y., Zhao, W.X.: Improving graph collaborative filtering with neighborhood-enriched contrastive learning. In: Proceedings of the ACM Web Conference 2022. pp. 2320–2329 (2022)
21. Lipton, Z.C., Berkowitz, J., Elkan, C.: A critical review of recurrent neural networks for sequence learning. arXiv preprint arXiv:1506.00019 (2015)
22. Luca, M.: Reviews, reputation, and revenue: The case of yelp. com. Com (March 15, 2016). Harvard Business School NOM Unit Working Paper (12-016) (2016)
23. Qiu, R., Huang, Z., Yin, H., Wang, Z.: Contrastive learning for representation degeneration problem in sequential recommendation. In: Proceedings of the fifteenth ACM international conference on web search and data mining. pp. 813–823 (2022)
24. Rendle, S., Freudenthaler, C., Gantner, Z., Schmidt-Thieme, L.: Bpr: Bayesian personalized ranking from implicit feedback. arXiv preprint arXiv:1205.2618 (2012)
25. Sun, F., Liu, J., Wu, J., Pei, C., Lin, X., Ou, W., Jiang, P.: Bert4rec: Sequential recommendation with bidirectional encoder representations from transformer. In: Proceedings of the 28th ACM international conference on information and knowledge management. pp. 1441–1450 (2019)
26. Vaswani, A., Shazeer, N., Parmar, N., Uszkoreit, J., Jones, L., Gomez, A.N., Kaiser, L., Polosukhin, I.: Attention is all you need. Advances in neural information processing systems **30** (2017)
27. Wang, C., Yu, Y., Ma, W., Zhang, M., Chen, C., Liu, Y., Ma, S.: Towards representation alignment and uniformity in collaborative filtering. In: Proceedings of the 28th ACM SIGKDD Conference on Knowledge Discovery and Data Mining. pp. 1816–1825 (2022)
28. Wang, T., Isola, P.: Understanding contrastive representation learning through alignment and uniformity on the hypersphere. In: International Conference on Machine Learning. pp. 9929–9939. PMLR (2020)
29. Wang, X., He, X., Cao, Y., Liu, M., Chua, T.S.: Kgat: Knowledge graph attention network for recommendation. In: Proceedings of the 25th ACM SIGKDD international conference on knowledge discovery & data mining. pp. 950–958 (2019)
30. Wang, X., He, X., Wang, M., Feng, F., Chua, T.S.: Neural graph collaborative filtering. In: Proceedings of the 42nd international ACM SIGIR conference on Research and development in Information Retrieval. pp. 165–174 (2019)
31. Wu, J., Wang, X., Feng, F., He, X., Chen, L., Lian, J., Xie, X.: Self-supervised graph learning for recommendation. In: Proceedings of the 44th international ACM SIGIR conference on research and development in information retrieval. pp. 726–735 (2021)
32. Wu, S., Tang, Y., Zhu, Y., Wang, L., Xie, X., Tan, T.: Session-based recommendation with graph neural networks. In: Proceedings of the AAAI conference on artificial intelligence. vol. 33, pp. 346–353 (2019)
33. Xia, L., Huang, C., Zhang, C.: Self-supervised hypergraph transformer for recommender systems. In: Proceedings of the 28th ACM SIGKDD Conference on Knowledge Discovery and Data Mining. pp. 2100–2109 (2022)

34. Xie, X., Sun, F., Liu, Z., Wu, S., Gao, J., Zhang, J., Ding, B., Cui, B.: Contrastive learning for sequential recommendation. In: 2022 IEEE 38th international conference on data engineering (ICDE). pp. 1259–1273. IEEE (2022)
35. Xiong, R., Yang, Y., He, D., Zheng, K., Zheng, S., Xing, C., Zhang, H., Lan, Y., Wang, L., Liu, T.: On layer normalization in the transformer architecture. In: International Conference on Machine Learning. pp. 10524–10533. PMLR (2020)
36. Xu, C., Zhao, P., Liu, Y., Sheng, V.S., Xu, J., Zhuang, F., Fang, J., Zhou, X.: Graph contextualized self-attention network for session-based recommendation. In: IJCAI. vol. 19, pp. 3940–3946 (2019)
37. Xu, Y., Wang, Z., Wang, Z., Guo, Y., Fan, R., Tian, H., Wang, X.: Simdcl: dropout-based simple graph contrastive learning for recommendation. Complex & Intelligent Systems pp. 1–13 (2023)
38. Yu, J., Yin, H., Xia, X., Chen, T., Cui, L., Nguyen, Q.V.H.: Are graph augmentations necessary? simple graph contrastive learning for recommendation. In: Proceedings of the 45th international ACM SIGIR conference on research and development in information retrieval. pp. 1294–1303 (2022)
39. Yu, T., Zhang, Z., Lan, C., Lu, Y., Chen, Z.: Mask-based latent reconstruction for reinforcement learning. Adv. Neural. Inf. Process. Syst. **35**, 25117–25131 (2022)
40. Zhao, W.X., Mu, S., Hou, Y., Lin, Z., Chen, Y., Pan, X., Li, K., Lu, Y., Wang, H., Tian, C., et al.: Recbole: Towards a unified, comprehensive and efficient framework for recommendation algorithms. In: proceedings of the 30th acm international conference on information & knowledge management. pp. 4653–4664 (2021)
41. Zhou, G., Mou, N., Fan, Y., Pi, Q., Bian, W., Zhou, C., Zhu, X., Gai, K.: Deep interest evolution network for click-through rate prediction. In: Proceedings of the AAAI conference on artificial intelligence. vol. 33, pp. 5941–5948 (2019)
42. Zhou, G., Zhu, X., Song, C., Fan, Y., Zhu, H., Ma, X., Yan, Y., Jin, J., Li, H., Gai, K.: Deep interest network for click-through rate prediction. In: Proceedings of the 24th ACM SIGKDD international conference on knowledge discovery & data mining. pp. 1059–1068 (2018)

Context-Augmented Contrastive Learning Method for Session-based Recommendation

Xianlan Sun[1], Xiangyun Gao[1], Subin Huang[1], Haibei Zhu[2], Chen Xu[3], Pingfu Chao[4], and Chao Kong[1(✉)]

[1] School of Computer and Information, Anhui Polytechnic University, Wuhu, China
{2220910106,2220920103}@stu.ahpu.edu.cn,
{subinhuang,kongchao}@ahpu.edu.cn
[2] School of Electrical and Computer Engineering, Georgia Institute of Technology, Atlanta, USA
hzhu389@gatech.edu
[3] School of Data Science and Engineering, East China Normal University, Shanghai, China
cxu@dase.ecnu.edu.cn
[4] School of Computer Science and Technology, Soochow University, Suzhou, China
pfchao@suda.edu.cn

Abstract. A session-based recommendation has become a hot research topic, which seeks to recommend the next item based on anonymous behavior sequences in a short time. While previous methods have made many efforts to address the complex information relationships between items, we contend that they still suffer from two inherent limitations: 1) they fail to consider the noisy preference information typically contained in user behavior sequences and 2) they are unaware of the importance of complex high-order relationships between non-adjacent items. In light of this, we contribute a novel solution named CCL (short for _Context-augmented Contrastive Learning_), which takes into account the joint effect of interest graph construction, context vectors, and contrastive learning. CCL decomposes session-based recommendation workflow into three steps. First, we adopt metric-based learning to reconstruct loose item sequences into tight item interest maps, making it easier to distinguish between the primary and secondary interests of users. Then, we propose adding a context vector to each session to provide a natural way to convey information beyond adjacent items. Finally, to improve the robustness of the model, we designed a contrastive self-supervised learning module as an auxiliary task to jointly learn the representation of items in the session. Extensive experiments have been conducted on two real-world datasets from different scenarios, demonstrating the superiority of CCL against several state-of-the-art methods.

Keywords: Session-based Recommendation · Graph Neural Network · Representation Learning · Contrastive Learning

1 Introduction

In the era marked by the rapid development of social media and the exponential growth of information, recommendation systems [1] play a crucial role in helping people alleviate information overload. However, traditional recommendation algorithms operate under the assumption that user's personal information and past activities are continuously recorded. In scenarios such as anonymous user login, these data become inaccessible, making it impossible to uncover their historical preferences and long-term interests. Consequently, this limitation leads to suboptimal recommendation performance. To address this challenge, session-based recommendation systems [2] have emerged. These systems, designed for anonymous or non-logged-in users, predict the next item for interaction based on the given user's anonymous sequential behavior. Their ability to capture dynamic user preferences and short-term interests has garnered widespread attention.

Currently, the latest advancements in session-based recommendation primarily focus on approaches based on deep learning [3]. Many of these methods employ graph structures [4] to model sessions, utilizing graph neural networks [5] to propagate information among adjacent items. These approaches have, to some extent, enhanced the performance of recommendation systems. However, user behavior sequences often contain noisy preference information that inadequately reflects users' actual preferences. For instance, users may accidentally click on unrelated items or items out of curiosity that do not align with their genuine preferences, and these interactions do not evolve into preferred items for future interactions. Therefore, such records may become noise in the user behavior history, disrupting their true interest in modeling. In addition, existing recommendation methods based on graph neural networks face challenges in modeling the relationships between non-adjacent items during the process of constructing a global graph. This process introduces noise and reduces the stability of the model. Due to the often overlooked complexity of high-order relationships between non-adjacent items [6], it has become necessary to conduct more complex modeling in the recommendation field.

To address the aforementioned issues, we propose a new solution. Firstly, we construct an interest graph by calculating the similarity between item pairs. By transforming loose item sequences into compact item diagrams, it is easier to distinguish between the primary and secondary interests of users. We constructed the current session graph (CSG) and global session graph (GSG) to learn item representations from both global and session levels. Secondly, we add a context vector to each session in the global graph, which provides a natural way to convey information beyond adjacent items. In addition, to obtain a more robust item representation, we generated an enhanced graph of GSG (AG-GSG). We maximize the consistency of context vectors for the same session in the original GSG graph and the enhanced graph AG-GSG for self-supervised contrastive learning. This model significantly improves the recommendation accuracy and enhances the robustness of the model to interaction noise by self-supervised tasks and recommendation tasks through joint optimization. The main contributions of this work are summarized as follows.

- We provide a novel solution called CCL, which integrates the construction of interest maps based on similarity and the transfer of information beyond adjacent items through context vectors into a unified framework.
 - We have designed a self-supervised contrastive learning module that can jointly optimize the contrastive self-supervised task and recommendation task, enhancing the model's robustness to interaction noise.
 - Extensive experiments are conducted to evaluate our solution under a wide spectrum of recommendation scenarios, demonstrating that our CCL model achieves substantial improvements over several competitive methods. Further studies verify the explainability of our model benefits brought by the session graph construction, non-adjacent item information propagation based on context vectors, and contrastive learning.

2 Related Work

Our work is highly relevant to two active research topics: 1) contrastive learning; and 2) session-based recommendation.

2.1 Contrastive Learning

In the past two years, with the significant success of contrastive learning (CL) [7] in computer vision (CV) [8] and natural language processing (NLP) [9], it has also been applied in recommendation systems. Comparative learning learns its representative features through self-supervised signals, thereby improving the performance and effectiveness of the model. MHCN [10] obtains a more comprehensive representation of users by constructing self-supervised signals from their social relationships. S^2-DHCN [11] constructs session data as a hypergraph and uses a hypergraph convolutional network to capture complex high-order relationships between items. COTREC [12] proposed a co-training mechanism for self-supervised graphs, which recursively utilizes different connectivity information from two views (item view and session view) to generate real samples, and then uses contrastive learning to supervise these samples.

2.2 Session-Based Recommendation

The preliminary exploration based on session recommendation mainly focuses on sequence modeling, where the Markov decision process [13] is the preferred technique in this stage. Rendle et al. [14] proposed a method that combines matrix factorization and first-order Markov chain to obtain long-term user preferences for recommendation. However, the Markov chain-based method only considers the dependency relationship between adjacent items. With the development of deep learning, recurrent neural networks are widely used in session recommendation [15] due to their excellent ability to model sequential data. GRU4RecR [16] is the first to apply RNN to session-based recommendation, using improved Gated Recurrent Units (GRU) to model interaction items. Although RNN can capture overall user preferences, it still has poor performance in capturing complex user preferences. NARM [17] captures the main preferences of users in the current session by modeling their sequential behavior using a hybrid encoder based on the attention mechanism. SASRec [18] models users' historical behavior information by using a self-attention mechanism to extract more valuable information.

Recently, graph neural networks have shown great potential in modeling complex item transformation patterns. Some large recommendation system data inherently possess the characteristics of graph topology, and graph neural networks can combine message propagation and aggregation mechanisms to capture graph topology information [19]. SR-GNN [20] is the first to propose transforming session sequences into session graphs and modeling complex transformation relationships between items through graph neural networks. TA-GNN [5] adaptively activates different user interests in a session by using target-aware attention. It learns different interest representations for different target items. CoKnow [21] can solve the problem of data sparsity and improve recommendation quality by selecting neighboring sessions at the label level. GCE-GNN [6] captures item transformation information more comprehensively by constructing session-level and global-level embeddings for items, respectively. MSGIFSR [22] model learns the interaction between continuous units from a high-level perspective by constructing a multi-granularity session graph.

Despite effectiveness, we contend that the session-based recommendation method mentioned above still suffers from two defects: 1) The current popular methods mostly model the behavior order of conversation sequences, ignoring that user behavior in long sequences reflects implicit and noisy preference signals. These records will deteriorate the modeling of their true interests; 2) The existing GNN-based recommendation methods make it difficult to model the relationship between non-adjacent items in the process of constructing a global graph and introduce noise information, which reduces the robustness of the model. Ignored the complex high-order relationships between non-adjacent items.

3 Problem Formulation

We first introduce the notations used in this paper and then formalize the session-based recommendation problem to be addressed.
Notations: Assuming $V = \{v_1, v_2, \cdots, v_m\}$ represents the set of items composed of all unique items in the session, and S represents all past sessions. Each anonymous session $s = \{v_{s,1}, v_{s,2}, \cdots, v_{s,n}\}$ consists of a series of items that interact in chronological order, where $v_{s,i} \in V$ represents a click by the user in session s. In a session-based recommendation model, this method generates a probability ranking list for all candidate items based on the interaction history sequence items of each user. The top k items in the ranking list will be selected for recommendation. Formally, the problem can be defined as:
Input: The adjacency matrices **A** , as well as the item embeddings **V**.
Output: A ranking list for all candidate items. The top k items in the ranking list will be selected for recommendation.

4 Proposed Model

In this section, we contribute a new model named CCL, which explores the construction of interest maps, modeling non-adjacent items based on context vectors, and contrastive learning. As illustrated in Fig. 1, there are four parts to our architecture: 1) Session Graph Construction; 2) Current Session Graph Embedding; 3) Global Session Graph Embedding; and 4) Model Prediction and Training.

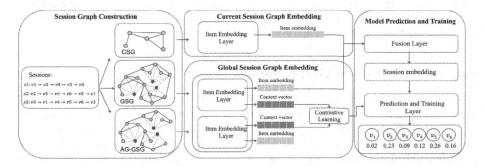

Fig. 1. An overview of the CCL model.

4.1 Session Graph Construction

To integrate and distinguish different types of preferences in the rich historical behavior of users, we can transform loose item sequences into tight item interest maps. The co-occurrence relationship between two items is a reasonable construction criterion, but the challenge is that the sparsity of the co-occurrence relationship is insufficient to generate a connected graph for each user. Therefore, we propose to calculate the similarity values between item pairs through cosine similarity, and automatically construct a graph structure for each interaction sequence to explore its interest distribution. The formula is $\alpha_{ij} = cos(\mathbf{W} \odot \mathbf{v}_i, \mathbf{W} \odot \mathbf{v}_j)$. Considering the introduction of noise, we only consider the most important connected node pairs by shielding those node pairs that are less than the non-negative threshold α. The core interest nodes have higher degrees than peripheral interest nodes because they connect more similar interests, and the higher the frequency of similar interests, the easier this modeling method is to distinguish between users' core and secondary interests. Therefore, we construct a session graph based on the similarity values of item pairs.

Considering that only considering the information within the session cannot better capture user preferences, we construct interest maps from both the current session and cross-session perspectives. The graph formed by modeling the sequence in the current session is defined as CSG=(V_g, E_g), similar to the current session graph modeling approach, we determine whether there is an edge relationship between two items by calculating the similarity of item pairs in the session sequence. At the same time, we add a global vector representation to each session in the graph and refer to it as the context vector. The context vector represents the session sequence as a whole and has bidirectional edges to all other nodes in the session, providing a natural way to transfer information between non-directly connected items. Therefore, for all historical session sequences of users, we define the global session graph as GCG=(V_g, E_g, C_g), where C_g represents the context vector of the session. Meanwhile, in order to improve the robustness of the model, we utilize a random edge dropout strategy to obtain the enhanced graph AG-GSG.

4.2 Current Session Graph Embedding

The Current Session Graph (CSG) aims to learn personalized item embeddings by modeling the current session. For each session, we embed each item into a unified embedding space and generate a d-dimensional embedding for each unique item in the session through an embedding layer, where the node vector represents the latent vector of the item. Graph neural networks can automatically extract features of session graphs while considering node connections, therefore, we use a gated graph neural network to update each item embedding in the current session graph CSG, iteratively updating the item node vector given the adjacency matrix $\mathbf{A}$ and item embedding vector $\mathbf{V}$.

$$
\begin{aligned}
\mathbf{a}_{s,i}^t &= \mathbf{A}_{i:}^s \left[\mathbf{V}_{s,i}^{t-1} \mathbf{H}_1, \mathbf{V}_{s,i}^{t-1} \mathbf{H}_2 \right]^T + \mathbf{b}, \\
\mathbf{z}_{s,i}^t &= \sigma \left(\mathbf{W}_z \mathbf{a}_{s,i}^t + \mathbf{U}_z \mathbf{v}_{s,i}^{t-1} \right), \\
\mathbf{r}_{s,i}^t &= \sigma \left(\mathbf{W}_r \mathbf{a}_{s,i}^t + \mathbf{U}_r \mathbf{v}_{s,i}^{t-1} \right), \\
\hat{\mathbf{v}}_i^t &= \tanh \left(\mathbf{W}_o \mathbf{a}_{s,i}^t + \mathbf{U}_o \left(\mathbf{r}_{s,i}^t \odot \mathbf{v}_{s,i}^{t-1} \right) \right), \\
\mathbf{v}_i^t &= \left(1 - \mathbf{z}_{s,i}^t \right) \odot \mathbf{v}_{s,i}^{t-1} + \mathbf{z}_{s,i}^t \odot \hat{\mathbf{v}}_i^t,
\end{aligned} \quad (1)
$$

where $\mathbf{H}_1, \mathbf{H}_2 \in \mathbb{R}^{d \times d}$ control the weight, $\mathbf{W}_z, \mathbf{W}_r, \mathbf{W}_o \in \mathbb{R}^{2d \times d}$ are the weight matrices, $\mathbf{z}_{s,i}^t$ is the reset gate, $\mathbf{r}_{s,i}^t$ is the update gate, t is the training step size, $\sigma(\cdot)$ is the sigmoid function, and $\odot$ is the element-wise multiplication operator.

Due to the presence of multiple layers of GNN, over-fitting and over-smoothing are prone to occur. We combine the output from the final layer of the module with the initial input to form the ultimate item representation.

$$
\begin{aligned}
\mathbf{g} &= \sigma(\mathbf{W}_s [\mathbf{v}_i^{s,0}; \mathbf{v}_i^{s,1}]), \\
\mathbf{v}_i^s &= \mathbf{g} \mathbf{v}_i^{s,0} + (1 - \mathbf{g}) \mathbf{v}_i^{s,1},
\end{aligned} \quad (2)
$$

where $\mathbf{W}_s \in \mathbb{R}^{2d \times d}$ is a learnable parameter.

4.3 Global Session Graph Embedding

Initialization. The GSG module aims to model complex high-order relationships between items through context vectors of sessions, to learn more powerful item embeddings. We designate the representation of the last item, denoted as $\mathbf{v}_m$, as the local embedding of session s, represented as $\mathbf{s}_l = \mathbf{v}_m$. Subsequently, we aggregate all node vectors within the session to generate a global preference embedding, denoted as $\mathbf{s}_g$. Finally, we use a soft-attention mechanism to learn their priorities and mix local embeddings with global embeddings $\mathbf{s}_l$ and $\mathbf{s}_g$ as follows.

$$\mathbf{a}_i = \mathbf{W}_0^T \sigma(\mathbf{W}_1 \mathbf{v}_m + \mathbf{W}_2 \mathbf{v}_i + \mathbf{b}),$$
$$\mathbf{s}_g = \sum_{i=1}^n \mathbf{a}_i \mathbf{v}_i, \qquad (3)$$

where $\mathbf{W}_0 \in \mathbb{R}^d$, $\mathbf{W}_1, \mathbf{W}_2 \in \mathbb{R}^{d \times d}$ are the learnable parameters that control the weight of the items, and $\mathbf{b} \in \mathbb{R}^d$ is the bias vector. Finally, we use a mixed embedding $\mathbf{s}_h = \mathbf{W}_3[\mathbf{s}_l; \mathbf{s}_g]$ as the corresponding initial context vector $\mathbf{c}^s$, i.e. $\mathbf{c}^s = \mathbf{s}_h$. This context vector combines the session's long-term and short-term interest preferences.

Node Updation. To learn the relationship between every two items in a conversation and the higher-order relationship between non-adjacent items in different conversations, we need to iteratively update the item embeddings and context vectors on the global session graph. Therefore, in the process of information dissemination, for each node in the graph, information is obtained from adjacent terms and context vectors. The first step is to ignore the context vector and then process the global session graph according to the current processing method of the session graph. For each node v_i at the layer in the graph, we update the representation of adjacent nodes from different sessions. Then, we add a dropout layer to alleviate overfitting. Due to the possibility of the item appearing in multiple sessions, each node can be connected to multiple context vectors. We calculate the similarity between node $\mathbf{v}_i$ and the context vectors $\mathbf{c}_{i,j}$ in layer l as attention weights and linearly combine them. Assuming the context vector of the session containing node v_i from the set $\mathbf{C}_i = [\mathbf{c}_{i,1}, \mathbf{c}_{i,2}, ...\mathbf{c}_{i,n}]$, where n is the number of sessions containing nodes v_i, the attention mechanism is as follows.

$$\alpha_{i,j}^l = \sigma\left(\frac{(\mathbf{W}_{q1} \mathbf{v}_i^{g,l})^T \mathbf{W}_{k1} \mathbf{c}_{i,j}^{l-1}}{\sqrt{d}}\right), \qquad (4)$$

where $\mathbf{W}_{q1}, \mathbf{W}_{k1} \in \mathbb{R}^{d \times d}$, $\mathbf{v}_i^{g,l}$ and $\mathbf{c}_{i,j}^{l-1}$ are the representations of node $\mathbf{v}_i$ in layer l, and the context vector representations of layer $l-1$. By performing a nonlinear mapping on vectors $\mathbf{v}_i^{g,l}$ and $\mathbf{c}_{i,j}^l$ to calculate the level priority β_i^l to balance the importance of the two vectors.

$$\mathbf{v}_i^{c,l} = \sum_{j=1}^m \alpha_{i,j}^l \mathbf{c}_j^l,$$
$$\beta_i^l = \sigma(\mathbf{W}_4[\mathbf{v}_i^{g,l}; \mathbf{v}_i^{c,l}]), \qquad (5)$$

where $\mathbf{v}_i^{g,l}$ is obtained from adjacent terms, $\mathbf{v}_i^{c,l}$ is obtained from the context vector at layer l, and $\mathbf{W}_4$ is a learnable parameter. Then, we integrate information from adjacent nodes and relevant context vectors,

$$\mathbf{v}_i^l = (1 - \beta_i^l)\mathbf{v}_i^{g,l} + \beta_i^l \mathbf{v}_i^{c,l}, \tag{6}$$

where $\mathbf{v}_i^l$ is the representation of the node at layer l. Finally, we aggregated the initial input and the output of the last layer to obtain the final item representation $\mathbf{v}_i^g$ by processing the current session graph.

Context Vector Updation. For each context vector in the global session graph, its information comes from the corresponding item information in the current session, as the information of the corresponding item varies in severity. So, we first assign different weights to each node v_i, as shown below.

$$\gamma_{j,i}^l = softmax\left(\frac{(\mathbf{W}_{k2}\mathbf{v}_i^l)^T(\mathbf{W}_{q2}\mathbf{c}_j^{l-1})}{\sqrt{d}}\right), \tag{7}$$

where $\mathbf{W}_{q2}, \mathbf{W}_{k2} \in \mathbb{R}^{d \times d}$ are trainable parameters, and $\gamma_{j,i}^l$ represents the importance of the ith item to the jth session in the lth layer. Finally, we linearly combine the item representations and aggregate the updated context vector $\mathbf{c}_j^{l'}$ and l-1 layer representations $\mathbf{c}_j^{l-1}$.

$$\begin{aligned}\mathbf{c}_j^{l'} &= \sum_{i=1}^n \gamma_{j,i}^l \mathbf{v}_i^l, \\ \varphi^l &= \sigma(\mathbf{W}_5[\mathbf{c}_j^{l'}; \mathbf{c}_j^{l-1}]), \\ \mathbf{c}_j^l &= \varphi^l \mathbf{c}_j^{l-1} + (1 - \varphi^l)\mathbf{c}_j^{l'},\end{aligned} \tag{8}$$

where $\mathbf{W}_5 \in \mathbb{R}^{d \times 2d}$ is a learnable parameter. Similarly, we use a highway network to combine the initial context vector with the output of the last layer to obtain the final representation $\mathbf{c}_j$.

Contrastive Learning. To address the issue of reducing model robustness due to the introduction of a large amount of noise information in constructing cross-session graphs. We have introduced a self-supervised contrastive learning module for auxiliary learning, we obtain the data augmentation graph AG-GSG of GSG by using the edge dropout method, we consider the same session in the original graph GSG and enhanced graph AG-GSG as positive pairs and the other 2(N-1) sessions in the two graphs as negative samples. For each session pair, since the context vector of each session can be considered as the overall representation of the current session, we can consider the updated context vector obtained from the session pair as a pair of positive samples. We use the InfoNCE loss of context vectors in two graphs as the optimization object defined below.

$$\mathcal{L}_{ssl}(\mathbf{c}_n, \mathbf{c}_n^{aug}) = -log\frac{exp(sim(\mathbf{c}_n, \mathbf{c}_n^{aug})\tau)}{\sum_{m=1}^{2N} exp(sim(\mathbf{c}_n, \mathbf{c}_m^{aug})/\tau)}, \tag{9}$$

where $sim(\cdot, \cdot)$ is the similarity function, τ is a hyperparameter that controls scaling. $\mathbf{c}_n^{aug}$ is the context vector of the AG-GSG graph.

4.4 Model Prediction and Training

Model Prediction. For each item v_j, they can obtain item representations from the CSG module and GSG module respectively. Therefore, the final representation of the item was obtained by merging information from different sources.

$$\mathbf{v}_j^{'} = \mathbf{v}_j^s + \mu \mathbf{v}_j^g, \tag{10}$$

where μ is a hyper-parameter to control the ratio of the representation learned from the CSG module. Meanwhile, the same method uses the initial session representation as the initialization context vector, using soft attention aggregation nodes to obtain the final session representation. After obtaining all item and session embeddings, we calculate the recommended score for each target item:

$$\hat{\mathbf{y}}_i = Softmax(\mathbf{s}_i^T \mathbf{v}_j^{'}). \tag{11}$$

Model Training. We used the cross-entropy of the prediction results and the ground truth labels $\mathbf{y}$ as the main loss defined in the following:

$$\mathcal{L}_{rec}(\hat{\mathbf{y}}) = -\sum_{i=1}^{m} \mathbf{y}_i \log\left(\hat{\mathbf{y}}_i\right) + (1 - \mathbf{y}_i) \log\left(1 - \hat{\mathbf{y}}_i\right), \tag{12}$$

where $\mathbf{y}$ denotes the one-hot encoding vector of the ground truth item.

Then, we combine recommendation loss with self-supervised contrastive learning loss for performance optimization, as shown below:

$$\mathcal{L} = \mathcal{L}_{rec} + \lambda \mathcal{L}_{ssl}, \tag{13}$$

where λ is a hyperparameter that controls the contrastive learning loss ratio.

5 Experiments

In this section, we perform experiments on two public datasets to demonstrate the effectiveness and rationality of our solution. Through empirical evaluation, we intend to answer three questions.
RQ1: How does the performance of CCL juxtapose against several contemporary state-of-the-art recommendation approaches?
RQ2: Can the introduction of context vectors in similarity modeling proposed in this article improve the model performance of existing methods?
RQ3: How do different hyperparameter settings impact the recommendation performance?

5.1 Experimental Setup

Data Description. We conducted experiments on two benchmark datasets, Diginetica and Tmall[1]. Diginetica comes from CIKM Cup 2016 and contains session information extracted from e-commerce search engine logs. Tmall comes from the IJCAI-15 competition, which includes shopping logs of anonymous users on the Tmall online shopping platform. Based on previous work, we filtered out sessions with a length of 1 and items with less than 5 occurrences. We will use the latest data as test data, and the remaining historical data will be used for training and validation.

Table 1. The descriptive statistics of datasets.

Datasets	All Clicks	Train Sessions	Test Sessions	Total Items	Avg. Length
Diginetica	982,961	719,470	68,977	43,097	5.12
Tmall	818,479	351,268	25,898	40,728	6.69

Evaluation Metrics. We adopted two metrics widely used in session-based recommendations: Precision@k and MRR@k To evaluate recommendation performance. Precision@K represents the accuracy of the recommendation, and MRR@k Indicates the position of the target item in the list of recommended items. Higher metrics indicate better ranking accuracy.

Baselines. We compare my method with the following baselines:

- FPMC [14]: This model captures users' long-term preferences for recommendation by combining matrix factorization and Markov chain.
- GRU4Rec [16]: This model is the first to apply RNN to the session-based recommendation, modeling user behavior sequences.
- NARM [17]: This model incorporates both RNN and attention mechanism into the session-based recommendation model to capture preferences.
- SR-GNN [20]: This model models session sequences as session graphs and uses graph neural networks for the first time to capture complex transformations between items.
- GCE-GNN [6]: This model captures local and global information by constructing two types of current session graphs and global session graphs.
- S^2-DNCN [11] : This model learns inter-session and intra-session information by constructing hypergraphs and uses self-supervised learning to provide supplementary information.
- COTREC [12] : This model constructs two views and selects and evolves pseudo labels as examples of information self-supervision.

---
[1] https://tianchi.aliyun.com/dataset/42.

- CORE [4]: This model designs an encoder that can embed linear combinations of input items into session embeddings.
- MGIR [23]: This model uses different aggregation layers to encode different relationships and generates enhanced session representations.
- MGS [24]: This model applies an iterative dual refinement mechanism to propagate information between the session graph and the mirror graph.

Hyperparameter Settings. We use a Gaussian distribution with a mean of 0 and a standard deviation of 0.1 to initialize the parameters. For regular settings, we set the embedding size and batch size to 256 and 1024, respectively. Each session is truncated within a maximum length of 20. We use Adam to optimize our model. The initial learning rate is $1e^{-3}$, the decay factor for the three periods is 0.1, and the L2 regularization is 10^{-5}.

Table 2. Recommendation Performance on Diginetica and Tmall.

Method	Diginetica		Tmall	
	Precision@20	MRR@20	Precision@20	MRR@20
FPMC	22.14%	6.66%	16.06%	7.32%
GRU4Rec	30.79%	8.22%	10.93%	5.89%
NARM	48.32%	16.00%	23.30%	10.70%
SR-GNN	51.26%	17.78%	27.57%	13.72%
GCE-GNN	54.22%	19.04%	33.42%	15.42%
S^2-DHCN	53.66%	18.51%	31.42%	15.05%
COTREC	54.18%	19.07%	36.35%	18.04%
CORE	52.84%	18.47%	39.16%	19.52%
MGIR	53.73%	18.77%	36.31%	17.42%
MGS	55.02%	19.14%	41.55%	18.39%
CCL	**55.42%****	**19.32%****	**41.94%****	**21.37%****

** indicates that the improvements are statistically significant for p<0.01 judged by paired t-test.

5.2 Performance Comparison with State-of-the-Arts (RQ1)

Overall Comparison. From the Table 2, we have four findings.

- The performance of the FPMC model is inferior to GRU4Reg and NARM. Both models then use RNN to model user sequences and capture user preferences, indicating that sequence correlation is crucial. The performance of NARM is superior to GRU4Rec, indicating that the attention mechanism can more accurately model the main interests of users.

- The performance of models based on graph neural networks is superior to traditional models such as NARM and GRU4Reg, demonstrating the effectiveness of graph neural networks in data modeling as they can capture transformation information between more complex items. In addition, GCE-GNN, S^2-DHCN, COTREC, MGIR, MGS, and other models capture local and global information by constructing two types of session graphs to obtain more accurate interest preferences. Indicating that only considering intra-session information to capture interest preferences is suboptimal.
- Models such as S^2-DHCN and COTREC use self-supervised modules to capture conversion information between more complex items by constructing comparison graphs, demonstrating the effectiveness of self-supervised contrastive learning in recommendation systems.
- We can see that our method achieved the best performance in terms of accuracy and MRR among all methods. These results demonstrate the rationality and effectiveness of my method in conversation-based recommendations.

5.3 Ablation Study (RQ2)

Table 3. Ablation study of key components of CCL.

Method	Diginetica		Tmall	
	Precision@20	MRR@20	Precision@20	MRR@20
CCL_{sd}	55.24%	19.02%	41.42%	20.89%
CCL_x	54.48%	18.77%	39.75%	19.91%
CCL_{xs}	55.26%	18.95%	41.63%	21.04%
CCL	**55.42%****	**19.32%****	**41.94%****	**21.37%****

** indicates that the improvements are statistically significant for p<0.01 judged by paired t-test.

To justify the rationality of our solution, we perform an ablation study in CCL. In this section, while keeping all other conditions the same, we designed three model variants to analyze the impact on model performance (RQ2): model CCL_{sd}, which removes the interest graph module built through similarity, the model CCL_x, which eliminates the context vector and contrastive learning module, and the model CCL_{xs}, which eliminates the contrastive learning module. The results are shown in Table 3. We noticed that the performance of the CCL model is better than that of the CCL_{sd} model, indicating that constructing interest maps through similarity can better eliminate the influence of noise and obtain user interests, which is effective in improving the performance of GNN-based session recommendations; We compared the variants of the CCL_x and CCL_{xs} models with the CCL model, indicating that constructing context vectors for sessions can help extract important behavioral information between non-adjacent items from session data, and through contrastive learning, session data can be better enhanced to achieve better recommendation results. Overall, the CCL model achieved the best performance on both datasets, indicating the necessity and rationality of considering the construction of interest maps through similarity and the propagation of information between non-adjacent items through context vectors.

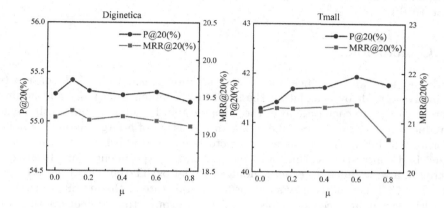

Fig. 2. The impact of CSG module ratio on algorithm performance.

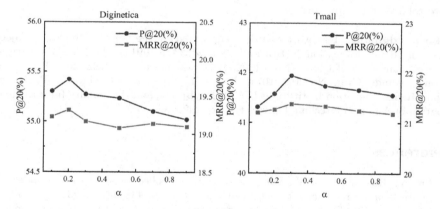

Fig. 3. The impact of similarity threshold on algorithm performance.

5.4 Hyperparameter Sensitivity (RQ3)

In this section, we investigated the influence of two important hyperparameters. For the sake of fairness, we keep all other parameters unchanged except for the measured parameters. To investigate the impact of the ratio of two modules, we tested a representative set of μ Value performance. From the results in Fig. 2, it can be seen that when the ratio μ in the Diginetica and Tmall datasets is set to 0.1 and 0.6, respectively, it can better aggregate item embeddings learned from global and session levels to improve the recommendation performance of the current session.

Meanwhile, we investigated the impact of similarity threshold size on performance when constructing interest maps. We tested on two datasets separately. The performance of this value is shown in Fig. 3. We found that the Diginetica dataset and Tmall dataset achieved optimal performance at $\alpha = 0.2$ and $\alpha = 0.3$ respectively.

6 Conclusions

In this article, we introduce an effective solution called CCL that constructs an interest graph by calculating the similarity between item pairs and adding context vectors to each session in the global graph. Finally, to address the issue of reduced model robustness due to the introduction of a large amount of noise information during the construction of cross-session graphs, we utilized a self-supervised contrastive learning module. Extensive experiments have been conducted in a wide range of recommendation scenarios. The empirical results indicate that CCL is significantly superior to several state-of-the-art methods.

Although great success has been achieved in converting project information between sessions, there are still many interesting issues to be explored in the research community. In the future, we hope to explore more efficient to establish more reliable recommendations.

Acknowledgment. This work has been supported by the Science Research Project of Anhui Higher Education Institutions (No. 2023AH050914), the National Natural Science Foundation of China (No. 61902001), the Science and Technology Project of Wuhu City under Grant No. 2023pt07 and 2023ly13, the Quality Engineering Project of Anhui Higher Education Institutions (No. 2023zybj018), and the Quality Improvement Program of Anhui Polytechnic University (No. 2022lzyybj02).

References

1. Liu, Y., Rao, Q., Pan, W., Ming, Z.: Variational collective graph autoencoder for multi-behavior recommendation. In: ICDM 2023. pp. 438–447 (2023)
2. Chen, Q., Guo, Z., Li, J., Li, G.: Knowledge-enhanced multi-view graph neural networks for session-based recommendation. In: SIGIR 2023. pp. 352–361 (2023)
3. Agrawal, N., Sirohi, A.K., Kumar, S., Jayadeva: No prejudice! fair federated graph neural networks for personalized recommendation. In: AAAI 2024. pp. 10775–10783 (2024)
4. Hou, Y., Hu, B., Zhang, Z., Zhao, W.X.: CORE: simple and effective session-based recommendation within consistent representation space. In: SIGIR 2022. pp. 1796–1801 (2022)
5. Yu, F., Zhu, Y., Liu, Q., Wu, S., Wang, L., Tan, T.: TAGNN: target attentive graph neural networks for session-based recommendation. In: SIGIR 2020. pp. 1921–1924 (2020)

6. Wang, Z., Wei, W., Cong, G., Li, X., Mao, X., Qiu, M.: Global context enhanced graph neural networks for session-based recommendation. In: SIGIR 2020. pp. 169–178 (2020)
7. Sun, G., Shen, Y., Zhou, S., Chen, X., Liu, H., Wu, C., Lei, C., Wei, X., Fang, F.: Self-supervised interest transfer network via prototypical contrastive learning for recommendation. In: AAAI 2023. pp. 4614–4622 (2023)
8. Datta, G., Liu, Z., Yin, Z., Sun, L., Jaiswal, A.R., Beerel, P.A.: Enabling ispless low-power computer vision. In: WACV 2023. pp. 2429–2438 (2023)
9. Schneider, P., Schopf, T., Vladika, J., Galkin, M., Simperl, E., Matthes, F.: A decade of knowledge graphs in natural language processing: A survey. In: IJCNLP 2022. pp. 601–614 (2022)
10. Yu, J., Yin, H., Li, J., Wang, Q., Hung, N.Q.V., Zhang, X.: Self-supervised multi-channel hypergraph convolutional network for social recommendation. In: WWW 2021. pp. 413–424 (2021)
11. Xia, X., Yin, H., Yu, J., Wang, Q., Cui, L., Zhang, X.: Self-supervised hypergraph convolutional networks for session-based recommendation. In: AAAI 2021. pp. 4503–4511 (2021)
12. Xia, X., Yin, H., Yu, J., Shao, Y., Cui, L.: Self-supervised graph co-training for session-based recommendation. In: CIKM 2021. pp. 2180–2190 (2021)
13. Ni, S., Hu, S., Li, L.: An intention-aware markov chain based method for top-k recommendation. IEEE Trans. Autom. Sci. Eng. **21**(1), 581–592 (2024)
14. Rendle, S., Freudenthaler, C., Schmidt-Thieme, L.: Factorizing personalized markov chains for next-basket recommendation. In: WWW 2010. pp. 811–820 (2010)
15. Salampasis, M., Siomos, T., Katsalis, A., Diamantaras, K.I., Christantonis, K., Delianidi, M., Karaveli, I.: Comparison of RNN and embeddings methods for next-item and last-basket session-based recommendations. In: ICMLC 2021. pp. 477–484 (2021)
16. Hidasi, B., Karatzoglou, A., Baltrunas, L., Tikk, D.: Session-based recommendations with recurrent neural networks. In: ICLR 2016 (2016)
17. Li, J., Ren, P., Chen, Z., Ren, Z., Lian, T., Ma, J.: Neural attentive session-based recommendation. In: CIKM 2017. pp. 1419–1428 (2017)
18. Kang, W., McAuley, J.J.: Self-attentive sequential recommendation. In: ICDM, 2018. pp. 197–206 (2018)
19. Chu, F., Jia, C.: Self-supervised global context graph neural network for session-based recommendation. PeerJ Comput. Sci. **8**, e1055 (2022)
20. Wu, S., Tang, Y., Zhu, Y., Wang, L., Xie, X., Tan, T.: Session-based recommendation with graph neural networks. In: AAAI 2019. pp. 346–353 (2019)
21. Liu, L., Wang, L., Lian, T.: Discovering proper neighbors to improve session-based recommendation. In: PKDD 2021. vol. 12975, pp. 353–369 (2021)
22. Guo, J., Yang, Y., Song, X., Zhang, Y., Wang, Y., Bai, J., Zhang, Y.: Learning multi-granularity consecutive user intent unit for session-based recommendation. In: WSDM 2022. pp. 343–352 (2022)
23. Han, Q., Zhang, C., Chen, R., Lai, R., Song, H., Li, L.: Multi-faceted global item relation learning for session-based recommendation. In: SIGIR 2022. pp. 1705–1715 (2022)
24. Lai, S., Meng, E., Zhang, F., Li, C., Wang, B., Sun, A.: An attribute-driven mirror graph network for session-based recommendation. In: SIGIR 2022. pp. 1674–1683 (2022)

Graph Contrastive Learning for Multi-behavior Recommendation

Haiying Li[1], Huihui Wang[1(✉)], Shunmei Meng[1], and Xingguo Chen[2]

[1] Nanjing University of Science and Technology, Nanjing, China
{haiyingli,huihuiwang,mengshunmei}@njust.edu.cn
[2] Nanjing University of Posts and Telecommunications, Nanjing, China
chenxg@njupt.edu.cn

Abstract. Traditional recommendation systems usually use the single type of user-item interaction behavior for recommendation. Multi-behavior recommendation explores multiple relationships between user different interaction (e.g., click, favorite) has been proposed to improve recommendation performance. Most existing multi-behavior recommendation methods are usually insufficient to capture heterogeneous information, which often emphasize the differences between multiple behaviors but ignore the common preferences of user-item interactions. In this paper, we propose a novel Graph Contrastive Learning for Multi-Behavior Recommendation (GCLMBR) framework which exploits the differences and dependencies between different user-item interaction behaviors. Firstly, we leverage a graph convolutional network (GCN) to learn the single-behavior representation from multi-behavior data, and integrate the learned single-behavior representation. Secondly, we apply an enhanced GCN with layer attention mechanism to encode global multi-behavior representation from the heterogeneous graph. The graph contrastive learning task is applied to enhance representation capability of the multi-behavior graph and single-behavior subgraphs. Finally, we adopt a multi-task learning strategy to jointly optimize the learning objectives, and predict the user-item interaction behaviors for recommendation. The empirical results on three real-world datasets show that our proposed GCLMBR outperforms various state-of-the-art baselines.

Keywords: Multi-Behavior Recommendation · Graph Contrastive Learning · Graph Convolutional Network

1 Introduction

With the rapid development of the internet, recommendation system has become an useful solution to address the information overload issue [7], and is widely used in shopping, music, social networks, etc. Traditional recommendation algorithms include content-based recommendation, collaborative filtering and hybrid recommendation [1]. However, traditional recommendation algorithms often suffer from the data sparsity and cold start problems. In recent years, deep learning models have been applied to effectively alleviate the above problems [29].

By combining deep learning and collaborative filtering technology, neural collaborative filtering (NCF) [10] and DeepFM [6] can learn rich user and item representations, and improve the accuracy and effectiveness of recommendation. Furthermore, auto-encoder addresses the cold-start problem better by learning the latent representation of users and items [2]. It has the capability to extract useful information from historical behavior data. Therefore, various methods based on auto-encoder (such as AutoRec [18], A-SAERec [30], etc.) have been proposed to improve the recommendation performance.

However, these methods only consider single interaction between users and items, ignore multi-behavior interactions (e.g. click, favorite, purchase) that can provide more information [24]. NMTR [5] captures the semantics of user behavior across various types by leveraging cascaded relationships, which fails to capture the relationship between users and items. Graph Neural Networks (GNNs) [26] have demonstrated outstanding performance across numerous domains. Therefore, various multi-behavior recommendation methods [20,22] based on GNNs have been proposed to effectively learn high-order information on user-item interactions. In particular, GNMR [22] and MB-GMN [23] divide the user interaction information into multiple subgraphs and perform graph convolutional network on these subgraphs to learn user/item representations. However, these methods mainly extract information from single-behavior subgraphs, and neglect the interactions between different behaviors, thus unable to comprehensively learn heterogeneous information. To fully capture heterogeneous information between behaviors, many approaches [3,12,15,21] have employed GNNs on heterogeneous graphs and integrated them with linear layers for recommendation. However, these methods still have limitations in effectively utilizing multi-behavior data. Firstly, they fail to model the dependencies among various behaviors. Secondly, they ignore the consistency between the user representations in single-behavior subgraphs and those in the heterogeneous graph.

To address this problem, we propose a novel model called Graph Contrastive Learning for Multi-Behavior Recommendation (GCLMBR). Firstly, GCLMBR constructs single-behavior subgraphs from user interactions under different behaviors, respectively. We adopt UltraGCN to reflect user preferences from each single-behavior subgraph. Secondly, in order to effectively extract the dependencies between user behaviors, we propose an enhanced GCN that uses the layer attention mechanism to extract dependencies between behaviors in each convolutional layer. Thirdly, we incorporate a graph contrastive learning task to capture the inherent similarity between user representations derived from single-behavior subgraphs and multi-behavior graph. Finally, we jointly optimize the recommendation model integrated the contrastive learning loss and Margin ranking loss for model training. The key contributions of this work are summarized as follows:

- We propose GCLMBR, a graph contrastive learning framework for multi-behavior recommendation. GCLMBR constructs user behavior graph and the employment of graph contrastive learning techniques, effectively captures the inherent patterns and preferences of user behaviors.

- We design the contrastive learning task to capture the commonalities between user's single-behavior representation and global multi-behavior representation.
- Extensive experiments are conducted on three real-world datasets. The results validate that GCLMBR outperforms several state-of-the-art methods.

2 Related Work

2.1 Graph-Based Recommendation

Graph Neural Networks (GNNs) have demonstrated excellent capabilities in exploring various types of correlated data [16,17]. For example, PinSage [27] integrates graph convolutional layers with random walks on the interaction graph to improve recommendation performance. STAR-GCN [28] combines stacked GCN encoder-decoders with intermediate supervision to improve prediction accuracy. In addition, NGCF [19] uses graph convolutional networks to capture higher-level user-item collaborative signal. To simplify GNNs architectures, LightGCN [9] simplifies the learning process of user and item embeddings by omitting feature transformation and non-linear activation. This makes it more suitable for recommendation tasks as it focuses on capturing essential interactions. In contrast to Light-GCN [9], UltraGCN [13] skips the information propagation of infinite layers and obtains embeddings by constraining the loss to approximate the limit of infinite layer graph convolution.

2.2 Multi-behavior Recommendation

In recent years, multi-behavior recommendation as a solution to address data sparsity, has attracted significant attention. R-GCN [15] proposes that the GNN framework can effectively handle multi-relationship networks and more accurately capture heterogeneous relationships between users and items. MB-GCN [12] utilizes graph convolutional networks to acquire user and item embeddings within a unified graph. GNMR [22] proposes the framework for categorizing multiple behaviors into different dimensions to models various user-item dependencies. GHCF [3] introduces a relation-aware propagation layer to embed joint representations of users/items and relationships into graph. Moreover, MB-GMN [23] learns behavior heterogeneity and interaction diversity are automatically leveraging meta-learning and GNN. MBRec [21] designs a graph-structured inter-relationship encoder to emphasize the interdependence between different behaviors to improve the accuracy of recommendation.

2.3 Contrastive Learning

Contrastive learning is a paradigm of self-supervised learning, is distinctive in its ability to be effectively trained even in the absence of labeled data. Thus it has been widely used in Computer Vision (CV) and Natural Language Process

(NLP) [4]. MoCo [8] leverages contrastive learning to implement unsupervised learning, and its performance in classification, detection, and segmentation tasks is comparable to or even surpasses that of supervised learning models. Furthermore, MMCLR [20] designs multiple contrastive learning tasks to better capture the commonalities and differences among different views. Therefore, we propose an innovative graph contrastive learning framework that can learn heterogeneous information in multiple behaviors using GNNs, meanwhile utilizing contrastive learning to capture commonalities between user representations to optimize the performance of recommendation.

3 Methodology

In this section, we propose a novel framework called Graph Contrastive Learning for Multi-Behavior Recommendation. The model architecture is illustrated in Fig. 1, which consisting of four components: (i) **Subgraph behavior encoder**, we utilize UltraGCN to learn the feature representations of nodes on a single-behavior graph, and then effectively aggregate all behavior information. (ii) **Multi-behavior encoder**, by combining the attention mechanism with graph convolutional networks, we employ the attention mechanism at each layer of graph convolution to aggregate different behavioral information, thus accurately capturing the dependencies among behaviors. (iii) **Graph contrastive learning**, which ensures the consistency between a user's behavior feature fusion representations and global multi-behavior representations. (iv) **Multi-task optimization**, model training using information from multiple behavioural profiles for recommendation and joint comparative learning.

3.1 Preliminaries

In our work, we define $U = \{u_1, u_2, \ldots, u_n\}$ represents the set of n users, $V = \{v_1, v_2, \ldots, v_m\}$ represents the set of m items, $R = \{r_1, r_2, \ldots, r_k\}$ represents the set of k behaviors. We use the three-dimensional tensor $B \in \mathbb{R}^{|U| \times |V| \times |R|}$ to represent user-item interaction information, where $|U|$, $|V|$ and $|R|$ are the total numbers of users, items and behaviors, respectively. $b_{i,j,k} \in B$, if $b_{i,j,k} = 1$ there exists behavior k between user i and item j, otherwise, $b_{i,j,k} = 0$.

3.2 Subgraph Behavior Encoder

In this subsection, our objective is to extract node feature information from the interactive subgraphs consisting of users and items across various behaviors. To capture the correlation between nodes, we employ UltraGCN [13] to learn the node embedding representations within each single-behavior subgraph. In each UltraGCN layer, we represent the user node embeddings as follows:

$$s_{u,r}^l = \sigma \left(\sum_{(v) \in N(u)} \beta_{u,v} s_{v,r}^{l-1} \cdot h_r \right) \quad (1)$$

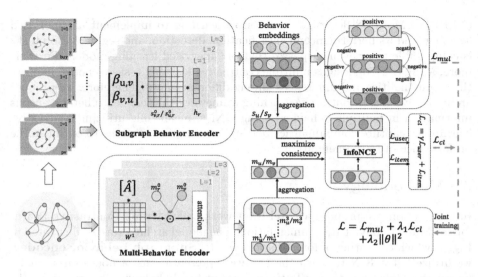

Fig. 1. An illustration of GCLMBR.

where $s_{u,r}^l \in \mathbb{R}^d$ denotes the embedding representation of user u on the behavior subgraph r at layer l, $s_{u,r}^0$ is randomly initialized using the ids of the user, $h_r \in \mathbb{R}^d$ denotes the embedding of the relation, and the initial values is randomly generated, σ is LeakyReLU. Inspired by [13], the use of $\beta_{u,v}$ for preventing node feature flattening, which is defined as $\beta_{u,v} = \frac{\sqrt{|N_u|+1}}{\sqrt{|N_v|+1}} \cdot \frac{1}{|N_u|}$.

To promote the inter-layer information exchange, we employ an aggregation mechanism to integrate the node embedding information from multiple layers. Meanwhile nodes have K subgraph feature representations, and these K features are aggregated to obtain the final embedding, the formula is expressed as follows:

$$s_{u,r} = \frac{1}{L+1} \sum_{l=0}^{L} s_{u,r}^l, s_u = \sum_{k=1}^{K} \hat{\alpha}_{u,k} s_{u,k} \quad (2)$$

where $\hat{\alpha}_{u,k}$ denotes the importance of behavior k to user u, i.e., the degree of influence on the user. In order to capture the varying importance of node representations across different behaviors, we employ a two-layer multi-layer perceptron (MLP) to derive corresponding weights. The formula for obtaining these weights is as: $\hat{\alpha}_{u,k} = softmax(W_2(W_1 s_{u,r} + b_1) + b_2)$, where $W_1 \in \mathbb{R}^d$, $W_2 \in \mathbb{R}^{(d \times d')}$ are trainable matrices, $b_1 \in \mathbb{R}$ and $b_2 \in \mathbb{R}^d$ are bias.

The embedding of items is similar as that of uses which has been introduced in Eq. (1–2).

3.3 Multi-behavior Encoder

After we have successfully derived embedding representations of nodes for individual behaviors, the behavior-to-behavior dependency information between behaviors is ignored. We uses an enhanced GCN with layer attention to capture heterogeneous information across different behaviors. The formula of extracting global multi-behavior representations can be defined:

$$m_{u,r}^l = \sigma \left(\sum_{(v,r) \in N(u)} \hat{A} W^{(l)} \emptyset \left(m_v^{(l-1)}, m_r^{(l-1)} \right) \right) \quad (3)$$

where $m_{u,r}^l \in \mathbb{R}^d$ denotes the embedding of user u on behavior r at layer l, $W^{(l)}$ is the learnable parameter of layer l, $\hat{A} = D^{-1}\tilde{A}$ is an asymmetric normalized adjacency matrix, D is a degree matrix of A, and $\tilde{A}$ is an adjacency matrix with a self-ring. $m_v^{(l-1)} \in \mathbb{R}^d$ and $m_r^{(l-1)} \in \mathbb{R}^d$ denote the embedding representation of item v and behavior r at the $(l-1)$ layer, respectively. $\emptyset(m_1, m_2) = m_1 \odot m_2$, the embedding representation of the behavior at each layer is updated as follows: $m_r^{(l)} = W_r^{(l)} m_r^{(l-1)}$.

In multi-behavior recommendation, to acknowledge the distinct effects of different behaviors on users, we leverage the attention mechanism to comprehend the relative importance of various behaviors, which can be expressed as follows:

$$\begin{aligned}\alpha_k^l &= softmax\left(\frac{QK^T}{\sqrt{d_k}}\right) V \\ &= softmax\left(\frac{m_{u,r_0}^l {m_{u,r_k}^l}^T}{\sqrt{d_k}}\right) m_{u,r_k}^l \end{aligned} \quad (4)$$

where α_k^l is the weight of the behavior k of the user at layer l, m_{u,r_0}^l denotes the target behavior. Meanwhile, in order to ensure that information can be exchanged between nodes in different layers, the node embedding information of each layer is aggregated to obtain the final user embedding, which is defined as follows:

$$m_u^l = \sum_{k=1}^{K} \alpha_k^l m_{u,r_k}^l, \; m_u = \sum_{l=0}^{L} \frac{1}{L+1} m_u^l \quad (5)$$

The embedding of items is similar as that of uses which has been introduced in Eq. (3–5).

3.4 Graph Contrastive Learning

Through single-behavior feature learning, the behavior aggregated representations (e.g., s_u and s_v) are aggregated from single-behavior subgraphs, there will be information loss compared with the global multi-behavior representations from the heterogeneous graph. Therefore, we introduce a graph contrastive learning to reduce the information loss, we use InfoNCE [14] loss to learn the consistency between s_u and m_u. which is formed as follows:

$$\mathcal{L}_{user} = \sum_{i \in U} -\log \frac{exp\left(\frac{sim(m_i, s_i)}{\tau}\right)}{\sum_{j \in U} exp\left(\frac{sim(m_i, s_j)}{\tau}\right)} \quad (6)$$

where s_i and m_i are user i feature fusion representation and global multi-behavior representation , respectively. $sim(m_i, s_i) = m_i \cdot s_i$ is a similarity function, and τ is the temperature parameter. To ensure the completeness of the information, we take the global multi-behavior representation as the anchor.

3.5 Multi-task Optimization

In real-world application scenarios, there is usually some correlation between users' behaviors. If we simply train on the target behavior, we tend to miss information about other behaviors, through multi-task learning, auxiliary behaviors are trained together, which not only shares data and features, but also effectively mitigates the cold-start problem and improves the generalization ability and effectiveness of the model. Therefore, we use multi-task learning to train the possibility of user-item interaction under different behaviors for assisting the target behavior, defined the loss function as below:

$$\mathcal{L}_{Mul} = \sum_{u \in U} \sum_{k=1}^{K} \sum_{(s,s') \in S_u} max\left(0, margin - \hat{y}_{u,s}^k + \hat{y}_{u,s'}^k\right) \tag{7}$$

where $\hat{y}_{u,s} = s_u^T W_k s_s$, $margin$ and W_k are adjustable hyperparameters, $(s, s') \in S_u$ is defined as the pair of positive and negative samples in the item.

The overall optimization objective function of the model is obtained by combining to loss functions of graph contrastive learning and multi-task learning:

$$\mathcal{L} = \mathcal{L}_{Mul} + \lambda_1(\gamma \mathcal{L}_{user} + \mathcal{L}_{item}) + \lambda_2 \|\Theta\|^2 \tag{8}$$

where λ_1 and λ_2 are the weight parameter, Θ is the hyperparameters in the model. The process is shown in Algorithm 1.

4 Experiments

In this section, to evaluate the performance of our GCLMBR framework, we conduct different experiments compared with other existing recommendation models on different datasets, and aim to answer the following research questions: (Q1) How well does our model GCLMBR compare to other baselines? (Q2) Are the components in our model useful for promoting recommendation? (Q3) How does our model perform with different hyper-parameters setup?

4.1 Experimental Setup

Datasets. We use three real-world datasets for our experiments: Beibei, Taobao and Yelp. Beibei dataset is a website for maternal and child products, where behaviors such as Page View, Cart, Purchase. Taobao dataset corresponds to the Taobao e-commerce platform, where behaviors such as Page View, Favorite, Cart, Purchase. Yelp dataset is a famous business review data in the United States, which contains information such as users' ratings of restaurants and enterprises. The statistical details of the datasets are summarized in Table 1:

Table 1. Statistics of the datasets.

Datasets	#user	#item	#interaction	types of interaction behaviors
Beibei	21716	7977	3338068	Page View, Cart, Purchase
Taobao	147894	99037	7658926	Page View, Favorite, Cart, Purchase
Yelp	19800	22734	1400036	Tip, Dislike, Neutral, Like

Evaluation Metrics. We use two metrics widely used for top-k recommendations, Hit Rate (HR) and Normalized Discounted Cumulative Gain (NDCG) to evaluate the model. To ensure the accuracy of the results, we divide the training set and the test set by the method of leave-one-out. At the same time, for each positive item, we sample 99 negative items (items that have no interaction with users) by random sampling.

Baselines. To evaluate the performance of our method, we use some classic and advanced single-behavior and multi-behavior recommendation algorithms as the baselines. The following:

- DMF [25]: DMF combines matrix factorization with neural network, and obtains users' rating prediction of items through the explicit feedback information of the interaction between users and items.
- NCF [10]: NCF expresses and expands the matrix factorization, and adopts multilayer perceptron to provide the model with nonlinear performance.
- AutoRec [18]: AutoRec combines the thought of autoencoder and collaborative filtering, and uses encoder and decoder to reconstruct the feature representation of users and items.
- R-GCN [15]: R-GCN is the earliest graph convolution network to deal with heterogeneous relationships, which considers the different influences of different relationships on nodes in the graph structure.
- GHCF [3]: GHCF is a state-of-the-art multi-relationship prediction algorithm based on non-sampling, which joint embeds nodes(users and items) and the representation of the relationship between nodes.
- GNMR [22]: It is a neighborhood-based multi-relationship recommendation algorithm that integrates the user's personalized preferences and the interactions between multiple relationships.
- MBGCN [12]: MBGCN is a state-of-the-art multi-relational GNNs model, which considers various relationships between users and items.
- MB-GMN [23]: MB-GMN is a state-of-the-art method which combines meta-learning with GNNS. By extraction of meta-knowledge of users and items, and learning the matching algorithm on the graph to make prediction.
- MORO [11]: MORO is a state-of-the-art recommendation algorithm based on multi-behaviors, which constructs multi-behavior graphs, embeds learning nodes and makes multi-behavior recommendations.

Parameter Settings. For all baselines, we use the code provided in the corresponding article and the optimal parameters setting. The model is implemented by pytorch, Adam is used as the optimizer, and the learning rate is initialized to 0.01. The model's training batch is set to 256, the depth of the graph convolutional network is select from $\{1, 2, 3, 4, 5\}$. The message loss rate of the dataset is set to be 0.2, which is used to prevent overfitting. The temperature parameter τ is 0.4 for Yelp, 0.5 for Beibei and 0.6 for Taobao. The training embedding dimension d is select from $\{32, 64, 128, 256\}$.

4.2 Performance Comparison(Q1)

The results of our model, GCLMBR, and other baselines are presented in Table 2. It can be observed that GCLMBR outperforms all baselines in terms of both HR@10 and NDCG@10 across all three datasets. For instance, when compared with MORO and MB-GMN on the Beibei dataset, GCLMBR exhibits improvements of 9.3% and 19.3% in HR@10, and 15.1% and 27.8% in NDCG@10, respectively. Similarly, on the Taobao dataset, GCLMBR achieves enhancements of 5.3% and 32.8% in HR@10, and 8.4% and 45.2% in NDCG@10 compared to MORO and MB-GMN, respectively. Moreover, we can observe that recommendation methods incorporating multi-behavior consistently outperform those considering only a single behavior (e.g., DMF, NCF, AutoRec). This finding indicates that integrating multi-behavior can significantly enhance recommendation outcomes, further underscoring the effectiveness of multi-behavior feature learning within our framework.

Table 2. Performance comparison of GCLMBR and other methods.

Datasets	Yelp		Beibei		Taobao	
Model	HR@10	NDCG@10	HR@10	NDCG@10	HR@10	NDCG@10
DMF [25]	0.756	0.485	0.597	0.336	0.305	0.189
NCF [10]	0.714	0.429	0.595	0.332	0.319	0.191
AutoRec [18]	0.765	0.472	0.607	0.341	0.313	0.190
R-GCN [15]	0.826	0.520	0.605	0.344	0.338	0.191
GHCF [3]	0.791	0.485	0.693	0.411	0.377	0.218
GNMR [22]	0.848	0.559	0.604	0.367	0.595	0.332
MBGCN [12]	0.779	0.465	0.642	0.376	0.369	0.222
MB-GMN [23]	0.852	0.567	0.691	0.410	0.491	0.301
MORO [11]	0.877	0.583	0.7544	0.455	0.619	0.403
GCLMBR	**0.886**	**0.596**	**0.8245**	**0.524**	**0.652**	**0.437**

Furthermore, we conduct experiments with varying values of N to assess the effectiveness of top-N recommendations. The experimental findings, as presented in Table 3 using the Beibei dataset, demonstrate that GCLMBR consistently outperforms other baselines across different N values. This indicates GCLMBR's superior ability in assigning higher scores to items of interest in the top-N list, attributed to its utilization of multi-behavior information and reduced information loss in representation aggregation through contrast enhancement.

Table 3. Comparison results on Beibei dataset with varying N value in terms of Hit@N and NDCG@N.

Model	@1		@3		@5		@7	
	Hit	NDCG	Hit	NDCG	Hit	NDCG	Hit	NDCG
NCF	0.123	0.123	0.317	0.232	0.447	0.283	0.53	0.315
AutoRec	0.128	0.128	0.321	0.236	0.456	0.291	0.54	0.322
R-GCN	0.134	0.134	0.323	0.242	0.453	0.295	0.535	0.323
GHCF	0.179	0.179	0.39	0.3	0.525	0.356	0.611	0.385
GNMR	0.168	0.168	0.336	0.265	0.436	0.307	0.504	0.328
MBGCN	0.167	0.167	0.374	0.284	0.498	0.337	0.541	0.322
MB-GMN	0.183	0.183	0.411	0.306	0.527	0.359	0.608	0.389
MORO	0.201	0.201	0.451	0.344	0.591	0.402	0.676	0.431
GCLMBR	**0.257**	**0.257**	**0.535**	**0.418**	**0.678**	**0.477**	**0.759**	**0.504**

4.3 Ablation Study(Q2)

In this subsection, we aim to validate the effectiveness of the components proposed in our work. We considered three variants of GCLMBR and design the following ablation experiments. A bar chart is employed to represent the results, as illustrated Fig. 2:

GCLMBR-MB. We denote 'MB' as representing the Multi-behavior encoder component. GCLMBR-MB indicates that this variant ignores global multi-behavior information and directly employs a single behavior for recommendation. Our observation reveals that GCLMBR outperforms GCLMBR-MB, suggesting the effectiveness of our global multi-behavior feature.

GCLMBR-CL. In this variant, we exclude the graph comparison learning and directly utilize the learned multi-behavior features and single-behavior features for recommendations. Compared with the GCLMBR model, the omission of graph contrast learning results in a decrease of 10.5% in HR@10 and 18.8% in NDCG@10 on the Beibei dataset. This indicates that employing graph comparison learning enhances the consistency of feature representation, thereby improving the accuracy of recommendations.

GCLMBR-MTR. Instead of utilizing multiple behaviors for the multi-task training prediction component, we utilize a single behavioral feature representation for recommendation prediction. On the Yelp dataset, this change results

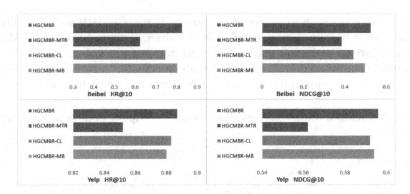

Fig. 2. Ablation study of GCLMBR framework on Beibei and Yelp data in terms of HR@10 and NDCG@10.

in a decrease of 4.1% and 6% in HR@10 and NDCG@10 respectively, while on the Taobao dataset, the decrease is more significant, with a reduction of 72.9% in HR@10 and 97.7% in NDCG@10. This highlights the importance of incorporating multiple behaviors for recommendation, which aids in learning the user's latent preferences and enhances recommendation performance.

Therefore, the removal of any component will lead to a significant decrease in performance, further confirming the positive impact of our components on enhancing the recommendation results.

4.4 Multi-behavior Recommendation Effect (Q3)

To investigate the impacts of hyperparameters on the recommendation performance in GCLMBR, we conduct a series of experiments. During each experiment, we focus solely on exploring the influence of a single hyperparameter, while maintaining the remaining hyperparameters at their default settings.

The Effect of Temperature Parameter τ. As mentioned previously, the temperature parameter τ plays a crucial role in regulating the consistency between the user's fused representation and the multi-behavior representation. The temperature parameter τ is searched from the set $\{0.2, 0.3, 0.4, 0.5, 0.6\}$, we compare the performance of GCLMBR and display the results in the left side of Fig. 3. Initially, as the τ increase, the model's effectiveness improves. However, as the τ continues to rise, the effectiveness of the model starts to gradually decline. It may be that too large τ results in overly smooth model outputs, which reduces its discriminative power across all samples. Therefore, the model may struggle to effectively differentiate between positive and negative samples, thereby causing a reduction in performance. Consequently, our analysis reveals that the optimal temperature parameter for the Yelp dataset is 0.4, for the Beibei dataset is 0.5, and for the Taobao dataset is 0.6.

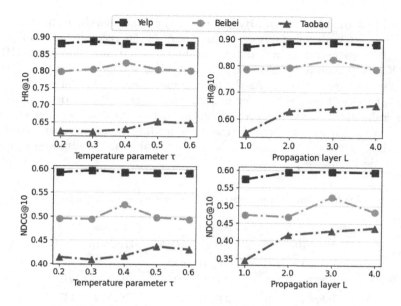

Fig. 3. Hyperparameter τ and L analysis of GCLMBR.

The Impact of Propagation Layers L. To investigate the impact of varying the propagation layers L on the recommendation performance of GCLMBR, we conduct a series of experiments with the propagation layers L of the graph neural network within the range of $\{1, 2, 3, 4, 5\}$. As clearly depicted on the right side of Fig. 3, our results indicate that for the Yelp and Beibei datasets, both HR@10 and NDCG@10 achieve optimal performance when the convolutional depth of the graph neural network is set to 3. Further increments in the convolutional depth lead to a decrease in performance, possibly due to the smoothing effect on feature embeddings. Conversely, for the Taobao dataset, optimal performance is observed when the convolution depth is set to 4.

Table 4. Impact of margin m of GCLMBR.

Datasets	Yelp		Beibei		Taobao	
m	HR@10	NDCG@10	HR@10	NDCG@10	HR@10	NDCG@10
0.5	0.8733	0.589	0.7946	0.493	0.6177	0.407
0.6	0.8817	0.591	0.797	0.495	0.6211	0.411
0.7	0.8869	0.596	0.8031	0.501	0.6188	0.412
0.8	0.8791	0.589	0.8002	0.4988	0.6522	0.437
0.9	0.8783	0.588	0.7952	0.493	0.6233	0.419

The Impact of Margin m**.** By adjusting the final margin form 0.5 to 0.9, we conduct an in-depth analysis of their impact on recommendation performance. As demonstrated in Table 4, our findings indicate that both HR@10 and NDCG@10 achieve optimal performance when the margin for the Yelp and Beibei datasets are set to 0.7. As the margin increases, the model's performance initially improves because a larger margin can enhance the model's robustness and generalization ability. However, when the margin exceeds 0.7, the model may become overly sensitive to noise in the data, leading to unnecessary losses during training, thereby impacting the model's generalization ability and causing a decline in performance.

Table 5. Impact of embedding dimensions d of GCLMBR.

Datasets	Yelp		Beibei		Taobao	
d	HR@10	NDCG@10	HR@10	NDCG@10	HR@10	NDCG@10
32	0.8583	0.55	0.6957	0.409	0.5798	0.366
64	0.877	0.586	0.7528	0.457	0.6122	0.399
128	0.8869	0.596	0.8031	0.501	0.6522	0.437
256	0.8761	0.587	0.8245	0.524	0.6251	0.416

The Effect of Embedding Dimensions d**.** We conduct experiments by varying the embedding dimensions d of the model within the set of $\{32, 64, 128, 256\}$. The experimental results, summarized in Table 5, reveal that optimal performance is observed for the Yelp and Taobao datasets when the embedding dimension is set to 128. Upon further increasing the embedding dimension, a decline in performance is typically observed. We hypothesize that the increased dimensionality of features may introduce unnecessary complexity, thereby compromising the effectiveness of the recommendation system.

5 Conclusions

In this paper, we propose a Graph Contrastive Learning for Multi-Behavior Recommendation(GCLMBR) framework. In GCLMBR, we firstly capture single behavior representations from single-behavior subgraphs using UltraGCN. Secondly, we employ enhance GCN with layer attention to encode the global multi-behavior representations. Thirdly, we employ graph contrastive learning to ensure the consistency between a user's behavior feature fusion representations and global multi-behavior representations. Finally, a multi-task learning framework is adopted to make full use of the multi-behavior data to predict different behaviors. Our framework achieves better recommendation performance when evaluated the on three real-world datasets compared to various baselines. Further model ablation studies showed the effectiveness of the components designed in the GCLMBR.

References

1. Aljunid, M.F., Huchaiah, M.D.: An efficient hybrid recommendation model based on collaborative filtering recommender systems. CAAI Transactions on Intelligence Technology **6**(4), 480–492 (2021)
2. Chae, D.K., Shin, J.A., Kim, S.W.: Adaptive autoencoders exploiting content preference for accurate recommendation. In: 2022 IEEE International Conference on Big Data and Smart Computing (BigComp). pp. 292–295. IEEE (2022)
3. Chen, C., Ma, W., Zhang, M., Wang, Z., He, X., Wang, C., Liu, Y., Ma, S.: Graph heterogeneous multi-relational recommendation. In: Proceedings of the AAAI Conference on Artificial Intelligence. vol. 35, pp. 3958–3966 (2021)
4. Chen, T., Kornblith, S., Norouzi, M., Hinton, G.: A simple framework for contrastive learning of visual representations (2020)
5. Gao, C., He, X., Gan, D., Chen, X., Feng, F., Li, Y., Chua, T.S., Jin, D.: Neural multi-task recommendation from multi-behavior data. In: 2019 IEEE 35th International Conference on Data Engineering (ICDE). pp. 1554–1557 (2019)
6. Guo, H., Tang, R., Ye, Y., Li, Z., He, X.: Deepfm: a factorization-machine based neural network for ctr prediction (2017)
7. Guo, Y., Wang, M., Li, X.: Application of an improved apriori algorithm in a mobile e-commerce recommendation system. Ind. Manag. Data Syst. **117**(2), 287–303 (2017)
8. He, K., Fan, H., Wu, Y., Xie, S., Girshick, R.: Momentum contrast for unsupervised visual representation learning (2020)
9. He, X., Deng, K., Wang, X., Li, Y., Zhang, Y., Wang, M.: Lightgcn: Simplifying and powering graph convolution network for recommendation. In: Proceedings of the 43rd International ACM SIGIR conference on research and development in Information Retrieval. pp. 639–648 (2020)
10. He, X., Liao, L., Zhang, H., Nie, L., Hu, X., Chua, T.S.: Neural collaborative filtering (2017)
11. Jiang, W., Duan, L., Ding, X., Chen, X.: Moro: A multi-behavior graph contrast network for recommendation. In: Asia-Pacific Web (APWeb) and Web-Age Information Management (WAIM) Joint International Conference on Web and Big Data. pp. 117–131. Springer (2022)
12. Jin, B., Gao, C., He, X., Jin, D., Li, Y.: Multi-behavior recommendation with graph convolutional networks. In: Proceedings of the 43rd International ACM SIGIR Conference on Research and Development in Information Retrieval. pp. 659–668 (2020)
13. Mao, K., Zhu, J., Xiao, X., Lu, B., Wang, Z., He, X.: Ultragcn: ultra simplification of graph convolutional networks for recommendation. In: Proceedings of the 30th ACM International Conference on Information & Knowledge Management. pp. 1253–1262 (2021)
14. Oord, A.v.d., Li, Y., Vinyals, O.: Representation learning with contrastive predictive coding. arXiv preprint arXiv:1807.03748 (2018)
15. Schlichtkrull, M., Kipf, T.N., Bloem, P., van den Berg, R., Titov, I., Welling, M.: Modeling Relational Data with Graph Convolutional Networks. In: Gangemi, A., Navigli, R., Vidal, M.-E., Hitzler, P., Troncy, R., Hollink, L., Tordai, A., Alam, M. (eds.) ESWC 2018. LNCS, vol. 10843, pp. 593–607. Springer, Cham (2018). https://doi.org/10.1007/978-3-319-93417-4_38
16. Song, J., Chang, C., Sun, F., Song, X., Jiang, P.: Ngat4rec: Neighbor-aware graph attention network for recommendation. arXiv preprint arXiv:2010.12256 (2020)

17. Veličković, P., Cucurull, G., Casanova, A., Romero, A., Lio, P., Bengio, Y.: Graph attention networks. arXiv preprint arXiv:1710.10903 (2017)
18. Wang, T.H., Hu, X., Jin, H., Song, Q., Han, X., Liu, Z.: Autorec: An automated recommender system. In: Proceedings of the 14th ACM Conference on Recommender Systems. pp. 582–584 (2020)
19. Wang, X., He, X., Wang, M., Feng, F., Chua, T.S.: Neural graph collaborative filtering. In: Proceedings of the 42nd international ACM SIGIR conference on Research and development in Information Retrieval. pp. 165–174 (2019)
20. Wu, Y., Xie, R., Zhu, Y., Ao, X., Chen, X., Zhang, X., Zhuang, F., Lin, L., He, Q.: Multi-view multi-behavior contrastive learning in recommendation. In: International Conference on Database Systems for Advanced Applications. pp. 166–182. Springer (2022)
21. Xia, L., Huang, C., Xu, Y., Dai, P., Bo, L.: Multi-behavior graph neural networks for recommender system. IEEE Transactions on Neural Networks and Learning Systems pp. 1–15 (2022). https://doi.org/10.1109/tnnls.2022.3204775, https://doi.org/10.1109%2Ftnnls.2022.3204775
22. Xia, L., Huang, C., Xu, Y., Dai, P., Lu, M., Bo, L.: Multi-behavior enhanced recommendation with cross-interaction collaborative relation modeling. In: 2021 IEEE 37th International Conference on Data Engineering (ICDE). pp. 1931–1936. IEEE (2021)
23. Xia, L., Xu, Y., Huang, C., Dai, P., Bo, L.: Graph meta network for multi-behavior recommendation. In: Proceedings of the 44th international ACM SIGIR conference on research and development in information retrieval. pp. 757–766 (2021)
24. Xuan, H., Liu, Y., Li, B., Yin, H.: Knowledge enhancement for contrastive multi-behavior recommendation. In: Proceedings of the Sixteenth ACM International Conference on Web Search and Data Mining. pp. 195–203 (2023)
25. Xue, H.J., Dai, X., Zhang, J., Huang, S., Chen, J.: Deep matrix factorization models for recommender systems. In: IJCAI. vol. 17, pp. 3203–3209. Melbourne, Australia (2017)
26. Yao, L., Mao, C., Luo, Y.: Graph convolutional networks for text classification. In: Proceedings of the AAAI conference on artificial intelligence. vol. 33, pp. 7370–7377 (2019)
27. Ying, R., He, R., Chen, K., Eksombatchai, P., Hamilton, W.L., Leskovec, J.: Graph convolutional neural networks for web-scale recommender systems. In: Proceedings of the 24th ACM SIGKDD international conference on knowledge discovery & data mining. pp. 974–983 (2018)
28. Zhang, J., Shi, X., Zhao, S., King, I.: Star-gcn: Stacked and reconstructed graph convolutional networks for recommender systems. arXiv preprint arXiv:1905.13129 (2019)
29. Zhang, S., Yao, L., Sun, A., Tay, Y.: Deep learning based recommender system: A survey and new perspectives. ACM computing surveys (CSUR) **52**(1), 1–38 (2019)
30. Zhang, Y., Zhao, C., Yuan, M., Chen, M., Liu, X.: Unifying attentive sparse autoencoder with neural collaborative filtering for recommendation. Intelligent Data Analysis **26**(4), 841–857 (2022)

UID-Net: Enhancing Click-Through Rate Prediction in Trigger-Induced Recommendation Through User Interest Decomposition

Jiazhen Lou[1], Zhao Li[2(✉)], Hong Wen[1], Jingsong Lv[2], Jing Zhang[3], Fuyu Lv[1], Zulong Chen[1], and Jia Wu[4]

[1] Alibaba Group, Hangzhou, China
{jane.ljz,fuyu.lfy,zulong.czl}@alibaba-inc.com
[2] Research Center for Data Hub and Security, Zhejiang Lab, Hangzhou, China
lzjoey@gmail.com
[3] The University of Sydney, Sydney, Australia
jing.zhang1@sydney.edu.au
[4] Macquarie University, Sydney, NSW, Australia
jia.wu@mq.edu.au

Abstract. This paper tackles the CTR prediction challenge within Trigger-Induced Recommendation (TIR) contexts. In TIR, users' current interests are exposed to some extent through clicked trigger items, which then lead to the recommendation of preference items. Discovering decomposed trigger-relevant and trigger-irrelevant user interests and adaptively fusing them for final CTR prediction remains a challenging and under-explored area in existing TIR-based CTR methods. To address this challenge, we introduce the User Interest Decomposition Network (UID-Net), an innovative model consisting of three key modules: Intent Extraction Module (IEM), Interest Decomposition Module (IDM), and Interest Fusing Module (IFM). IEM initially decomposes each behavior's embedding vector into trigger-relevant and trigger-irrelevant component embeddings, forming two component sequences used to extract corresponding interests. IDM employs self-supervised contrastive learning to further distinguish these interests, resulting in more discriminative representations. Finally, IFM predicts a fusion weight that adaptively combines trigger-relevant and trigger-irrelevant interests for accurate CTR prediction. Extensive offline-to-online experiments showcase the superiority of UID-Net to the state-of-the-art (SOTA) models.

Keywords: Trigger-Induced Recommendation · CTR Prediction · Recommender System · User Interest Decomposition · Contrastive Learning

---

J. Lou and Z. Li—These authors contributed equally to this work.

1 Introduction

Trigger-Induced Recommendation (TIR), a method that recommends follow-up items according to the clicked trigger item just now, has become a staple feature on numerous prominent e-commerce platforms. This approach significantly enhances the shopping experience for users online, thereby substantially increasing Item Page Views (IPV), as documented in [12].

Consider Fliggy[1], one of China's leading online travel platforms (OTPs), as an illustrative example [15]. The platform's homepage is initially populated with various function-specific portals, each featuring a personalized recommendation, as illustrated in Fig. 1(a). The "Daily Promotion" portal may offer items at discounted prices. Upon a user's selection of a portal, they are directed to a new page known as the Item Feeds Flow Page (IFFP). Here, the item that was just clicked, referred to as the trigger item, is prominently displayed first, followed by other related recommendations, as shown in Fig. 1(b). Obviously, the IFFP scenario is a kind of typical TIR scenarios. Similar TIR applications are evident in other e-commerce applications, such as Taobao's "Youhaohuo" module, eBay's "More Similar Items" section, and Shoppee's "Flash Deals" feature.

To enhance user experience within TIR contexts, one of the most important challenge is how to improve the Click-Through Rate (CTR), specifically targeting the estimation of the probability of users engaging with recommended items.

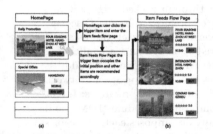
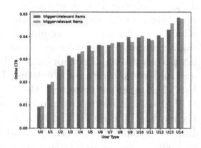

Fig. 1. An illustration of the online TIR scenarios on Fliggy app.

Fig. 2. The online CTR performance of different groups of users on trigger-relevant and trigger-irrelevant items.

In the following, we examine the discrepancies between traditional CTR modeling methods and TIR-based methods. Traditional CTR models like DIN [24] or DIEN [23] primarily focus on capturing users' dynamic interests across various item categories. However, TIR presents unique challenges:

Firstly, TIR scenarios expose users' current interests through trigger items, necessitating the incorporation of both historical user behaviors and explicit exploitation of trigger items to extract relevant interests.

Secondly, TIR aims not only to recommend trigger-relevant items but also trigger-irrelevant items, to explore users' potential varied interests. Figure 2

---
[1] https://www.fliggy.com/.

demonstrates empirical evidence from real-world recommender system, showing that for some user groups, such as *U5* or *U9*, the CTR performance on trigger-irrelevant items is superior, indicating a desire for diverse recommendations.

It is crucial for CTR prediction in TIR scenarios to accurately extract users' trigger-relevant and trigger-irrelevant interests and exploit trigger items for adaptive recommendations. To date, no previous studies have addressed this problem explicitly. Traditional CTR methods, such as PNN [8], DeepFM [1], DCN [17], DIN [24], DIEN [23], are ineffective in TIR settings due to their lack of awareness of users' current interests induced by trigger items. Recently, trigger-based methods like DIHN [12] have been developed for TIR scenarios, demonstrating effectiveness through trigger exploitation. However, these methods often overemphasize trigger-relevant interests, overlooking potentially interesting trigger-irrelevant items and leading to unsuitable recommendations.

To tackle this issue, we propose a novel model dubbed User Interest Decomposition Network (UID-Net) for CTR prediction within TIR contexts, which comprises three key modules namely the Intent Extraction Module (IEM), Interest Decomposition Module (IDM), and Interest Fusing Module (IFM).

Concretely, Our primary contributions are outlined below:

- We introduce UID-Net, a novel CTR model tailored for TIR scenarios. This model is designed to capture more nuanced expressions of user interests, thereby enhancing the accuracy of CTR predictions. The effectiveness of UID-Net is attributed to the integration of three meticulously crafted modules.
- We leverage the idea of self-supervise contrastive learning to effectively learn discriminative trigger-relevant and trigger-irrelevant interest representations from their behaviors and adaptively fuse the two kinds of interests concerning corresponding trigger and target items.
- We have performed extensive experiments on public and industrial datasets. The experimental results confirm that the proposed UID-Net can consistently achieve superior performance over other representative methods. It is notable that UID-Net has been deployed in the *Daily Promotion* module of the Fliggy app and delivers significant improvement, *i.e.*, contributing up to 8.96% gains on CTR, which verifies its value in real-world applications.

The structure of this paper is as follows: Sect. 2 reviews related literature. Section 3 details our proposed methodology, succeeded by comprehensive experimental results and their analysis in Sect. 4. We conclude the paper in Sect. 5.

2 Related Work

In this section, related studies are reviewed in the subsequent two dimensions.

Sequential Recommendation: Sequential Recommendation [18] targets to recommend next items with the highest users clicking probability by leveraging their historical item sequences. Early works on sequential recommendation implicitly model users' state dynamics and derive the next coming behaviors based on Markov chain [10]. Recently, as the development of deep learning methodologies in various research fields [2], RNN, CNN, Capsule, and Transformer are deployed in recommender systems to discover users' multiple interests from users' behavior sequences [3,13,14]. However, the existing methods are ineffective in TIR scenarios, due to the ignorance of explicitly modeling users' current interests exposed by trigger items.

Trigger-Induced Recommendation: Different from the traditional recommendation, the TIR has been rarely investigated. Recently, *relevant recommendation* [4,21] is a more related topic. For example, R3S [21] proposed a hybrid strategy to jointly fuse the information of the triplet of user, trigger, and item. However, it ignores the impact of users' historical behaviors, as well as the difference between trigger-relevant and trigger-irrelevant interests. The most relevant work to ours is the DIHN [11], which is exactly designed for TIR scenarios. It firstly derives users' preferences on trigger items, followed by utilizing it to fuse the embeddings of the trigger and target items and further reveal users' interests at the moment. Despite being effective, DIHN implicitly assumes that the user has an explicit preference induced by the trigger item and thus over-emphasizes the role of trigger-relevant interests, resulting in the potentially interested trigger-irrelevant items being overlooked, which, by contrast, can be exactly captured by our UID-Net.

3 Methodology

To capture both the trigger-relevant and trigger-irrelevant interests for recommendation in TIR scenarios, a novel model dubbed UID-Net for CTR prediction is proposed and elaborated in this section, with the overall architecture depicted in Fig. 3. It comprises the following four key components. 1) Feature Representation Layer (FRL): This layer performs dimension reduction by transforming high-dimensional sparse embeddings into low-dimensional dense embeddings with fixed-length; 2) Interest Extraction Module (IEM): This module extracts trigger-relevant and trigger-irrelevant interests concerning corresponding trigger and target items; 3) Interest Decomposition Module (IDM): This module ensures that the extracted interests are closer to their corresponding proxies than the opposing ones; 4) Interest Fusing Module (IFM): This module adaptively fuses the trigger-relevant and trigger-irrelevant interests.

3.1 Feature Representation Layer

Five feature groups are considered, namely *user profile* u_p, *user sequential behaviors* u_b, *trigger item* t, *target item* i and *context features* c, where $u_b = \{i_1, i_2, ...i_{|T|}\}$, $|T|$ is the length of the behavior sequence and i_t is the t-th interacted item. In industrial recommender systems, each feature group is typically multi-field, with fine-grained features represented by high-dimensional sparse one-hot encodings. For instance, in the *sex* field of the group u_p, the *male* might be encoded as $[1, 0]^T$. Meanwhile, assuming that u_p, u_b, t, i, c are concatenations of these one-hot features from their respective fields, they need to be transformed

UID-Net: Enhancing CTR Prediction Through User Interest Decomposition 53

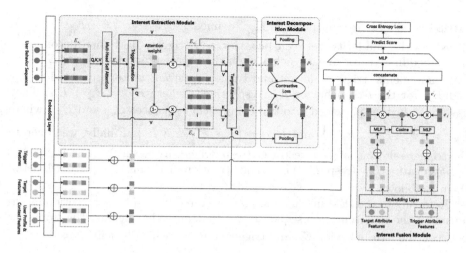

Fig. 3. The overview architecture of UID-Net.

into low-dimensional dense embeddings: E_{u_p}, E_{u_b}, E_t, E_i, and E_c, respectively. Specifically, $E_{u_b} = \{E_{i_1}, E_{i_2}, ...E_{i_{|T|}}\}$, where E_{i_t} represents the embedding vector of the t-th interacted item. Now, Our objective is to forecast the likelihood of user u clicking on item i, contingent upon E_{u_p}, E_{u_b}, E_t, E_i, and E_c. This is formulted as $\hat{y}_{u,i} = \mathcal{F}(E_{u_p}, E_{u_b}, E_t, E_i, E_c; \theta)$, where $\mathcal{F}$ is the learning function that is always implemented as a neural network with learnable parameters θ.

3.2 Interest Extraction Module

Generally, users could exhibit multiple kinds of interests from their historical behaviors, *i.e.*, users can simultaneously be interested in items from varied categories such as *clothing*, *electronics*, and *food*. However, in the context of TIR scenarios, the trigger items have explicitly exposed users' current interests to some extent. In other words, irrelevant behaviors with the trigger items could make limited contributions to the prediction of users clicking target items. One naive solution is to follow the hard-search paradigm proposed in SIM [7], which involves selecting behaviors from the trigger item's category, forming a sub-behavior sequence, and applying an attention mechanism to model the dynamic interests of users in response to various target items. We contend that not only behaviors from the trigger item's category but also those from different categories are valuable, as they can mutually collaborate with each other to reveal users' accurate interests. However, it is challenging to evaluate the effects of these two behavioral types on interest extraction while considering both the trigger and target items. This challenge motivates us to propose the Interest Extraction Module (IEM).

To captures dependencies between representation pairs within a sequence, IEM first processes the original sequence E_{u_b} using Multi-Head Self-Attention

(MHSA) [16], further outputting $E_{u'_b}$, defined as $E_{u'_b} = \left\{ E_{i'_1}, E_{i'_2}, ... E_{i'_{|T|}} \right\}$. Based on the similarity between behavior and trigger, IEM then decomposes the embedding vector of each behavior in the representation sequence $E_{u'_b}$ into two components: trigger-relevant and trigger-irrelevant vectors, denoted as $E_{i_t^r}$ and $E_{i_t^f}$ for the t-th behavior $E_{i'_t}$, respectively, where $E_{i'_t} = E_{i_t^r} + E_{i_t^f}$. They have been further packaged into two component sequences $E_{u_b^r}$ and $E_{u_b^f}$, where $E_{u_b^r} = \left\{ E_{i_1^r}, E_{i_2^r}, ... E_{i_{|T|}^r} \right\}$ and $E_{u_b^f} = \left\{ E_{i_1^f}, E_{i_2^f}, ... E_{i_{|T|}^f} \right\}$. Finally, we generate the trigger-relevant interest vector e_r (resp. trigger-irrelevant interest vector e_f) by aggregating $E_{u_b^r}$ (resp. $E_{u_b^f}$) based on attention mechanism [16]. We will detail them as follows.

Referring to the IEM module in Fig. 3, we directly adopt the attention mechanism on the representation sequence $E_{u'_b}$ to adaptively calculate the similarity between the t-th behavior $E_{i'_t}$ and trigger item E_t, which is defined as:

$$w_t = \text{a}(E_{i'_t} || E_t || E_{i'_t} - E_t), \tag{1}$$

where w_t is the attention weight, $||$ is the concatenation operation, a(.) is a three-layer feed-forward network. The output w_t, which ranges from 0 to 1, is obtained by using a *Sigmoid* activation function [6] after the last layer. Now, given w_t and $E_{i'_t}$, the trigger-relevant component vector $E_{i_t^r}$ can be calculated via an element-wise multiplication of w_t and $E_{i'_t}$, defined as: $E_{i_t^r} = w_t E_{i'_t}$. Accordingly, the trigger-irrelevant component vector $E_{i_t^f}$ is as follows: $E_{i_t^f} = (1 - w_t) E_{i'_t}$.

In this way, the two component sequences $E_{u_b^r}$ and $E_{u_b^f}$ can be naturally packaged. Given $E_{u_b^r}$ and $E_{u_b^f}$, we then apply the attention mechanism [16] for target items to adaptively generate users' trigger-relevant and trigger-irrelevant interest representations. Taking the aggregated process of sequence $E_{u_b^r}$ as an example, the attention score (*i.e.*, α_j^r) of the j-th behavior (*i.e.*, $E_{i_j^r}$) with respect to target item (*i.e.*, E_i) can be obtained as follows:

$$\alpha_j^{r'} = f(E_{i_j^r} || E_i || E_{i_j^r} - E_i), \tag{2}$$

$$\alpha_j^r = \frac{exp(\alpha_j^{r'})}{\sum_{j=1}^{|T|} exp(\alpha_j^{r'})}, \tag{3}$$

where $f(.)$ is a three-layer feed-forward network with *Relu* [6] as the activation function, and $||$ is the concatenation operation. Therefore, the final trigger-relevant interest representation e_r can be obtained in a weighted-sum aggregation manner, which is mathematically defined as: $e_r = \sum_{j=1}^{|T|} \alpha_j^r E_j^r$. Following the same way, the final trigger-irrelevant interest representation e_f can be obtained as: $e_f = \sum_{j=1}^{|T|} \alpha_j^f E_j^f$, where α_j^f is the weight of the j-th behavior (*i.e.*, E_j^f) in component sequence $E_{u_b^f}$ in line with the target item (*i.e.*, E_i).

Intuitively, the trigger-relevant and trigger-irrelevant interests, namely e_r and e_f can explain the recommended target items. e_r indicates users are interested not only in the trigger item but also in the target item, suggesting that users are initially attracted by the trigger item and are eager to view more trigger-relevant items. While e_f represents that users are probably not so interested in the trigger item and enter the feed page casually, where more trigger-irrelevant target items are preferred and required to be displayed to them.

3.3 Interest Decomposition Module

In this section, referred to the self-supervised contrastive learning [5], we propose an Interest Decomposition Module (IDM) to supervise the learning of both interest representations e_r and e_f. This module ensures that each interest representation is more similar to its corresponding proxy than to the opposite proxy in a contrastive manner. The *proxy*, extracted from the corresponding sequence, serve as the pseudo label for interest representation. We will detail it as follows.

First, we use the notation p_r (resp. p_f) to represent the *proxy* of component sequence $E_{u_b^r}$ (resp. $E_{u_b^f}$), defined as the mean pooling representation of the corresponding sequence itself. Mathematically, they are calculated as follows:

$$p_r = mean(E_{u_b}^r) = \frac{1}{|T|} \sum_{j=1}^{|T|} E_j^r, \tag{4}$$

$$p_f = mean(E_{u_b}^f) = \frac{1}{|T|} \sum_{j=1}^{|T|} E_j^f, \tag{5}$$

where $mean(.)$ denotes the mean pooling operation. Actually, according to our experiment results, the mean pooling operation is proven to be very effective.

Next, given the proxies p_r and p_f, they are further used to supervise the disentanglement of trigger-relevant interest e_r and trigger-irrelevant interest e_f. Since these two types of interests may overlap to some extent, our goal is to align the learned representations of each interest closely with their respective proxies, distinguishing them from opposing proxies. This approach avoids strictly forcing the interest representations to be dissimilar to the opposite ones. Referring to Fig. 3, we formally define two groups of contrastive tasks:

$$similarity(e_r, p_r) > similarity(e_r, p_f), \tag{6}$$

$$similarity(e_f, p_f) > similarity(e_f, p_r), \tag{7}$$

where $similarity(.,.)$ denotes the inner product of both embedding representations. Equation 6 (resp. Eq. 7) supervises the learning of trigger-irrelevant interest e_r (resp. trigger-irrelevant interest e_f). For example, meeting the requirement in Eq. 6 implies encouraging the obtained e_r to be more similar to its proxy p_r than the opposite proxy p_f, while meeting the requirement in Eq. 7 implies encouraging the obtained e_f to be more similar with its proxy p_f than the opposite proxy p_r. In a nutshell, the self-supervision of both interests can provide more reliable disentangled representations. Finally, the loss formulated by Bayesian Personalized Ranking (BPR) [9] is adopted to supervise the learning of Eq. 6 and 7:

$$Loss_{con} = g(e_r, p_r, p_f) + g(e_f, p_f, p_r), \tag{8}$$

where g is the BPR loss, defined as:

$$g(e_r, p_r, p_f) = softplus(dot(e_r, p_f) - dot(e_r, p_r)), \tag{9}$$

where *softplus(.)* [6] represents a typical activation function and *dot(.,.)* is the operation of inner product.

3.4 Interest Fusion Module

Given the two interest representations e_r and e_f, we figure out that both representations contribute to the prediction of target items. However, how to evaluate the importance of both representations with respect to different target items is not trivial and challenging. One intuitive solution is to concatenate them to form a new representation vector and feed it to follow-up modules for final CTR prediction. However, it ignores an intuitive fact that the more similar trigger and target items are, the more important role the trigger-relevant interest representation will act, and vice versa. It motivates us to propose an Interest Fusion Module (IFM) to dynamically fuse the trigger-relevant and trigger-irrelevant interests with respect to different target items.

Specifically, referring to Fig. 3, we firstly elaborately select several attribute features for both the trigger and target items, *e.g.*, *item id, category id, historically accumulated ctr, etc*. Without loss of generality, assuming the concatenation results of these attribute features of trigger item (resp. target item) are denoted as $E_t^{'}$ (resp. $E_i^{'}$), they are fed into parameters-specific multi-layer perceptron network to obtain the representation vector $V_t^{'}$ (resp. $V_i^{'}$), defined as $V_t^{'} = MLP(E_t^{'})$ and $V_i^{'} = MLP(E_i^{'})$, where $MLP(.)$ consists of a three-layer neural network architecture that employs the Rectified Linear Unit (ReLU) activation function, as referenced in [6]. Besides, $V_t^{'}$ and $V_i^{'}$ have the same dimension d. We will investigate its impact on the final performance in Sect. 4.4. Furthermore, the similarity score (*i.e.*, w) between trigger and target items can be defined as: $w = cosine(V_t^{'}, V_i^{'})$, where $cosine(.,.)$ represents the cosine similarity of two embedding vectors. Now, given w, the trigger-relevant and trigger-irrelevant interests can be further adaptively refined:

$$e_r^{'} = we_r, \tag{10}$$

$$e_f^{'} = (1-w)e_f, \tag{11}$$

where $e_r^{'}$ (resp. $e_f^{'}$) represents the refined interest representation of e_r (resp. e_f), *i.e.*, the more similar trigger and target items are, the more important role the trigger-relevant interest representation will act, and vice versa. Then, we use the concatenation operation as the fusing strategy, i.e.,

$$e_{rf} = [e_r^{'} || e_f^{'}], \tag{12}$$

where e_{rf} is the fusion representation. We will investigate the influence of different fusion strategies in Sect. 4.4.

Finally, the dense representation vectors are integrated with the raw context features, expressed as $x = [E_{u_p}||E_t||E_i||E_c||e_{rf}]$. This composite vector is then input into fully connected layers designed to capture higher-order feature interactions. Subsequently, a *Sigmoid* activation function is applied to derive the predicted probability $\hat{y}$ of the target item receiving a click. The objective of this module is encapsulated by the negative log-likelihood loss function, defined as:

$$Loss_{main} = -\frac{1}{|N|} \sum_{(x,y) \in S} (y\log\hat{y} + (1-y)\log(1-\hat{y})), \qquad (13)$$

where S represents the training dataset comprising a total of $|N|$ instances, and $y \in \{0, 1\}$ signifies the ground truth label, indicating user engagement with the target items through clicks or the lack thereof. Finally, the total loss to supervise the training of the proposed UID-Net is defined as:

$$Loss_{tot} = Loss_{main} + \alpha Loss_{con}, \qquad (14)$$

where the hyperparameter α balances the two loss functions. Its effect on the model's performance will be explored in Sect. 4.4.

4 Empirical Evaluations

This section presents comprehensive experiments designed to validate the efficacy of the proposed UID-Net model, while addressing research inquires as below:

- RQ1: How about the performance of UID-Net versus other models?
- RQ2: What is the individual impacts of UID-Net components?
- RQ3: What is the influence of key hyper-parameters on UID-Net?
- RQ4: How does UID-Net stack up against rivals in real-world online recommender systems?

4.1 Experimental Setting

Dataset Description. Evaluations are executed on a real-world industrial dataset and a public dataset.

- **Fliggy-Dataset**: To our best knowledge, no existing public datasets are precisely tailored for the task of TIR prediction, especially considering whether trigger items are explicitly declared. To fill the gap, an offline dataset, called *Fliggy-Dataset*, is collected from users' traffic logs of the *Daily Promotion* module of Fliggy[2] app from 2022-11-15 to 2022-12-20. The offline dataset encompasses 0.28 million users and 0.21 million travel items recorded over a span of 36 consecutive days. It is meticulously partitioned into non-overlapping training and testing subsets; the training subset comprises data from the initial 35 days, while the testing subset includes data from the remaining day. For further specifics, please refer to Table 1.

---

[2] https://www.fliggy.com/.

Table 1. Statistics of the offline datasets.

Dataset	Fliggy	Public
Users	276,799	873,630
Items	210,014	769,228
Total	3,680,963	17,796,987
Training	3,550,618	15,545,566
Testing	130,345	2,251,421

Table 2. Settings of Hyper-parameters.

Hyper-parameters	Choice
Decay rate	0.96
No. of layers in MLP	4
Dim of layers in MLP	[512,256,128,32]
Batch size	2048
Learning rate	0.0005

- **Public-Dataset**: The *Public-Dataset*, sourced from the Alimama Dataset[3], encompasses a collection of 26 million samples, capturing interactions among 1 million users and 0.8 million advertisements over an 8-day period. To address the scarcity of trigger items in TIR scenarios, the dataset is adapted accordingly. Specifically, for each instance, the most recent advertisement clicked by a user within a 4-hour window is designated as the sample's trigger item. Based on the manual rule, the samples without triggers are omitted. Besides, a user's latest 50 behaviors are selected as his (her) corresponding original behavior sequence. Similarly, data within the first seven days are the training data, and the rest are the testing data. More details can be found in Table 1.

Implementation Details. We have implemented all the comparative models using TensorFlow, employing the Adam optimizer alongside an exponential decay learning rate schedule. For instance, the initial learning rate is designated as 0.0005, with a decay rate of 0.96. Comprehensive details regarding the hyper-parameter configurations are delineated in Table 2.

Evaluation Metrics. To thoroughly assess the performance of our proposed model alongside its competitors, we have adopted two well-established metrics prevalent in recommender and advertising systems: Area Under the Curve (AUC) [24] and Relative Improvement (RI) [11]. AUC is indicative of the ranking efficacy of CTR methods, while RI serves as a direct comparative metric, illustrating the percentage improvement of a target model over a baseline model. The RI is mathematically defined as $RI = \left(\frac{AUC(target)-0.5}{AUC(base)-0.5} - 1\right) \times 100\%$.

4.2 Competitors

To substantiate the effectiveness of UID-Net, we have conducted comparative analyses with a selection of SOTA methods, categorized into two distinct groups. Group 1 encompasses the methodologies of **DeepFM** [1], **DIN** [24], **DIEN** [23], and **MIAN** [22]. These models do not explicitly account for the dynamic evolution of users' current interests as influenced by the trigger items. On the contrary,

---
[3] https://tianchi.aliyun.com/dataset/dataDetail?dataId=56.

Group 2 explicitly exploit the trigger items to reveal users' interests, including **DIN++**, **DIEN++**, **R3S** [21], **R3S++** and **DIHN** [12], where several approaches have been adapted from existing CTR models, with enhancements that enable them to effectively leverage the influence of trigger items.

4.3 Offline Experimental Results (RQ1)

The offline evaluation outcomes for select methods on both the Fliggy-Dataset and the Public-Dataset can be found in Table 3. Each method was executed five times, and the final results reflect the mean performance across these trials.

Table 3. Comparison results of all the competitors, where RI represents the relative improvement over DeepFM.

Group	Model	Fliggy-Dataset		Public-Dataset	
		AUC	RI	AUC	RI
Group 1	DeepFM	0.6540	0.00%	0.6191	0.00%
	DIN	0.6594	3.51%	0.6243	4.37%
	DIEN	0.6662	7.92%	0.6260	5.79%
	MIAN	0.6679	9.03%	0.6282	7.64%
Group 2	R3S	0.6646	6.88%	0.6229	3.19%
	DIN++	0.6668	8.31%	0.6274	6.97%
	DIEN++	0.6688	9.58%	0.6282	7.64%
	R3S++	0.6707	10.84%	0.6283	7.72%
	DIHN	0.6723	11.88%	0.6297	8.90%
	UID-Net	**0.7133**	**38.51%**	**0.6314**	**10.33%**

Results in Group 1. Examining the results of all competitors within Group 1, the following observations can be made: 1) The sequential modeling methods, such as DIN, DIEN, and MIAN, can achieve superior performance over non-sequential representative method DeepFM on both datasets, e.g., DIN obtains AUC RI gains of 3.51% and 4.37% over DeepFM on the Fliggy-Dataset and Public-Dataset, respectively, which demonstrates that discovering users' interests from their behaviors is very critical. 2) The proposed UID-Net explicitly discovers users' current interests from trigger items ignored by the above-mentioned methods, and thus achieves the most superior performance among all the evaluated methods. For instance, compared with the best competitor MIAN in this group, the average AUC RI gains of UID-Net are 27.04% and 2.50% on the two datasets, respectively, which indeed denotes a significant increment for online CTR improvement [19].

Results in Group 2. In Group 2, methods such as DIN++, DIEN++, and R3S++ incorporate an attention mechanism that operates on both individual target items and trigger items, thereby revealing users' current interests.

Our observations are as follows: 1) R3S obtains the worst performance in this group, due to the ignorance of users' historical behaviors, implying the benefit of revealing users' interests from their behaviors. 2) R3S++ achieves superior performance over DIN++ (or DIEN++), *i.e.*, 2.34% (or 1.16%) and 0.71% (or 0.08%) AUC improvements on the Fliggy-Dataset and Public-Dataset, respectively, since R3S++ considers the critical semantic relevance between trigger items and corresponding targets ignored by DIN++ or DIEN++. 3) As for DIHN, it utilizes labeled data to supervise the fusing process of the trigger and target items, further benefiting the extraction of users' interests, resulting in better AUC over R3S++, *i.e.*, 0.94% and 1.09% on the two datasets, respectively. 4) In comparison with DIHN, which overlooks the potentially interested trigger-irrelevant items, UID-Net can tackle this issue owing to the IEM module and thus delivers the best performance. 5) The majority of competitors in Group 2 can obtain significant improvements over those in Group 1, which indeed demonstrates the necessity of explicitly exploiting the trigger items to discover users' current interests in TIR scenarios.

4.4 Ablation Study (RQ2,RQ3)

To delve into the efficacy of UID-Net's individual components, we present ablation studies conducted specifically on the Fliggy-Dataset.

The Effectiveness of IFM. To evaluate the impact of IFM, UID-Net are compared with two groups of variants namely *UID-Net w/o IFM* and *UID-Net w/ DIHN_UIN* in Table 4, which represent removing IFM and replacing the fusing weight with the weight delivered by the User Intent Network (UIN) module of DIHN [12] respectively. Compared with UID-Net, the performance of both groups are reduced by 9.80% and 7.64% respectively. The former confirms the importance of IFM. And the latter shows DIHN over-emphasizes the role of trigger-relevant interests, while UID-Net captures both interests.

The Effectiveness of IDM. To elucidate the impact of the IDM on the supervision of trigger-relevant and trigger-irrelevant interests, we have conducted an experiment by excluding it from UID-Net, yielding a modified version termed *UID-Net w/o IDM*. As depicted in Table 4, the performance of *UID-Net w/o IDM* declines by 17.67% in Relative Improvement (RI) when juxtaposed with the original UID-Net. This reduction underscores the IDM's pivotal role in guiding the network to acquire distinctive representations of user interests.

The Influence of α. To investigate the influence of the hyper-parameter α in Eq. 14, five groups of experiments are conducted. As illustrated in Table 5, UID-Net achieves optimal performance with α set to 0.1. When α increases or decreases to be larger or smaller than 0.1, the performance is always getting more worse accordingly. It implies that a too-large α probably dominates the main CTR task and a too-small α will weaken the role of self-supervision in the IDM module. Therefore, α is set to 0.1 by default.

Table 4. Ablation study of the IFM module and IDM module.

Module	Model	AUC	RI
IFM module	UID-Net	**0.7133**	0.00
	UID-Net w/o IFM	0.6924	−9.80%
	UID-Net w/ DIHN_UIN	0.6970	−7.64%
IDM module	UID-Net	**0.7133**	0.00
	UID-Net w/o IDM	0.6756	−17.67%

The Influence of the Fusing Strategy in IFM. Considering the refined trigger-relevant and trigger-irrelevant interests, denoted as *i.e.*, $e_r^{'}$ and $e_f^{'}$, we have experimented with two alternative fusing strategies, *i.e.*, *Mean Pooling* and *Max Pooling*. The outcomes of these experiments are delineated in Table 5. It is evident that, the original concatenation strategy in UID-Net achieves better performance than *Mean Pooling* and *Max Pooling* operations. We suspect it is because the concatenation operation can provide more information than the other two operations for end-to-end learning.

Table 5. The influence of the hyper-parameter α, the fusing strategy in IFM and the dimension d in IFM.

influencing factors	values	AUC	RI
α	0	0.6756	−17.67%
	0.01	0.7035	−4.59%
	0.1	**0.7133**	0.00
	0.2	0.6859	−12.85%
	0.5	0.6772	−16.92%
Model	UID-Net	**0.7133**	0.00
	UID-Net w/ Mean Pooling	0.7010	−5.77%
	UID-Net w/ Max Pooling	0.7038	−4.45%
Dimension d	8	0.6960	−8.11%
	16	0.6909	−10.50%
	32	**0.7133**	0.00
	64	0.6894	−11.20%
	92	0.6819	−14.72%

The Influence of d in IFM. We delve into the impact of the vector dimensions d for $V_t^{'}$ and $V_i^{'}$ within the IFM Module. We conduct five groups of experiments by setting different dimensions, *e.g.*, 8, 16, 32, 64, and 92. As shown in Table 5, when d takes 32, the model obtains the best performance. We guess vectors in a

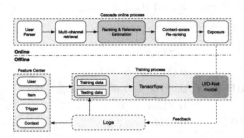

Fig. 4. The online workflow of UID-Net deployed at Fliggy.

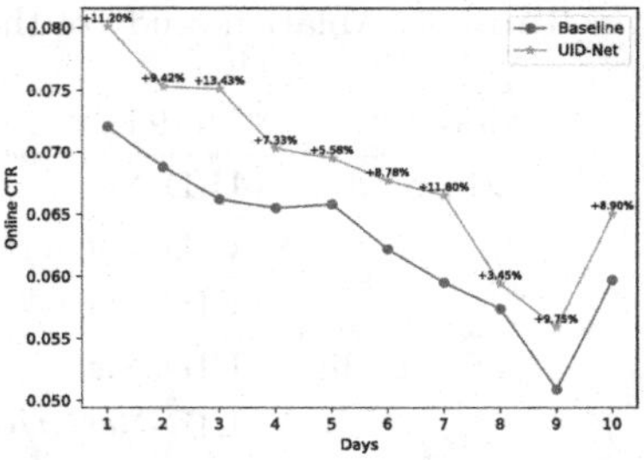

Fig. 5. Online A/B testing results.

large dimension probably contain noise information, while a too-small dimension may lose useful information.

4.5 Online A/B Testing (RQ4)

To further attest to the efficacy of UID-Net, we have initiated an online A/B testing deployment. The comprehensive workflow is outlined in Fig. 4, which encompasses three main sequential phases [20] namely *Retrieval*, *Rank*, and *Exposure* after user identification, *i.e.*, *User Parser*. In the *Retrieval* phase, recommender system recalls a large number of relevant items from all candidates, *e.g.*, tens of thousands of items, based on users' historical behavior preferences. Then, they are delivered to *Rank* phase, where they are ranked according to CTR, Conversion Rate (CVR) [20] or other ranking metrics. Moreover, hundreds of top-scored items are delivered to *Exposure* phase, where users can directly click or purchase at the terminal. Our proposed UID-Net model, which is trained in an offline manner with corresponding scenario feedback logs and features from *Feature Center*, works in *Rank* phase for online real-time serving. Online A/B testing is conducted from 2022-12-30 to 2023-01-08, and the results are depicted in Fig. 5, where the baseline model is the best offline competitor DIHN [12]. As shown, UID-Net consistently obtains improvement over the baseline. On average, it contributes up to 8.96% CTR promotion, demonstrating its significant business value. Now, UID-Net has been deployed as a large-scaled online service for the main traffic.

5 Conclusion

In this study, a novel approach dubbed UID-Net is devised to extract users' decomposed trigger-relevant and trigger-irrelevant interests in TIR scenarios. It fulfills this goal through three critical modules namely Intent Extraction Module, Interest Decomposition Module, and Interest Fusing Module. They aim to corporately capture users' trigger-relevant and trigger-irrelevant interests in regard to the specific target and trigger items in a self-supervised contrastive manner.

UID-Net not only obtains great improvement over the SOTA models on offline datasets but also achieves 8.96% increase in CTR on the real-world online application, thereby affirming its exceptional performance in TIR scenarios. How to investigate other ways (*e.g.*, graph learning) to explore users' dynamic interests in TIR scenarios is an interesting topic and will be studied in the future.

Acknowledgements. This work was supported in part by National Key Research and Development Program of China 2023YFB4502400.

References

1. Guo, H., Tang, R., Ye, Y., Li, Z., He, X.: Deepfm: a factorization-machine based neural network for ctr prediction. arXiv preprint arXiv:1703.04247 (2017)
2. Guo, J., He, H., He, T., Lausen, L., Li, M., Lin, H., Shi, X., Wang, C., Xie, J., Zha, S., et al.: Gluoncv and gluonnlp: Deep learning in computer vision and natural language processing. J. Mach. Learn. Res. **21**(23), 1–7 (2020)
3. Li, C., Liu, Z., Wu, M., Xu, Y., Zhao, H., Huang, P., Kang, G., Chen, Q., Li, W., Lee, D.L.: Multi-interest network with dynamic routing for recommendation at tmall. In: Proceedings of the 28th ACM International Conference on Information and Knowledge Management. pp. 2615–2623 (2019)
4. Lin, Z., Wang, H., Mao, J., Zhao, W.X., Wang, C., Jiang, P., Wen, J.R.: Feature-aware diversified re-ranking with disentangled representations for relevant recommendation. In: Proceedings of the 28th ACM SIGKDD Conference on Knowledge Discovery and Data Mining. pp. 3327–3335 (2022)
5. Ma, J., Zhou, C., Yang, H., Cui, P., Wang, X., Zhu, W.: Disentangled self-supervision in sequential recommenders. In: Proceedings of the 26th ACM SIGKDD International Conference on Knowledge Discovery & Data Mining. pp. 483–491 (2020)
6. Nwankpa, C., Ijomah, W., Gachagan, A., Marshall, S.: Activation functions: Comparison of trends in practice and research for deep learning. arXiv preprint arXiv:1811.03378 (2018)
7. Pi, Q., Zhou, G., Zhang, Y., Wang, Z., Ren, L., Fan, Y., Zhu, X., Gai, K.: Search-based user interest modeling with lifelong sequential behavior data for click-through rate prediction. In: Proceedings of the 29th ACM International Conference on Information & Knowledge Management. pp. 2685–2692 (2020)
8. Qu, Y., Cai, H., Ren, K., Zhang, W., Yu, Y., Wen, Y., Wang, J.: Product-based neural networks for user response prediction. In: 2016 IEEE 16th International Conference on Data Mining (ICDM). pp. 1149–1154. IEEE (2016)
9. Rendle, S., Freudenthaler, C., Gantner, Z., Schmidt-Thieme, L.: Bpr: Bayesian personalized ranking from implicit feedback. arXiv preprint arXiv:1205.2618 (2012)
10. Rendle, S., Freudenthaler, C., Schmidt-Thieme, L.: Factorizing personalized markov chains for next-basket recommendation. In: Proceedings of the 19th international conference on World wide web. pp. 811–820 (2010)
11. Shen, Q., Tao, W., Zhang, J., Wen, H., Chen, Z., Lu, Q.: Sar-net: A scenario-aware ranking network for personalized fair recommendation in hundreds of travel scenarios. In: Proceedings of the 30th ACM International Conference on Information & Knowledge Management. pp. 4094–4103 (2021)
12. Shen, Q., Wen, H., Tao, W., Zhang, J., Lv, F., Chen, Z., Li, Z.: Deep interest highlight network for click-through rate prediction in trigger-induced recommendation. In: Proceedings of the ACM Web Conference 2022. pp. 422–430 (2022)

13. Sun, F., Liu, J., Wu, J., Pei, C., Lin, X., Ou, W., Jiang, P.: Bert4rec: Sequential recommendation with bidirectional encoder representations from transformer. In: Proceedings of the 28th ACM international conference on information and knowledge management. pp. 1441–1450 (2019)
14. Tang, J., Wang, K.: Personalized top-n sequential recommendation via convolutional sequence embedding. In: Proceedings of the Eleventh ACM International Conference on Web Search and Data Mining. pp. 565–573 (2018)
15. Tao, W., Fu, Z.H., Li, L., Chen, Z., Wen, H., Liu, Y., Shen, Q., Chen, P.: A dual channel intent evolution network for predicting period-aware travel intentions at fliggy. In: Proceedings of the 31st ACM International Conference on Information & Knowledge Management. pp. 3524–3533 (2022)
16. Vaswani, A., Shazeer, N., Parmar, N., Uszkoreit, J., Jones, L., Gomez, A.N., Kaiser, Ł., Polosukhin, I.: Attention is all you need. Advances in neural information processing systems **30** (2017)
17. Wang, R., Fu, B., Fu, G., Wang, M.: Deep & cross network for ad click predictions. In: Proceedings of the ADKDD'17, pp. 1–7 (2017)
18. Wang, S., Hu, L., Wang, Y., Cao, L., Sheng, Q.Z., Orgun, M.: Sequential recommender systems: challenges, progress and prospects. arXiv preprint arXiv:2001.04830 (2019)
19. Wen, H., Zhang, J., Lin, Q., Yang, K., Huang, P.: Multi-level deep cascade trees for conversion rate prediction in recommendation system. In: Proceedings of the AAAI Conference on Artificial Intelligence. vol. 33, pp. 338–345 (2019)
20. Wen, H., Zhang, J., Wang, Y., Lv, F., Bao, W., Lin, Q., Yang, K.: Entire space multi-task modeling via post-click behavior decomposition for conversion rate prediction. In: Proceedings of the 43rd International ACM SIGIR conference on research and development in Information Retrieval. pp. 2377–2386 (2020)
21. Xie, R., Wang, R., Zhang, S., Yang, Z., Xia, F., Lin, L.: Real-time relevant recommendation suggestion. In: Proceedings of the 14th ACM International Conference on Web Search and Data Mining. pp. 112–120 (2021)
22. Zhang, K., Qian, H., Cui, Q., Liu, Q., Li, L., Zhou, J., Ma, J., Chen, E.: Multi-interactive attention network for fine-grained feature learning in ctr prediction. In: Proceedings of the 14th ACM International Conference on Web Search and Data Mining. pp. 984–992 (2021)
23. Zhou, G., Mou, N., Fan, Y., Pi, Q., Bian, W., Zhou, C., Zhu, X., Gai, K.: Deep interest evolution network for click-through rate prediction. In: Proceedings of the AAAI Conference on Artificial Intelligence (2019)
24. Zhou, G., Zhu, X., Song, C., Fan, Y., Zhu, H., Ma, X., Yan, Y., Jin, J., Li, H., Gai, K.: Deep interest network for click-through rate prediction. In: Proceedings of the 24th ACM SIGKDD International Conference on Knowledge Discovery & Data Mining. pp. 1059–1068. ACM (2018)

PDC-FRS: Privacy-Preserving Data Contribution for Federated Recommender System

Chaoqun Yang[1], Wei Yuan[2], Liang Qu[2], and Thanh Tam Nguyen[1](✉)

[1] Griffith University, Gold Coast, Australia
t.nguyen19@griffith.edu.au
[2] The University of Queensland, Brisbane, Australia

Abstract. Federated recommender systems (FedRecs) have emerged as a popular research direction for protecting users' privacy in on-device recommendations. In FedRecs, users keep their data locally and only contribute their local collaborative information by uploading model parameters to a central server. While this rigid framework protects users' raw data during training, it severely compromises the recommendation model's performance due to the following reasons: (1) Due to the power law distribution nature of user behavior data, individual users have few data points to train a recommendation model, resulting in uploaded model updates that may be far from optimal; (2) As each user's uploaded parameters are learned from local data, which lacks global collaborative information, relying solely on parameter aggregation methods such as FedAvg to fuse global collaborative information may be suboptimal. To bridge this performance gap, we propose a novel federated recommendation framework, PDC-FRS. Specifically, we design a privacy-preserving data contribution mechanism that allows users to share their data with a differential privacy guarantee. Based on the shared but perturbed data, an auxiliary model is trained in parallel with the original federated recommendation process. This auxiliary model enhances FedRec by augmenting each user's local dataset and integrating global collaborative information. To demonstrate the effectiveness of PDC-FRS, we conduct extensive experiments on two widely used recommendation datasets. The empirical results showcase the superiority of PDC-FRScompared to baseline methods.

Keywords: Recommender System · Federated Learning · Privacy Protection

1 Introduction

Nowadays, the demand for recommender systems has dramatically increased in most online services (e.g., e-commerce [5] and social media [25]) due to their success in alleviating information overload. The remarkable personal recommendation ability of these systems mainly originates from discovering and mining latent user-item relationships from massive user data, such as user profiles and historical behaviors [14]. Conventionally, this personal user data is collected on a central server, and a recommendation model is trained using it [8], which carries high risks of privacy leakage. With growing awareness of privacy and the recent release of privacy protection regulations, these traditional centralized recommender model training paradigms have become harder to implement [24].

Federated learning [34] is a privacy-preserving decentralized training scheme that collaboratively trains a model without sharing clients' sensitive raw data. Consequently, many works investigate the combination of federated learning and recommender systems, known as federated recommender systems (FedRecs) [1,26]. In FedRecs, users/clients[1] manage their own data locally while a central server coordinates the training process. Specifically, the central server distributes recommendation model parameters to clients. Upon receiving the recommendation model, clients train it on their local data and send the updated model parameters back to the central server. The central server then aggregates these parameters to integrate each client's collaborative information.

Although the generic federated recommendation framework conceals users' raw data and thus protects their privacy, it may only achieve suboptimal performance for the following reasons. Firstly, since user-item interaction data follows a power law distribution [33], most users have few data points to train a recommendation model effectively. Consequently, the locally trained parameters will be far from optimal when sent to the central server. Furthermore, the only way to integrate collaborative information from different clients is by aggregating the uploaded parameters, which may be less effective.

To address the above problems, recent works have broken the rigid FedRec learning protocol by allowing users to control the portion of data they want to share [2,18]. In this approach, users can contribute their collaborative information not only by uploading the locally trained parameters but also by sharing their raw data. Unfortunately, we argue that these methods pose high privacy risks since users directly disclose their real and clean data to the central server. For example, when the exposed data is sufficient, a curious central server can easily infer users' remaining data using graph completion techniques [6].

Local differential privacy (LDP) has become the gold standard for providing protection guarantees in federated recommender systems [12,20,36]. These works apply LDP by transforming users' local model parameters into a noisy version before uploading them to the central server. Inspired by this idea, we attempt to leverage LDP to provide a privacy guarantee for user data contributions. Nevertheless, there are significant challenges in achieving differentially private user data contributions. Firstly, current LDP methods in FedRecs are not directly applicable to user data since they are not typical numerical values like model parameters. Additionally, how to effectively leverage perturbed user data in the FedRec learning process remains underexplored.

---

[1] In this paper, we focus on cross-user federated recommendation, where each client represents one user. Therefore, the concepts of user and client are interchangeable.

In this paper, we propose PDC-FRS, the first federated recommendation framework that enables user data contribution with privacy guarantees. Specifically, PDC-FRSutilizes the exponential mechanism to achieve LDP on user-shared data. Then, PDC-FRSsimultaneously trains an auxiliary model on the users' uploaded data. This auxiliary model enhances FedRecs in two ways. First, it acts as a data augmentor to enrich each client's local dataset. Additionally, the auxiliary model parameters are fused into FedRecs via contrastive learning to provide global collaborative information. To demonstrate the effectiveness of PDC-FRS, we conduct extensive experiments on two datasets. The experimental results indicate the superiority of our method compared to several federated recommendation baselines.

To sum up, the main contributions of this paper are as follows:

- To the best of our knowledge, we are the first to investigate user data contribution with a privacy-preserving mechanism in a federated recommender system (PDC-FRS), allowing users to share their data with differential privacy guarantees.
- In PDC-FRS, we design an exponential mechanism-based privacy protection for user data and provide two methods for using the perturbed data to enhance federated recommendation performance.
- We conduct extensive experiments on public real-world recommendation datasets to validate the effectiveness of PDC-FRS. The experimental results demonstrate the promising performance of our proposed methods.

2 Related Work

2.1 Centralized Recommender System

Recommender systems aim to provide personalized recommendations to users based on their historical data (e.g., purchases and browses) [14,38]. Generally, these systems can be classified into matrix factorization (MF)-based methods [6,10], deep learning-based methods [7,9], and graph neural network (GNN)-based methods [8,21]. MF methods decompose the user-item rating matrix into two lower-dimensional vectors representing user and item preferences. Deep learning-based systems utilize deep learning models to discover nonlinear relationships between users and items. Recently, GNN-based recommendation has gained popularity as GNN models can effectively capture high-order user-item interactions, treating user-item interactions as a graph structure. However, all these methods are traditionally trained in a centralized manner, which raises privacy concerns.

2.2 Federated Recommender System

Federated recommender systems (FedRecs) have garnered significant attention recently due to their privacy protection capabilities [22,26]. Ammad et al. [1]

were pioneers in designing the first federated recommendation framework, spawning subsequent research in this area [19]. These studies can be categorized into four perspectives based on their objectives: effectiveness [15,20,28,30], efficiency [13,17], privacy [29,31,36], and security [27,32,35]. For instance, Wu et al. [23] and Yuan et al. [30] applied advanced neural networks and contrastive learning in FedRecs to enhance recommendation performance. Muhammad et al. [17] investigated efficient training methods for FedRecs with rapid convergence, while Zhang et al. [36] explored sensitive attribute privacy in FedRecs. However, these works operate within the original federated recommendation framework where users can only contribute collaborative information by uploading model parameters, which may limit effectiveness. Anelli et al. [2] and Qu et al. [18] highlighted this limitation and proposed FedRecs frameworks based on user-governed data contribution. Nonetheless, in their approaches, parts of users' clean data are directly transmitted to the central server, lacking privacy guarantees and posing significant privacy risks. In response, our paper introduces a privacy-preserving user data contribution framework for FedRecs.

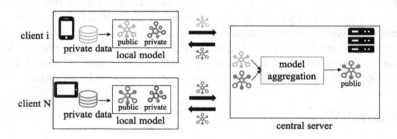

Fig. 1. Overview of a typical federated recommender system

3 Preliminaries

3.1 Federated Recommendation Framework

Let $\mathcal{U} = \{u_i\}_{i=1}^{|\mathcal{U}|}$ and $\mathcal{V} = \{v_j\}_{j=1}^{|\mathcal{V}|}$ represent the set of users and items in FedRec, where $|\mathcal{U}|$ and $|\mathcal{V}|$ are the number of users and items, respectively. $\mathcal{D}_{u_i}$ is user u_i's raw private dataset, which contains a set of triples (u_i, v_j, r_{ij}) representing its interaction status. $r_{ij} = 1$ indicates u_i has interacted with item v_j while $r_{ij} = 0$ means that v_j is still in non-interacted item pool[2]. In traditional FedRec, $\mathcal{D}_{u_i}$ is always stored in the user's local device and cannot be used by any other participants. The goal of FedRec is to train a recommender model that can predict user preference tendency for non-interacted items using these distributed private datasets.

---

[2] In this paper, we mainly focus on recommendation with implicit feedback.

To achieve this goal, most FedRec systems follow the learning protocol shown in Fig. 1. A central server acts as a coordinator to manage the entire training process, with the recommendation model divided into public and private parameters. Private parameters typically refer to user embeddings **U**, while public parameters primarily consist of item embeddings **V**. Clients engage in collaborative learning by transmitting public parameters. Specifically, there are four steps iteratively executed during the FedRec learning process. At first, the central server selects a group of clients $\mathcal{U}_{t-1}$ to participate and disperses public parameters $\mathbf{V}^{t-1}$ to them. Then, the client $u_i \in \mathcal{U}_{t-1}$ utilizes the received public parameters $\mathbf{V}^{t-1}$ and its corresponding local parameters $\mathbf{u}_i$ to train the recommendation model on their local datasets $\mathcal{D}_{u_i}$ using certain objective function, such as:

$$\mathcal{L}^{original} = -\sum\nolimits_{(u_i, v_j, r_{ij}) \in \mathcal{D}_{u_i}} r_{ij} \log \hat{r}_{ij} + (1 - r_{ij}) \log(1 - \hat{r}_{ij}) \qquad (1)$$

After training, the clients send the updated public parameters $\mathbf{V}^{t-1}_{u_i}$ back to the central server. Finally, the central server aggregates these uploaded parameters $\{\mathbf{V}^{t-1}_{u_i}\}_{u_i \in \mathcal{U}_{t-1}}$ to get a new version of public parameters $\mathbf{V}^t$. It is evident that clients' collaborative information can only be conveyed through the public parameters $\mathbf{V}^{t-1}_{u_i}$ via aggregation, which may be less effective.

3.2 Base Recommendation Model

Federated recommendation framework is compatible with most deep learning based recommender systems. In this paper, without loss of generality, we adopt neural collaborative filtering (NCF) [9] as the experimental base models, which are widely used in this research area [1,28,35]. Based on matrix factorization, NCF leverages a multi-layer perceptron (MLP) and the concatenation of user and item embeddings to predict a user's preference score for an item:

$$\hat{r}_{ij} = \sigma(\mathbf{h}^\top MLP([\mathbf{u}_i, \mathbf{v}_j])) \qquad (2)$$

4 Methodology

In this section, we present the technical details of our PDC-FRS. We first introduce the general framework of PDC-FRSin Sect. 4.1, and then discuss the specific component designs. Specifically, we describe how to achieve privacy-preserving data contribution in Sect. 4.2. In Sect. 4.3, we introduce the details of utilizing an auxiliary model to enhance FedRec performance.

4.1 General Framework of PDC-FRS

The traditional federated recommender system depicted in Fig. 1 relies solely on public parameters to convey clients' collaborative information. This approach has limitations because (1) each client's uploaded parameters vary in quality

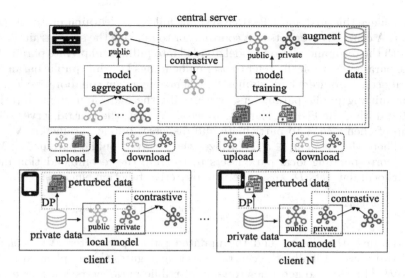

Fig. 2. Overview of PDC-FRS. The yellow dotted box highlights the privacy-preserving data contribution while the red dotted box depicts the methods of using data contribution to enhance federated recommendation, which are the two most important parts of PDC-FRS. (Color figure online)

and (2) the uploaded parameters are trained only on local data, lacking global information.

In this work, we propose PDC-FRSto enhance FedRec's performance by allowing users to contribute to the system using data in a privacy-preserving manner. Revealing users' clean data to the central server, as done in [2,18], would significantly compromise FedRec's privacy-preserving capability. Therefore, the first challenge for PDC-FRSis "how to share user data in a privacy-preserving manner?" As shown in Fig. 2, we employ a differential privacy method to transform users' private data into a noisy version before uploading it. However, this perturbation reduces the utility of the uploaded data, presenting another challenge for PDC-FRS: "how to leverage these noisy data to enhance federated recommendation performance?" In this paper, we first train an auxiliary model on the user-uploaded data and then explore two ways to enhance FedRec: data augmentation and contrastive learning. In the following two subsections, we detail how we address these two challenges.

4.2 Exponential Mechanism-Based Data Contribution

Local Differential Privacy (LDP) [3,37], a variant of Differential Privacy (DP) [16], provides a rigorous mathematical privacy guarantee and has thus become the *de facto* standard for privacy protection in decentralized settings. Consequently, in PDC-FRS, we design a data contribution mechanism that satisfies LDP.

Definition 1: Local Differential Privacy (LDP). Let $\mathcal{A}(\cdot)$ be a randomized function, for example, a perturbation algorithm or machine learning model, and $\epsilon > 0$ is the privacy budget. The function $\mathcal{A}(\cdot)$ is said to provide ϵ-Local different privacy if for any two input data $x_1, x_2 \in \mathcal{X}$ and for any possible output $y \in \mathcal{Y}$ satisfied the following inequality:

$$Pr[\mathcal{A}(x_1) = y] \leq exp(\epsilon) Pr[\mathcal{A}(x_2) = y] \tag{3}$$

where $Pr[\cdot]$ is the probability and ϵ controls the trade-off between data utility and privacy. Intuitively, Eq. 3 implies that based on the output of algorithm $\mathcal{A}(\cdot)$, the adversarial attacker cannot accurately infer the original data is x_1 or x_2, since their output probability difference of value y is bounded by $exp(\epsilon)$.

To achieve LDP, each user needs to perturb its protected objectives before sending it to the collector. For example, in traditional FedRec, the protected objective is model public parameters $\mathbf{V}$. A user u_i will locally transform $\mathbf{V}$ into a noisy version $\mathbf{V}'$ using a perturbation algorithm such as Laplace mechanism before sending it to the central server [36].

In PDC-FRS, the user will upload their private dataset $\mathcal{D}_{u_i} = \{(u_i, v_j, r_{ij})\}_{v_j \in \mathcal{V}}$ to the central server. As we mainly focus on the recommendation with implicit feedback, the actual protection objective is user u_i's interacted item set $\mathcal{V}_{u_i}^+$. Therefore, before sending data contribution, the user needs to perturb $\mathcal{V}_{u_i}^+$ to a noisy version $\mathcal{V}_{u_i}^{+'}$. As $\mathcal{V}_{u_i}^+$ is a set, in this paper, we adopt the exponential mechanism as the perturbation algorithm to protect user data.

Definition 2: Exponential Mechanism. The exponential mechanism $\mathcal{A}$ is said to preserve ϵ-LDP if for any input $x \in \mathcal{X}$, the probability of any output $y \in \mathcal{Y}$ is:

$$Pr[\mathcal{A}(x) = y] = \frac{exp(\frac{\epsilon \rho(x,y)}{2\Delta})}{\sum_{y' \in \mathcal{Y}} exp(\frac{\epsilon \rho(x,y')}{2\Delta})} \tag{4}$$

where ϵ is the privacy budget, $\rho(\cdot)$ is the rating function, and Δ is the sensitivity of $\rho(\cdot)$.

In PDC-FRS, we utilize the similarity between two items v_i and v_j as the rating function to sample a new item set $\mathcal{V}_{u_i}^{+'}$ based on $\mathcal{V}_{u_i}^+$, which can protect privacy while keep data utility, since replacing v_i with similar item v_j can still keep user's expected collaborative information to a large extent:

$$Pr[\mathcal{A}(v_i) = v_j] = \frac{exp(\frac{\epsilon sim(v_i, v_j)}{2\Delta})}{\sum_{v_k \in \mathcal{V}} exp(\frac{\epsilon sim(v_i, v_k)}{2\Delta})} \tag{5}$$

However, how to calculate the similarity score is not trivial. A simple method is to directly compute based on the item embedding table $\mathbf{V}$. Unfortunately, in the first few rounds, the item embedding table is from scratch and therefore contains less useful information, making the similarity scores low-quality. Considering that in most cases, items contain content information (e.g., title), and this information can represent items to some extent. Without loss of generality, in this paper, we calculate the similarity based on item titles using pre-trained word embeddings.

4.3 Enhance Federated Recommendation with Data Contribution

After the central server received all users privacy-preserving data contributions $\{\mathcal{D}'_{u_i}\}_{u_i \in \mathcal{U}}$, the central server constructs an auxiliary model $\mathbf{M}_{aux} = \{\mathbf{U}_{aux}, \mathbf{V}_{aux}\}$ and will train the model on the perturbed dataset. Specifically, before a global training round t starts, the central server updates $\mathbf{M}_{aux}^{t-1}$ to $\mathbf{M}_{aux}^{t}$ on dataset $\{\mathcal{D}'_{u_i}\}_{u_i \in \mathcal{U}}$ using the same learning hyper-parameters as the clients' local training. Then, the auxiliary model $\mathbf{M}_{aux}^{t}$ is used to improve federated recommendation from two perspectives: data augmentation and contrastive learning.

Auxiliary Model as Data Augmentor. As we mentioned in Sect. 1, most clients lack training data to support the effective updating of the recommendation model. As $\mathbf{M}_{aux}^{t}$ is trained on a massive dataset that contains multiple users' data, $\mathbf{M}_{aux}^{t}$ can learn some broader collaborative information. Therefore, we utilize $\mathbf{M}_{aux}^{t}$ as the data augmentor to generate an augmented dataset $\widetilde{\mathcal{D}}_{u_i}^{t} = \{(u_i, v_j, \widetilde{r}_{ij}^{t})\}_{v_j \in \mathcal{V}}$ for each user:

$$\widetilde{r}_{ij}^{t} = f_{rec}(\mathbf{M}_{aux}^{t}|(u_i, v_j)) \qquad (6)$$

where f_{rec} is the recommendation algorithm. However, directly transfer $\widetilde{\mathcal{D}}_{u_i}^{t}$ is ineffective since (1) the dataset is very large and (2) $\mathbf{M}_{aux}^{t}$ is trained on perturbed data thus not all predictions are helpful. To solve this problem, we select a subset $\widetilde{\mathcal{V}}_{u_i}$ based on the ranking of $\widetilde{r}_{ij}^{t}$. Intuitively, the higher $\widetilde{r}_{ij}^{t}$ score indicates that the model has higher confidence in predicting the user will prefer the item v_j. Finally, each client u_i's local training objective is transformed from Eq. 1 to:

$$\mathcal{L}^{rec} = -\sum\nolimits_{(u_i, v_j, r_{ij}) \in \mathcal{D}_{u_i} \cup \widetilde{\mathcal{D}}_{u_i}^{t}} r_{ij} \log \hat{r}_{ij} + (1 - r_{ij}) \log(1 - \hat{r}_{ij}) \qquad (7)$$

Auxiliary Model as Contrastive View. As discussed in Sect. 1, the other problem of FedRec is that users' collaborative information is only fused via parameter aggregation, such as FedAvg. In PDC-FRS, we enhance the knowledge fusion across users using contrastive learning. Specifically, since the auxiliary model $\mathbf{M}_{aux}^{t}$ is trained on the broader dataset that is perturbed by replacing items with their similar items in possibility, we treat the $\mathbf{M}_{aux}^{t}$ as contrastive views. As shown in Fig. 2, after aggregation, the central server obtains the public parameter trained by clients $\mathbf{V}^{t}$. Then, the central server conducts contrastive learning by treating the counterpart item embeddings in $\mathbf{V}^{t}$ and $\mathbf{V}_{aux}^{t}$ as positive views while others as negative views:

$$\mathcal{L}^{itemCL} = \beta \sum_{v_i \in \mathcal{V}} -log \frac{exp(sim(\mathbf{v}_i, \mathbf{v}_i^{aux})/\tau)}{\sum_{v_j \in \mathcal{V}} exp(sim(\mathbf{v}_i, \mathbf{v}_j^{aux})/\tau)} \qquad (8)$$

where β controls the strengths of $\mathcal{L}^{itemCL}$ and τ is the temperature.

Besides, PDC-FRS sends the auxiliary model's user embeddings $\mathbf{U}_{aux}^t$ to clients for user embedding collaborative learning:

$$\mathcal{L}^{userCL} = -log\frac{exp(sim(\mathbf{u}_i, \mathbf{u}_i^{aux})/\tau)}{\sum_{u_j \in \mathcal{U}} exp(sim(\mathbf{u}_i, \mathbf{u}_j^{aux})/\tau)} \quad (9)$$

Therefore, the client u_i's local training objective is transformed from Eq. 7 to:

$$\mathcal{L} = \mathcal{L}^{rec} + \lambda \mathcal{L}^{userCL} \quad (10)$$

where λ controls the strengths of $\mathcal{L}^{userCL}$. Algorithm 1 summarizes PDC-FRS.

Algorithm 1 The pseudo-code for PDC-FRS.

Input: global round T; learning rate lr, ...
Output: well-trained model $\mathbf{M}^T = \{\mathbf{U}^T, \mathbf{V}^T\}$
1: each client initializes private parameter $\mathbf{u}_i^0$
2: client generates perturbed data $\mathcal{D}'_{u_i}$ using Eq. 5 and uploads to the server
3: server initializes model $\mathbf{V}^0$, $\mathbf{M}_{aux}^0$, and receives clients' data $\{\mathcal{D}'_{u_i}\}_{u_i \in \mathcal{U}}$
4: **for** each round t =0, ..., $T-1$ **do**
5:     // execute on server sides
6:     $\mathbf{M}_{aux}^{t+1} \leftarrow$ update model $\mathbf{M}_{aux}^t$ on $\{\mathcal{D}'_{u_i}\}_{u_i \in \mathcal{U}}$
7:     **for** $u_i \in \mathcal{U}$ **in parallel do**
8:         $\widetilde{\mathcal{D}}_{u_i}^t \leftarrow$ generate augmented data using $\mathbf{M}_{aux}^{t+1}$ with Eq. 6
9:         // execute on client sides
10:         $\mathbf{V}_{u_i}^{t+1} \leftarrow$ CLIENTTRAIN$(\mathbf{V}^t, \widetilde{\mathcal{D}}_{u_i}^t, \mathbf{U}_{aux}^{t+1})$
11:     **end for**
12:     // execute on central server
13:     $\mathbf{V}^{l+1} \leftarrow$ aggregate received client model parameters $\{\mathbf{V}_{u_i}^{t+1}\}_{u_i \in \mathcal{U}}$
14:     update $\mathbf{V}^{l+1}$ using Eq. 8
15: **end for**
16: **function** CLIENTTRAIN$(\mathbf{V}^t, \widetilde{\mathcal{D}}_{u_i}^t, \mathbf{U}_{aux}^{t+1})$
17:     $\mathbf{u}_i^{t+1}, \mathbf{V}_{u_i}^{t+1} \leftarrow$ update local model with Eq. 10
18:     **return** $\mathbf{V}_{u_i}^{t+1}$
19: **end function**

---

5 Experiments

In this work, we explore the following research questions (RQs):

- **RQ1.** How is the performance of PDC-FRS compared to the base FedRecs?
- **RQ2.** What is the impact of privacy-preserving mechanisms on data contribution?
- **RQ3.** What is the impact of data augmentation module?
- **RQ4.** What is the impact of contrastive learning module?

5.1 Datasets

In this work, we conduct experiments on two widely used recommendation datasets, MovieLens-1M[3] (ML) and Amazon Office Products (AZ)[4], which are collected from different platforms in various scenarios. ML contains 6,040 users and 3706 movies with 1,000,000 user rating records. AZ includes 53,258 reviews among 4905 users and 2421 office products. Following most recommendations with implicit feedback data preprocessing [9,29,35], we convert all interacted items' ratings into 1 and randomly sample negative items from non-interacted item pools with ratio 1 : 4. Besides, 80% data are used for training while the remaining data are for testing.

5.2 Evaluation Metrics

We leverage Recall@K and NDCG@K to evaluate the recommendation performance, and K is set to 20. We treat all non-interacted items as the candidate set to avoid sampling-based evaluation bias [11].

5.3 Baselines

To show the effectiveness of PDC-FRS, we employ the following baselines:

- **FCF** [1]. It is the first work that combines federated learning with collaborative filtering.
- **FedMF** [4]. It combines federated learning with matrix factorization and utilizes a homomorphic encryption technique to ensure user privacy.
- **FedNCF**. This is the system that removes the data contribution of PDC-FRS. The details of it are described in Sect. 3.
- **FedNCF+Aug**. We employ privacy-preserving data contribution to FedNCF but only utilize the data with the data augmentation method proposed in Sect. 4.3.
- **FedNCF+CL**. We employ privacy-preserving data contribution to FedNCF with a contrastive learning method proposed in Sect. 4.3.

5.4 Implementation Details

Following [9,18,28], the dimensions of user and item embeddings in the NCF model are set to 32. Three feedforward layers with sizes 32, 16, and 8 are used to capture user and item collaborative information. The maximum number of global training rounds is 20, and all clients are ensured to participate during a global round. Specifically, at the start of a global round, a shuffled order of clients is generated. Then, a batch of 256 clients is sequentially selected to train the model. The local training epoch for each client is 5. The optimizer is Adam with 0.001 learning rate. The values of ϵ, α, β, and λ are set to 5, 30, 0.5, and 0.5, respectively, and we will explore their influences in the following sections.

---

[3] https://grouplens.org/datasets/movielens/1m/.
[4] https://jmcauley.ucsd.edu/data/amazon/.

Table 1. Comparison between PDC-FRS and baselines. "Aug" is short for data augmentation and "CL" is abbreviated for contrastive learning.

	ML		AZ	
	Recall@20	NDCG@20	Recall@20	NDCG@20
FCF	0.01286	0.02142	0.00724	0.00326
FedMF	0.01420	0.02273	0.00763	0.00338
FedNCF	0.01529	0.02355	0.00875	0.00385
FedNCF+Aug	0.01580	0.02400	0.01341	**0.00670**
FedNCF+CL	0.01547	0.02377	0.00910	0.00368
FedNCF+Aug+CL (PDC-FRS)	**0.01679**	**0.02557**	**0.01387**	0.00663

5.5 Effectiveness of PDC-FRS (RQ1)

Table 1 presents the comparison results of PDC-FRS with baselines on the ML and AZ datasets. Overall, PDC-FRS achieves superior performance among these baselines across most metrics on both datasets. By comparing FedNCF with FedNCF+Aug and FedNCF+CL, we observe that incorporating data contributions through either our proposed augmentation method or the contrastive learning approach enhances the base FedRec performance. Furthermore, the influence of data augmentation is relatively better than that of contrastive learning, indicating that clients in FedRec suffer from a severe data scarcity problem. Finally, the combination of augmentation and contrastive learning, as implemented in PDC-FRS, further improves performance, showcasing its effectiveness. Specifically, PDC-FRS improves FedNCF's Recall@20 scores from 0.0152 to 0.0167 on ML and from 0.008 to 0.0138 on AZ.

5.6 Impact of Privacy-Preserving Data Contribution (RQ2)

One major contribution of this paper is the proposal of a privacy-preserving data contribution method. In this section, we investigate the relationship between utility and privacy protection. The privacy-preserving ability of PDC-FRS is controlled by ϵ. Theoretically, privacy protection is negatively correlated with ϵ. In our case, a small ϵ adds more randomness to item perturbation, while a larger ϵ makes the exponential mechanism tend to keep the item unchanged. Figure 3 shows the performance trends on the ML and AZ datasets. Generally, as ϵ increases, the model performance first rises to a peak and then gradually declines. This phenomenon is caused by two factors. Firstly, when ϵ is too small, the uploaded user data becomes random, making it difficult for the system to learn useful information. Secondly, when ϵ is too large, the uploaded data closely resembles the original dataset. Therefore, when using the auxiliary model as a data augmenter, the generated data will be too similar to the client's original data, resulting in ineffectiveness.

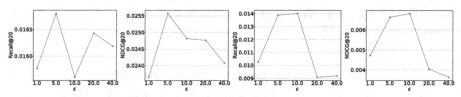

(a) Performance Trend on ML with ϵ. (b) Performance Trend on AZ with ϵ.

Fig. 3. The performance trend with privacy controller ϵ.

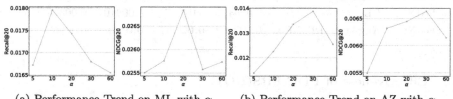

(a) Performance Trend on ML with α. (b) Performance Trend on AZ with α.

Fig. 4. The performance trend with augmentation dataset size α.

5.7 Impact of Data Augmentation (RQ3)

Utilizing the auxiliary model $\mathbf{M}_{aux}$ to augment clients' local datasets is an effective way to leverage users' noisy data contributions, as demonstrated in Table 1. In this subsection, we further investigate the influence of the augmented dataset size, α, on model performance. As depicted in Fig. 4, we explore augmented dataset sizes ranging from 5 to 60 on the ML and AZ datasets. The performance trends on these two datasets are similar: model performance increases to a peak and then decreases as the augmented dataset size continues to grow. This indicates that an appropriate amount of data augmentation can improve model performance, however, if the augmentation set becomes too large, it will impede the model's ability to learn from the original dataset.

5.8 Impact of Contrastive Learning (RQ4)

Treating the auxiliary model $\mathbf{M}_{aux}$ as a contrastive learning view is a novel method to leverage user data contributions. PDC-FRSincludes two contrastive learning tasks, $\mathcal{L}^{userCL}$ and $\mathcal{L}^{itemCL}$, corresponding to user and item embeddings, respectively. These two tasks are controlled by factors λ and β. Figure 5 shows the performance changes with varying λ. PDC-FRSachieves the best performance when $\lambda = 0.05$ on both the ML and AZ datasets. Figure 6 presents the impact of β. Similar to λ, on the ML dataset, PDC-FRSobtains the best performance when $\beta = 0.05$, while on the AZ dataset, the best performance is achieved when $\beta = 0.1$. Overall, based on these two figures, we find that too small factors limit the influence of contrastive learning, while too large factors make contrastive learning overwhelming, impeding the model learning process.

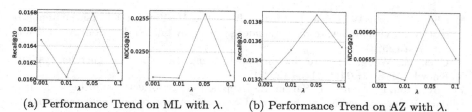

(a) Performance Trend on ML with λ. (b) Performance Trend on AZ with λ.

Fig. 5. The performance trend with user contrastive learning $\mathcal{L}^{userCL}$ factor λ.

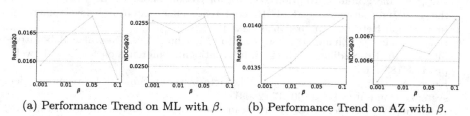

(a) Performance Trend on ML with β. (b) Performance Trend on AZ with β.

Fig. 6. The performance trend with item contrastive learning $\mathcal{L}^{itemCL}$ factor β.

6 Conclusion and Future Work

In this paper, we propose a novel federated recommendation framework, PDC-FRS, which enables users to contribute their data to enhance FedRec performance in a privacy-preserving manner. Specifically, PDC-FRSemploys the exponential mechanism to protect users' data before uploading it to the central server. To effectively utilize these noisy data contributions, an auxiliary model is trained upon them, acting as a data augmentor and providing contrastive learning views during FedRec's training. Extensive experiments on two datasets demonstrate the effectiveness of PDC-FRS.

References

1. Ammad-Ud-Din, M., Ivannikova, E., Khan, S.A., Oyomno, W., Fu, Q., Tan, K.E., Flanagan, A.: Federated collaborative filtering for privacy-preserving personalized recommendation system. arXiv preprint arXiv:1901.09888 (2019)
2. Anelli, V.W., Deldjoo, Y., Di Noia, T., Ferrara, A., Narducci, F.: FedeRank: User Controlled Feedback with Federated Recommender Systems. In: Hiemstra, D., Moens, M.-F., Mothe, J., Perego, R., Potthast, M., Sebastiani, F. (eds.) ECIR 2021. LNCS, vol. 12656, pp. 32–47. Springer, Cham (2021). https://doi.org/10.1007/978-3-030-72113-8_3
3. Arachchige, P.C.M., Bertok, P., Khalil, I., Liu, D., Camtepe, S., Atiquzzaman, M.: Local differential privacy for deep learning. IEEE Internet Things J. **7**(7), 5827–5842 (2019)
4. Chai, D., Wang, L., Chen, K., Yang, Q.: Secure federated matrix factorization. IEEE Intell. Syst. **36**(5), 11–20 (2020)

5. Chen, L., Yuan, W., Chen, T., Ye, G., Hung, N.Q.V., Yin, H.: Adversarial item promotion on visually-aware recommender systems by guided diffusion. ACM Trans. Inf, Syst (2024)
6. Chen, Z., Wang, S.: A review on matrix completion for recommender systems. Knowl. Inf. Syst. **64**(1), 1–34 (2022)
7. Covington, P., Adams, J., Sargin, E.: Deep neural networks for youtube recommendations. In: Proceedings of the 10th ACM conference on recommender systems. pp. 191–198 (2016)
8. He, X., Deng, K., Wang, X., Li, Y., Zhang, Y., Wang, M.: Lightgcn: Simplifying and powering graph convolution network for recommendation. In: Proceedings of the 43rd International ACM SIGIR conference on research and development in Information Retrieval. pp. 639–648 (2020)
9. He, X., Liao, L., Zhang, H., Nie, L., Hu, X., Chua, T.S.: Neural collaborative filtering. In: Proceedings of the 26th international conference on world wide web. pp. 173–182 (2017)
10. Koren, Y., Bell, R., Volinsky, C.: Matrix factorization techniques for recommender systems. Computer **42**(8), 30–37 (2009)
11. Krichene, W., Rendle, S.: On sampled metrics for item recommendation. In: Proceedings of the 26th ACM SIGKDD international conference on knowledge discovery & data mining. pp. 1748–1757 (2020)
12. Li, T., Song, L., Fragouli, C.: Federated recommendation system via differential privacy. In: 2020 IEEE international symposium on information theory (ISIT). pp. 2592–2597. IEEE (2020)
13. Liang, F., Pan, W., Ming, Z.: Fedrec++: Lossless federated recommendation with explicit feedback. In: Proceedings of the AAAI conference on artificial intelligence. vol. 35, pp. 4224–4231 (2021)
14. Lu, J., Wu, D., Mao, M., Wang, W., Zhang, G.: Recommender system application developments: a survey. Decis. Support Syst. **74**, 12–32 (2015)
15. Luo, S., Xiao, Y., Song, L.: Personalized federated recommendation via joint representation learning, user clustering, and model adaptation. In: Proceedings of the 31st ACM international conference on information & knowledge management. pp. 4289–4293 (2022)
16. McSherry, F., Talwar, K.: Mechanism design via differential privacy. In: 48th Annual IEEE Symposium on Foundations of Computer Science (FOCS'07). pp. 94–103. IEEE (2007)
17. Muhammad, K., Wang, Q., O'Reilly-Morgan, D., Tragos, E., Smyth, B., Hurley, N., Geraci, J., Lawlor, A.: Fedfast: Going beyond average for faster training of federated recommender systems. In: Proceedings of the 26th ACM SIGKDD International Conference on Knowledge Discovery & Data Mining. pp. 1234–1242 (2020)
18. Qu, L., Yuan, W., Zheng, R., Cui, L., Shi, Y., Yin, H.: Towards personalized privacy: User-governed data contribution for federated recommendation. arXiv preprint arXiv:2401.17630 (2024)
19. Sun, Z., Xu, Y., Liu, Y., He, W., Jiang, Y., Wu, F., Cui, L.: A survey on federated recommendation systems. arXiv preprint arXiv:2301.00767 (2022)
20. Wang, Q., Yin, H., Chen, T., Yu, J., Zhou, A., Zhang, X.: Fast-adapting and privacy-preserving federated recommender system. VLDB J. **31**(5), 877–896 (2022)
21. Wang, X., He, X., Wang, M., Feng, F., Chua, T.S.: Neural graph collaborative filtering. In: Proceedings of the 42nd international ACM SIGIR conference on Research and development in Information Retrieval. pp. 165–174 (2019)
22. Wang, Z., Yu, J., Gao, M., Yuan, W., Ye, G., Sadiq, S., Yin, H.: Poisoning attacks and defenses in recommender systems: A survey (2024)

23. Wu, C., Wu, F., Cao, Y., Huang, Y., Xie, X.: Fedgnn: Federated graph neural network for privacy-preserving recommendation. arXiv preprint arXiv:2102.04925 (2021)
24. Yang, L., Tan, B., Zheng, V.W., Chen, K., Yang, Q.: Federated recommendation systems. Federated Learning: Privacy and Incentive pp. 225–239 (2020)
25. Yin, H., Cui, B.: Spatio-temporal recommendation in social media. Springer (2016)
26. Yin, H., Qu, L., Chen, T., Yuan, W., Zheng, R., Long, J., Xia, X., Shi, Y., Zhang, C.: On-device recommender systems: A comprehensive survey. arXiv preprint arXiv:2401.11441 (2024)
27. Yuan, W., Nguyen, Q.V.H., He, T., Chen, L., Yin, H.: Manipulating federated recommender systems: Poisoning with synthetic users and its countermeasures. arXiv preprint arXiv:2304.03054 (2023)
28. Yuan, W., Qu, L., Cui, L., Tong, Y., Zhou, X., Yin, H.: Hetefedrec: Federated recommender systems with model heterogeneity. arXiv preprint arXiv:2307.12810 (2023)
29. Yuan, W., Yang, C., Nguyen, Q.V.H., Cui, L., He, T., Yin, H.: Interaction-level membership inference attack against federated recommender systems. In: Proceedings of the ACM Web Conference 2023. pp. 1053–1062 (2023)
30. Yuan, W., Yang, C., Qu, L., Ye, G., Nguyen, Q.V.H., Yin, H.: Robust federated contrastive recommender system against model poisoning attack. arXiv preprint arXiv:2403.20107 (2024)
31. Yuan, W., Yin, H., Wu, F., Zhang, S., He, T., Wang, H.: Federated unlearning for on-device recommendation. In: Proceedings of the Sixteenth ACM International Conference on Web Search and Data Mining. pp. 393–401 (2023)
32. Yuan, W., Yuan, S., Zheng, K., Nguyen, Q.V.H., Yin, H.: Manipulating visually-aware federated recommender systems and its countermeasures. arXiv preprint arXiv:2305.08183 (2023)
33. Zaier, Z., Godin, R., Faucher, L.: Evaluating recommender systems. In: 2008 International Conference on Automated Solutions for Cross Media Content and Multi-Channel Distribution. pp. 211–217. IEEE (2008)
34. Zhang, C., Xie, Y., Bai, H., Yu, B., Li, W., Gao, Y.: A survey on federated learning. Knowl.-Based Syst. **216**, 106775 (2021)
35. Zhang, S., Yin, H., Chen, T., Huang, Z., Nguyen, Q.V.H., Cui, L.: Pipattack: Poisoning federated recommender systems for manipulating item promotion. In: Proceedings of the Fifteenth ACM International Conference on Web Search and Data Mining. pp. 1415–1423 (2022)
36. Zhang, S., Yuan, W., Yin, H.: Comprehensive privacy analysis on federated recommender system against attribute inference attacks. IEEE Transactions on Knowledge and Data Engineering (2023)
37. Zhang, Y., Ye, Q., Chen, R., Hu, H., Han, Q.: Trajectory data collection with local differential privacy. Proceedings of the VLDB Endowment **16**(10), 2591–2604 (2023)
38. Zheng, R., Qu, L., Cui, B., Shi, Y., Yin, H.: Automl for deep recommender systems: A survey. ACM Transactions on Information Systems **41**(4), 1–38 (2023)

Explicit and Implicit Counterfactual Data Augmentation for Sequential Recommendation

Zhouying Xu[1], Xuejun Liu[1(✉)], Zhuoya Xing[1], Jiasheng Cao[2], Tao He[1], and Xiaoyang Huang[1]

[1] Nanjing Tech University, Nanjing 211816, China
xjliu@njtech.edu.cn
[2] University of Birmingham, Birmingham B15 2TT, UK
jxc1481@student.bham.ac.uk

Abstract. Existing counterfactual data augmentation methods for sequential recommendation only consider users' implicit feedback for generating augmented counterfactual samples, while the explicit feedback is ignored. Therefore, we propose an **E**xplicit and **I**mplicit **C**ounterfactual data **A**ugmentation algorithm for **S**equential **R**ecommendation (EI-CASR) to address this issue. Our algorithm takes into account both explicit and implicit feedback information of users. By learning the logical inverse (NOT) operation, neural logical reasoning can model explicit feedback in sequential learning, thus making it possible to conduct counterfactual reasoning over users' explicit feedback to generate explicit counterfactual samples for data augmentation. At the same time, the implicit sampler generates implicit counterfactual samples by replacing historical items of user interaction. Two sets of augmented training samples, together with the original training samples, can help improve the recommendation performance by generating synthetic data to cover the unexplored input space. Experimental results on three public datasets demonstrate that EI-CASR significantly improves the performance of sequential recommendation tasks and effectively addresses the data sparsity problem commonly encountered in sequential recommendations.

Keywords: Sequential recommendation · Neural logic reasoning · Data sparsity · Counterfactual data augmentation

1 Introduction

In today's era of information explosion, people are faced with massive information and choices. How to effectively recommend personalized content for users has become a major challenge in Internet services. Traditional recommender systems often only consider the relationship between users and items, ignoring the timing of information, while sequential recommender systems make up for this deficiency and better reflect the evolution user behavior. Sequential recommender system is based on the user's historical behavior sequence, using temporal information and sequential relationships, which can more accurately capture the user's interest evolution and behavioral pattern changes, so as to provide users with more personalized recommendation services. It has been successfully applied in many fields such as e-commerce, news recommendation, and video streaming [7].

However, one of the main challenges facing sequential recommendation tasks is the problem of data sparsity [2]. Users do not interact with enough items so that an accurate user profile cannot be constructed. This makes it difficult to provide personalized recommendations to users, especially for new or inactive users.

To meet this challenge, researchers have explored various data augmentation techniques [12,15,19]. In recent years, some researchers have proposed the use of counterfactual inference methods to enhance sequential recommendations [6]. They attempt to model user behavior through counterfactual analysis to infer how users would act in different scenarios that do not occur in reality. Despite the success of counterfactual data augmentation in improving many recommendation models, a major problem with existing counterfactual data augmentation methods is that they only consider users' implicit feedback (e.g., click or purchase information) to generate augmented counterfactual samples, ignoring explicit feedback such as users' like or dislike preference on the items. However, explicit feedback contains rich user preference information, which helps to understand user behaviors and make accurate predictions.

To address this problem, we propose an explicit and implicit counterfactual data augmentation method for sequential recommendation, called EI-CASR. Unlike the existing counterfactual data augmentation methods, this method can use the neural logical reasoning method to model the explicit feedback in the sequence, so that the counterfactual reasoning can be performed on the users' explicit like/dislike feedback to generate explicit counterfactual samples. Even if the user's interaction history is the same, user's counterfactual preferences on the historical items can create informative counterfactual samples to augment the recommendation model. The main contributions of this paper are summarized as follows:

- To the best of our knowledge, this is the first work to combine explicit counterfactual data augmentation method and implicit counterfactual data augmentation method in the sequential recommendation task.
- We explore the joint ability of logical and counterfactual reasoning, which are two important types of reasoning abilities for machine learning.
- We conduct extensive experiments on three real-world datasets to verify the effectiveness of our approach on different models.

2 Related Work

2.1 Sequential Recommendation

Sequential recommendation (SR) aims to accurately characterize users' dynamic interests by modeling their past behavior sequences [13,14]. For sequential recommendation, early work is usually based on Markov chains [10] for modeling.

FPMC [20] combines the advantages of Markov chains and matrix decomposition, fusing sequence patterns and users' general interests. With the rapid development of deep learning, many deep recommendation models have been developed, such as methods based on convolutional neural networks [23] and recurrent neural networks [11]. As Transformer shows great potential in the field of natural language processing, many related models based on attention mechanism have been developed, such as SASRec [13], BERT4Rec [22], S^3Rec [29], etc. LSAN [14] reduces the model parameters by temporal context-aware embedding and dual-attention network, and realizes a lightweight model. ASRep [17] further mitigates the data sparsity problem by augmenting short sequences with a pretrained transformer on modified user behavior sequences. Although effective, they do not explicitly address the data sparsity problem.

2.2 Counterfactual Data Augmentation

Compared to non-sequential models, sequential models require higher-quality sequential data for training because the output is directly affected by the input sequence [6,24,25]. However, the sparsity of the actual data may hinder the performance of the sequential recommendation models. To mitigate this problem, researchers have been exploring data augmentation of counterfactual thinking to enhance datasets and models. The existing counterfactual data augmentation process uses alternatives to replace the past behaviors and generate counterfactual samples. Based on this idea, counterfactual data augmentation has led to some important achievements in recommender systems [6,24–27].

However, existing counterfactual data augmentation methods for recommender systems only consider users' implicit feedback and cannot deal with the explicit feedback for counterfactual sample generation. In fact, explicit information is also very useful in recommendation because explicit and implicit feedback show user preference from different perspectives. Therefore, considering both types of feedback simultaneously can improve the performance of sequential modeling [5]. For example, neural logic reasoning (NLR) [21], neural collaborative reasoning (NCR) [4], and graph collaborative reasoning (GCR) [3] consider users' like and dislike feedback for sequential modeling.

3 Notations and Background Knowledge

In this section, we present the important notations used in this paper and an approach to modeling explicit feedback for sequential recommendation.

Let the user set $U = \{u_1, u_2, \cdots, u_{|U|}\}$ and the item set $V = \{v_1, v_2, \cdots, v_{|V|}\}$. The user u_i interacted with a sequence of historical items $H_i = \{v_1^i, v_2^i, \cdots, v_n^i\}$, and the feedback corresponding to these interacted historical items are $B_i = \{b_1^i, b_2^i, \cdots, b_n^i\}$. If the user likes the item, let $b_t^i = 1$; if the user dislikes the item, let $b_t^i = 0$. The sequential recommendation model f predicts the ranking score r_{ij} of the user u_i to the item v_j based on H_i and B_i:

$$r_{ij} = f(H_i, B_i, v_j) \tag{1}$$

The recommendation model predicts user preferences and generates a recommendation list by ranking candidate items in descending order of ranking score r_{ij}. The sequential recommendation model can predict the user's recent behavior based on the user's interaction history. Sometimes, we can not only get the user's interaction items from the dataset, but also know whether they like each item. However, most of the existing sequential recommendation models only consider the interaction history for recommendation, i.e., only implicit feedback information is used to train the model and predict, which may lose some useful information and affect the accuracy of prediction.

One approach to modeling explicit feedback for sequential recommendation is neural collaborative reasoning (NCR) [4], which models explicit feedback for sequential recommendation by training the neural logical NOT operator. By applying the NOT operation to the historical items, the model can distinguish the positive and negative feedback of the user on the historical items. Specifically, NCR encodes a user-item interaction into an interaction event vector:

$$e_u^v = W_2 \phi \left(W_1 \begin{bmatrix} u \\ v \end{bmatrix} + b_1 \right) + b_2 \qquad (2)$$

where u and v are the latent embedding vectors of the user and the item, respectively. W_1, W_2 and b_1, b_2 are the weighted matrices and bias vectors to be learned. e_u^v is the encoded event vector representing the interaction between the user u and the item v, and $\phi(\cdot)$ is the activation function. In this paper, it is a rectified linear unit (ReLU). By introducing neural logical modules: AND ($\wedge$), OR ($\vee$), NOT ($\neg$), NCR can convert sequential data into logical expressions. Suppose that the user u has negative feedback on item v_1 and positive feedback on other items, then the explicit reasoning expression is:

$$\neg e_u^{v_1} \wedge e_u^{v_2} \wedge \cdots \wedge e_u^{v_n} \to e_u^{v_n+1} \qquad (3)$$

where $\{e_u^{v_1}, e_u^{v_2}, \cdots, e_u^{v_n}\}$ represents the user's historical event, and $e_u^{v_n+1}$ is the next item that the user interacts with. In order to consider explicit feedback information, e_u^v is used to indicate that user u has a positive interaction with item v, while $\neg e_u^v$ is used to indicate that user u has a negative interaction with item v. Ideally, the sequential recommendation process can predict $e_u^{v_n+1}$ based on $\{e_u^{v_1}, e_u^{v_2}, \cdots, e_u^{v_n}\}$. According to the definition of material implication, the above expression is equivalent to:

$$\left(\neg\neg e_u^{v_1} \vee \neg e_u^{v_2} \vee \cdots \vee \neg e_u^{v_n} \right) \vee e_u^{v_n+1} \qquad (4)$$

The recommendation score of candidate item v_{n+1} is calculated based on the similarity between the logical expression and the constant True (T) vector. According to the score, the model will decide whether to recommend the item (if the expression is close to True) or not (if the expression is close to False).

4 Method

4.1 Overview

Users' explicit feedback is difficult to quantify and model in previous works, which is an important factor affecting the recommendation results. To address this problem, we propose an explicit and implicit counterfactual data augmentation method for sequential recommendation (EI-CASR). The overall architecture of the proposed EI-CASR method is illustrated in Fig. 1.

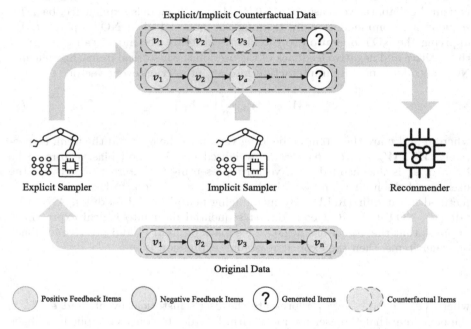

Fig. 1. The overall architecture of the proposed EI-CASR method.

Our approach consists of two samplers $\mathcal{S}$ and a sequential recommendation model $\mathcal{A}$. The explicit sampler generates explicit counterfactual data from the original data by intervening with user's explicit feedback, while the implicit sampler generates implicit counterfactual data by replacing the historical items of user interaction. The sequential recommendation model (i.e., the anchor model) is responsible for providing the final recommendation list. Firstly, both $\mathcal{A}$ and $\mathcal{S}$ in our model are pre-trained based on the original dataset. Then, the anchor model $\mathcal{A}$ is re-optimized using the explicit and implicit counterfactual data generated by the sampler. Eventually, the re-optimized anchor model will provide the final recommendation list to the user.

4.2 Explicit Sampler

In order to generate explicit counterfactual data, we use NCR as a sampler for counterfactual reasoning. This is because NCR can consider the counterfactual of explicit feedback with the help of logical negations ($\neg$). The first step of the explicit sampler is to decide which explicit feedback of user u_i should be changed. We use a binary vector $\boldsymbol{\Delta}_i = \{0,1\}^{|B_i|}$ to represent the change to the explicit feedback, whose vector size is equal to the size of the user's explicit feedback vector $\boldsymbol{B}_i$. Then, we apply the intervention on $\boldsymbol{B}_i$:

$$\boldsymbol{B}_i^* = (1 - \boldsymbol{B}_i) \odot \boldsymbol{\Delta}_i + \boldsymbol{B}_i \odot (1 - \boldsymbol{\Delta}_i) \tag{5}$$

For each $\delta_t \in \boldsymbol{\Delta}_i$, if $\delta_t = 1$, the corresponding feedback is changed; otherwise, the feedback remains unchanged. For example, if $\boldsymbol{B}_i = [0,1,1]$, $\boldsymbol{\Delta}_i = [1,1,0]$, this means that the user's feedback on the first and second items should be changed, so $\boldsymbol{B}_i^* = [1,0,1]$.

It has been demonstrated in previous work [8] that the samples with larger loss can usually provide more knowledge to widen the model's experience and improve the model performance. In the explicit sampler, we learn the explicit counterfactual sequence by maximizing the loss of the anchor model. For a given historical item $\boldsymbol{H}_i$ and feedback $\boldsymbol{B}_i$, we optimize the following objective:

$$\begin{aligned} \max\nolimits_{\boldsymbol{\Delta}_i \in \{0,1\}^{|B_i|}} & -\log \mathcal{A}\left(\hat{v}_{n+1} \mid \boldsymbol{H}_i, \boldsymbol{B}_i^*\right) \\ \text{s.t.} \quad & \hat{v}_{n+1} = \operatorname{argmax}\nolimits_{v \in I} \mathcal{S}\left(v \mid \boldsymbol{H}_i, \boldsymbol{B}_i^*\right) \\ & \|\boldsymbol{\Delta}_i\|_0 \leq \lambda \end{aligned} \tag{6}$$

where $\hat{v}_{n+1}$ represents the next item that the user may interact with under the action of the intervened feedback vector $\boldsymbol{B}_i^*$. I is a set of items in the dataset, which can be the whole item set V or another set involving prior knowledge. $\|\boldsymbol{\Delta}_i\|_0$ is the zero-norm of the intervention vector $\boldsymbol{\Delta}_i$ that represents the amount of changed feedback. In this equation, the objective is to maximize the loss of the anchor model. The first constraint states that the sequence is generated based on $\mathcal{S}$, and the second constraint controls the number of changes in the user's explicit feedback through λ.

In practice, the main challenge is that objective (6) is not differentiable. In order to solve this problem, we relax $\boldsymbol{\Delta}_i$ into a real-valued vector, and relax the l_0-norm $\|\boldsymbol{\Delta}_i\|_0$ into l_1-norm $\|\boldsymbol{\Delta}_i\|_1$, so that the intervention vector $\boldsymbol{\Delta}_i$ can be learned. This method has been proved to be effective in prior research and helps to reduce the number of changed explicit feedback in the sequence. In order to learn $\boldsymbol{\Delta}_i$, we relax objective (6) to the following differentiable target:

$$\max_{\boldsymbol{\Delta}_i} -\sum_{m=1}^{|I|} \frac{\exp\left(\tau \mathcal{S}\left(m \mid \boldsymbol{H}_i, \boldsymbol{B}_i^*\right)\right)}{\sum_{n=1}^{|I|} \exp\left(\tau \mathcal{S}\left(n \mid \boldsymbol{H}_i, \boldsymbol{B}_i^*\right)\right)} \log \mathcal{A}\left(m \mid \boldsymbol{H}_i, \boldsymbol{B}_i^*\right) - \alpha \|\boldsymbol{\Delta}_i\|_1 \tag{7}$$

where α is a tuning parameter. $-\log \mathcal{A}\left(m \mid \boldsymbol{H}_i, \boldsymbol{B}_i^*\right)$ denotes the loss of the sample when m is the current item. $\frac{\exp(\tau \mathcal{S}(m|\boldsymbol{H}_i,\boldsymbol{B}_i^*))}{\sum_{n=1}^{|I|} \exp(\tau \mathcal{S}(n|\boldsymbol{H}_i,\boldsymbol{B}_i^*))}$ imposes a softmax layer on $\mathcal{S}$, which represents the probability of interacting with item m. By multiplying them, the first term of Eq. (7) is actually calculating the expectation of the anchor model loss. τ is a temperature parameter used to tunes the softness of the item distribution.

As mentioned above, the sampler $\mathcal{S}$ is implemented with NCR to accommodate explicit feedback in the sequence. Specifically, for each user-item interaction event e in the history H_i, assume that $\boldsymbol{e}$ is the event vector corresponding to the event, and the corresponding value of the event in the intervention vector is δ_e, then the intervened event vector $\boldsymbol{e}^*$ is:

$$\boldsymbol{e}^* = \neg \boldsymbol{e} \cdot \delta_e + \boldsymbol{e} \cdot (1 - \delta_e) \qquad (8)$$

These intervened event vectors constitute the counterfactual history $\{\boldsymbol{H}_i, \boldsymbol{B}_i^*\}$ for calculating the item ranking score, as shown in Eq. (7). After optimization, the values in the learned intervention vector $\boldsymbol{\Delta}_i$ may not be completely equal to 0 or 1. Therefore, a threshold is used to binarize $\boldsymbol{\Delta}_i$ as the final output. In the experiments, we set the threshold to 0.5, i.e., for elements greater than 0.5 in $\boldsymbol{\Delta}_i$, we set them to 1, otherwise, we set them to 0. Finally, the binarized intervention vector $\boldsymbol{\Delta}_i$ is used to generate the new next item $\hat{v}_{n+1}$ for the explicit counterfactual sample:

$$\hat{v}_{n+1} = \text{argmax}_{v \in I}\, \mathcal{S}\left(v \mid \boldsymbol{H}_i, \boldsymbol{B}_i^*\right) \qquad (9)$$

4.3 Implicit Sampler

In typical classification problems, the input feature space can be divided into many subspaces according to different output labels. The boundaries between different input subspaces are called decision boundaries. For samples close to the decision boundaries, the labels can be easily changed even if the input features change little. As demonstrated in the previous work [1], these decision boundary samples are usually discriminative in revealing the underlying data patterns, and training based on them may improve model performance. Our implicit sampler is based on this principle. We generate the counterfactual sequence by "minimally" changing the user's historical items, such that the next item of user interaction can be "exactly" altered. Formally, assuming that $\boldsymbol{v} \in \mathbb{R}^D$ is the embedding of item v, we measure the change of user behavior based on the embedding space. For a given real sequence $(\{v_1, v_2, \cdots, v_n\}, v_{n+1})$ and index d, we first replace v_d with a random item v_a. Then, the replaced sequence $H_i^* = \{v_1, \cdots, v_{d-1}, v_a, v_{d+1}, \cdots, v_n\}$ is brought into the following objective for optimization:

$$\min_{v_a \in C} \|\boldsymbol{v}_a - \boldsymbol{v}_d\|_2^2 \\ \text{s.t.} \quad v_{n+1} \neq \text{argmax}_{v \in I}\, \mathcal{S}\left(v \mid \boldsymbol{H}_i^*, \boldsymbol{B}_i\right) \qquad (10)$$

where $\boldsymbol{v}_a$ and $\boldsymbol{v}_d$ are the embeddings of the replaced and original items. C represents the item set for replacement, which can be specified as V or other sets involving prior knowledge. The objective of the optimization is to minimize the distance between the item embeddings before and after the replacement (i.e., minimally changing the user's historical items). The constraint ensures that the current item is no longer v_{n+1}. By combining them, we hope to change the historical item in a minimal way so that the current item can be exactly altered. Once we get v_a, the final implicit counterfactual sequence is $(H_i^*, B_i, \hat{v}_{n+1})$, where $\hat{v}_{n+1}$ can be derived by:

$$\hat{v}_{n+1} = \operatorname{argmax}_{v \in I} \mathcal{S}(v \mid H_i^*, B_i) \tag{11}$$

To make this objective easier to handle, we introduce a "virtual" item $\widetilde{v_a}$, whose embedding is $\widetilde{v_a} = v_d + \Delta$, where Δ is a continuous vector representing the distance between $\widetilde{v_a}$ and v_d. Our general idea is first learning $\widetilde{v_a}$ in a differentiable way, and then projecting $\widetilde{v_a}$ to the nearest real item. In order to learn $\widetilde{v_a}$, the relaxed objective is:

$$\min_{\Delta \in \mathbb{R}^D} \|\Delta\|_2^2 + \beta \mathcal{S}(v_{n+1} \mid v_1, \cdots, v_{d-1}, v_d + \Delta, v_{d+1}, \cdots, v_n, B_i) \tag{12}$$

where Δ is the only learnable parameter. The first term aims to minimize the distance between the virtual and real items. The second item penalizes the probability of interacting with the real current item v_{n+1}, in other words, v_{n+1} is not expected to be predicted by $\mathcal{S}$. Therefore, this term plays a similar role to the constraint in Eq. (10). β is the tuning parameter to balance different objectives. Once we get $\widetilde{v_a}$, the substitution v_a will be derived by the following projection method:

$$v_a = \operatorname{argmin}_{v_a \in C} \|\widetilde{v_a} - v_a\|_2^2 \tag{13}$$

4.4 Reduce Noisy Samples

As mentioned above, for the generated explicit and implicit counterfactual data, we select the next interaction item based on the sampler. However, since the sampler model is not completely accurate, it may generate inaccurate predictions. The accurate and inaccurate counterfactual samples generated by the sampler will be used to re-optimize the anchor model $\mathcal{A}$. Therefore, if we do not set some constraints to reduce the number of inaccurate counterfactual samples, the performance of the re-optimized anchor model may be affected. Inspired by [24], we set a confidence parameter $\kappa \in [0, 1)$ to alleviate this problem. Only when $\mathcal{S}(\hat{v}_{n+1} \mid H_i, B_i^*) > \kappa$ or $\mathcal{S}(\hat{v}_{n+1} \mid H_i^*, B_i) > \kappa$, do we accept the generated counterfactual data. This means that only when the sampler has sufficient confidence in the counterfactual sample will we accept it. Otherwise, the model will discard the sample.

5 Experiments

In this section, we conduct experiments on three real-world datasets and compare the results of the original sequential recommendation model without data augmentation, models with implicit counterfactual data augmentation, and models with our EI-CASR framework.

5.1 Experimental Settings

Datasets. We evaluate our proposed method on three real-world public datasets. The MovieLens-100K [9] (ML100K) dataset stores users' preference for various movies. The Amazon [18] dataset is a large collection of product reviews and ratings from the Amazon online platform. We select two of the datasets, i.e., Movies & TV and Electronics, for experiments. The basic statistics of the datasets are summarized in Table 1.

Table 1. Statistics of the datasets.

Dataset	Users	Items	Interactions	Density
ML100K	943	1,682	100,000	6.30%
Movies & TV	123,961	50,053	1,697,533	0.027%
Electronics	192,404	63,002	1,689,188	0.014%

We regard 1–3 ratings as negative feedback, label 0, 4–5 ratings as positive feedback, label 1. We use the positive LeaveOne-Out [4,28] method to create the training, validation, and testing dataset. For each user, the last positive interaction and its subsequent negative interactions are put into the testing set, and the last but one positive interaction and its subsequent negative interactions are put into the validation set. Then, we put all the remaining interactions into the training set. If a user has less than 5 interactions, we put all the interactions into the training set to avoid cold start.

Metrics. In the experiments, we use Hit Ratio (HR) and Normalized Discounted Cumulative Gain (NDCG) to evaluate the recommendation performance. Hit Ratio is more focused on whether the item exists in the recommendation list, while NDCG is more focused on the ranking of the recommendation sequence. In this paper, HR@5, HR@10, NDCG@5, and NDCG@10 are used to evaluate the model. For each user-item pair in the validation set and the test set, we randomly select 100 irrelevant items and rank all these 101 items for recommendation.

Baselines. We use the following baseline models for comparison to demonstrate the effectiveness of EI-CASR.

- **GRU4Rec** [11] is a sequential recommendation model based on Recurrent Neural Networks (RNN).
- **STAMP** [16] is a sequential recommendation model based on attention mechanism, which can capture users' long-term and short-term preferences for recommendation.
- **SASRec** [13] is a sequential recommendation model based on self-attention mechanism.

Table 2. Comparison of experimental results.

Dataset	Method	HR@5	HR@10	NDCG@5	NDCG@10
ML100K	GRU4Rec	0.5014	0.6719	0.3401	0.4037
	CASR-GRU4Rec	0.5082	0.6796	0.3488	0.4120
	EICASR-GRU4Rec	**0.5191**	**0.6874**	**0.3639**	**0.4266**
	STAMP	0.5028	0.6651	0.3422	0.4017
	CASR-STAMP	0.5105	0.6768	0.3516	0.4059
	EICASR-STAMP	**0.5263**	**0.6917**	**0.3624**	**0.4150**
	SASRec	0.5077	0.6783	0.3476	0.4115
	CASR-SASRec	0.5182	0.6855	0.3569	0.4152
	EICASR-SASRec	**0.5271**	**0.6998**	**0.3732**	**0.4264**
	NCR	0.5140	0.6801	0.3586	0.4122
	CASR-NCR	0.5184	0.6892	0.3617	0.4193
	EICASR-NCR	**0.5323**	**0.7026**	**0.3749**	**0.4316**
Movies & TV	GRU4Rec	0.5378	0.6610	0.4111	0.4315
	CASR-GRU4Rec	0.5421	0.6714	0.4146	0.4539
	EICASR-GRU4Rec	**0.5535**	**0.6903**	**0.4258**	**0.4624**
	STAMP	0.5217	0.6568	0.4060	0.4274
	CASR-STAMP	0.5320	0.6609	0.4123	0.4456
	EICASR-STAMP	**0.5501**	**0.6839**	**0.4337**	**0.4615**
	SASRec	0.5430	0.6671	0.4125	0.4559
	CASR-SASRec	0.5488	0.6729	0.4204	0.4613
	EICASR-SASRec	**0.5612**	**0.6958**	**0.4350**	**0.4771**
	NCR	0.5511	0.6726	0.4155	0.4563
	CASR-NCR	0.5548	0.6822	0.4177	0.4581
	EICASR-NCR	**0.5643**	**0.6970**	**0.4336**	**0.4694**
Electronics	GRU4Rec	0.4325	0.5536	0.3123	0.3547
	CASR-GRU4Rec	0.4472	0.5600	0.3265	0.3691
	EICASR-GRU4Rec	**0.4718**	**0.5779**	**0.3423**	**0.3834**
	STAMP	0.4119	0.5420	0.3014	0.3406
	CASR-STAMP	0.4246	0.5531	0.3073	0.3488
	EICASR-STAMP	**0.4497**	**0.5675**	**0.3238**	**0.3612**
	SASRec	0.4392	0.5581	0.3229	0.3573
	CASR-SASRec	0.4505	0.5747	0.3355	0.3658
	EICASR-SASRec	**0.4680**	**0.5896**	**0.3542**	**0.3881**
	NCR	0.4406	0.5575	0.3317	0.3662
	CASR-NCR	0.4513	0.5698	0.3389	0.3744
	EICASR-NCR	**0.4652**	**0.5863**	**0.3505**	**0.3932**

- **NCR** [4] is a sequential recommendation model based on neural logical reasoning, which captures the logical relationship between user-item interactions for recommendation.
- **CASR** [24] is a state-of-the-art implicit counterfactual data augmentation method for sequential modeling.

CASR and EI-CASR can be applied as data augmentation methods to the GRU4Rec, STAMP, SASRec and NCR models.

Implementation. In the experiments, all methods use the ReLU function as the activation function, the embedding size is set to 64, and the methods are optimized using mini-batch with a batch size of 128. Grid search is performed for hyper-parameters: search for learning rate in [0.0001, 0.001, 0.01, 0.1], search for regularization parameters α and β in [10^{-3}, 10^{-2}, 10^{-1}, 1, 10^1, 10^2, 10^3]. The confidence parameter κ ranges from 0 to 1. In the following experiments, we will discuss the influence of different confidence parameter κ and index d. We tune each model's parameters to its own best performance on the validation set.

5.2 Performance Comparison

Table 2 shows the results of the four baseline models on the datasets without data augmentation, with data augmentation using CASR, and with data augmentation using EI-CASR. If "X" represents one of the four sequential recommendation models, then "CASR-X" represents the results of model X on the dataset enhanced by CASR, and "EICASR-X" represents the results of model X on the dataset enhanced by EI-CASR. We show the best results in bold.

The results show that compared with the original model and the model enhanced by implicit counterfactual data, Our EI-CASR framework achieves significantly better performance than the baseline methods on all of these three datasets. Specifically, the hit ratio (HR) and normalized discounted cumulative gain (NDCG) of EI-CASR are on average 2.8% and 3.8% higher than those of CASR, and 4.5% and 6.1% higher than those without data augmentation. Compared with the original model, EI-CASR can obtain better results based on the generated explicit and implicit counterfactual data, which alleviates the data scarcity and encodes informative samples into the training dataset. Compared with the implicit counterfactual data augmentation method CASR, EI-CASR utilizes explicit feedback, so the counterfactual data generated by EI-CASR are more effective.

5.3 Influence of the Iteration Number

Since the re-optimized anchor model $\mathcal{A}'$ achieves better performance than the original anchor model $\mathcal{A}$, will the performance be better if we use the re-optimized anchor model to generate a new set of augmented data and optimize $\mathcal{A}'$ again to $\mathcal{A}''$. We use NCR as an example of the anchor model to test this idea. Unfortunately, as shown in Fig. 2, the performance on HR and NDCG decreases with the increase of the iteration times. The reason is that the sampler is not completely accurate, and the generated counterfactual samples may contain noise, which is learned into the anchor model. Therefore, multiple rounds of augmentation and re-optimization may propagate this noise and thus degrade the performance. This means that data augmentation cannot infinitely improve model performance.

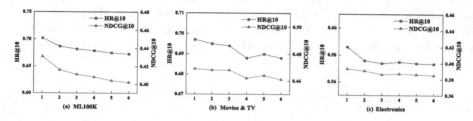

Fig. 2. Influence of the iteration number.

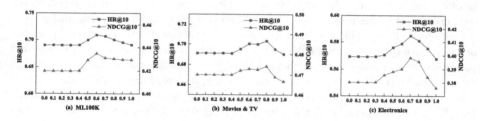

Fig. 3. Influence of the confidence parameter κ.

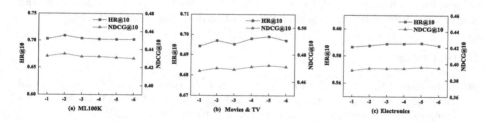

Fig. 4. Influence of the index d.

5.4 Hyper-Parameter Analysis

Influence of the Confidence Parameter κ. During the generation of the counterfactual data, we set a confidence parameter κ. Only when the ranking score of the data is greater than κ, do we accept the generated counterfactual data. We tested different values of κ on all three datasets, and the results are shown in Fig. 3. From the figure, we can see that when κ is set around 0.7, our framework has the best performance. When κ is very small, it does not change the performance of the framework because all the generated data can pass the constraint of κ. When κ is too large, the performance will decrease because only a few counterfactual data can pass the constraint, so we cannot obtain enough counterfactual data to re-optimize the anchor model.

Influence of the Intervention Index d. In order to investigate the influence of the user recent behaviors, we tune d in the range of $[-1, -2, -3, -4, -5, -6]$, where the negative value represents the index from back to front. The experimental results are shown in Fig. 4. We can see that in most cases, the performance fluctuates little as d varies, and slightly better performance is usually obtained when d falls in the range of $[-2, -5]$. This shows that in practice, a moderate d may be a better choice.

6 Conclusion

In this paper, we propose an explicit and implicit counterfactual data augmentation method for sequential recommendation task, which considers not only implicit feedback but also explicit feedback to improve sequential recommendation models. This method constructs two counterfactual samplers and a recommender. The explicit sampler generates explicit counterfactual data from the original data by intervening with user's explicit feedback, while the implicit sampler generates implicit counterfactual data by replacing the historical items of user interaction. The recommender uses the explicit and implicit counterfactual data generated by the sampler for re-optimization, and the re-optimized recommender will provide the user with the final recommendation list. In addition, the approach explores the joint ability of logical and counterfactual reasoning, which are two important types of reasoning abilities for machine learning. Experimental results on three public datasets demonstrate that the recommendation model can achieve better results when trained on the data generated by EI-CASR.

References

1. Abbasnejad, E., Teney, D., Parvaneh, A., Shi, J., Hengel, A.v.d.: Counterfactual vision and language learning. In: Proceedings of the IEEE/CVF conference on computer vision and pattern recognition. pp. 10044–10054 (2020)
2. Bian, S., Zhao, W.X., Wang, J., Wen, J.R.: A relevant and diverse retrieval-enhanced data augmentation framework for sequential recommendation. In: Proceedings of the 31st ACM International Conference on Information & Knowledge Management. pp. 2923–2932 (2022)
3. Chen, H., Li, Y., Shi, S., Liu, S., Zhu, H., Zhang, Y.: Graph collaborative reasoning. In: Proceedings of the Fifteenth ACM International Conference on Web Search and Data Mining. pp. 75–84 (2022)
4. Chen, H., Shi, S., Li, Y., Zhang, Y.: Neural collaborative reasoning. In: Proceedings of the Web Conference 2021. pp. 1516–1527 (2021)
5. Chen, S., Peng, Y.: Matrix factorization for recommendation with explicit and implicit feedback. Knowl.-Based Syst. **158**, 109–117 (2018)

6. Chen, X., Wang, Z., Xu, H., Zhang, J., Zhang, Y., Zhao, W.X., Wen, J.R.: Data augmented sequential recommendation based on counterfactual thinking. IEEE Transactions on Knowledge and Data Engineering (2022)
7. Cheng, M., Liu, Z., Liu, Q., Ge, S., Chen, E.: Towards automatic discovering of deep hybrid network architecture for sequential recommendation. In: Proceedings of the ACM Web Conference 2022. pp. 1923–1932 (2022)
8. Fu, T.-J., Wang, X.E., Peterson, M.F., Grafton, S.T., Eckstein, M.P., Wang, W.Y.: Counterfactual Vision-and-Language Navigation via Adversarial Path Sampler. In: Vedaldi, A., Bischof, H., Brox, T., Frahm, J.-M. (eds.) ECCV 2020. LNCS, vol. 12351, pp. 71–86. Springer, Cham (2020). https://doi.org/10.1007/978-3-030-58539-6_5
9. Harper, F.M., Konstan, J.A.: The movielens datasets: History and context. Acm transactions on interactive intelligent systems (tiis) **5**(4), 1–19 (2015)
10. He, R., McAuley, J.: Fusing similarity models with markov chains for sparse sequential recommendation. In: 2016 IEEE 16th international conference on data mining (ICDM). pp. 191–200. IEEE (2016)
11. Hidasi, B., Karatzoglou, A., Baltrunas, L., Tikk, D.: Session-based recommendations with recurrent neural networks. arXiv preprint arXiv:1511.06939 (2015)
12. Jiang, S., Chu, Y., Wang, Z., Ma, T., Wang, H., Lu, W., Zang, T., Wang, B.: Explainable text classification via attentive and targeted mixing data augmentation. In: International Joint Conference on Artificial Intelligence (2023)
13. Kang, W.C., McAuley, J.: Self-attentive sequential recommendation. In: 2018 IEEE international conference on data mining (ICDM). pp. 197–206. IEEE (2018)
14. Li, Y., Chen, T., Zhang, P.F., Yin, H.: Lightweight self-attentive sequential recommendation. In: Proceedings of the 30th ACM International Conference on Information & Knowledge Management. pp. 967–977 (2021)
15. Li, Y., Luo, Y., Zhang, Z., Sadiq, S., Cui, P.: Context-aware attention-based data augmentation for poi recommendation. In: 2019 IEEE 35th International Conference on Data Engineering Workshops (ICDEW). pp. 177–184. IEEE (2019)
16. Liu, Q., Zeng, Y., Mokhosi, R., Zhang, H.: Stamp: short-term attention/memory priority model for session-based recommendation. In: Proceedings of the 24th ACM SIGKDD international conference on knowledge discovery & data mining. pp. 1831–1839 (2018)
17. Liu, Z., Fan, Z., Wang, Y., Yu, P.S.: Augmenting sequential recommendation with pseudo-prior items via reversely pre-training transformer. In: Proceedings of the 44th international ACM SIGIR conference on Research and development in information retrieval. pp. 1608–1612 (2021)
18. Ni, J., Li, J., McAuley, J.: Justifying recommendations using distantly-labeled reviews and fine-grained aspects. In: Proceedings of the 2019 conference on empirical methods in natural language processing and the 9th international joint conference on natural language processing (EMNLP-IJCNLP). pp. 188–197 (2019)
19. Ni, S., Zhou, W., Wen, J., Hu, L., Qiao, S.: Enhancing sequential recommendation with contrastive generative adversarial network. Information Processing & Management **60**(3), 103331 (2023)
20. Rendle, S., Freudenthaler, C., Schmidt-Thieme, L.: Factorizing personalized markov chains for next-basket recommendation. In: Proceedings of the 19th international conference on World wide web. pp. 811–820 (2010)
21. Shi, S., Chen, H., Ma, W., Mao, J., Zhang, M., Zhang, Y.: Neural logic reasoning. In: Proceedings of the 29th ACM International Conference on Information & Knowledge Management. pp. 1365–1374 (2020)

22. Sun, F., Liu, J., Wu, J., Pei, C., Lin, X., Ou, W., Jiang, P.: Bert4rec: Sequential recommendation with bidirectional encoder representations from transformer. In: Proceedings of the 28th ACM international conference on information and knowledge management. pp. 1441–1450 (2019)
23. Tang, J., Wang, K.: Personalized top-n sequential recommendation via convolutional sequence embedding. In: Proceedings of the eleventh ACM international conference on web search and data mining. pp. 565–573 (2018)
24. Wang, Z., Zhang, J., Xu, H., Chen, X., Zhang, Y., Zhao, W.X., Wen, J.R.: Counterfactual data-augmented sequential recommendation. In: Proceedings of the 44th international ACM SIGIR conference on research and development in information retrieval. pp. 347–356 (2021)
25. Xiong, K., Ye, W., Chen, X., Zhang, Y., Zhao, W.X., Hu, B., Zhang, Z., Zhou, J.: Counterfactual review-based recommendation. In: Proceedings of the 30th ACM International Conference on Information & Knowledge Management. pp. 2231–2240 (2021)
26. Yang, M., Dai, Q., Dong, Z., Chen, X., He, X., Wang, J.: Top-n recommendation with counterfactual user preference simulation. In: Proceedings of the 30th ACM International Conference on Information & Knowledge Management. pp. 2342–2351 (2021)
27. Zhang, S., Yao, D., Zhao, Z., Chua, T.S., Wu, F.: Causerec: Counterfactual user sequence synthesis for sequential recommendation. In: Proceedings of the 44th International ACM SIGIR Conference on Research and Development in Information Retrieval. pp. 367–377 (2021)
28. Zhao, W.X., Chen, J., Wang, P., Gu, Q., Wen, J.R.: Revisiting alternative experimental settings for evaluating top-n item recommendation algorithms. In: Proceedings of the 29th ACM International Conference on Information & Knowledge Management. pp. 2329–2332 (2020)
29. Zhou, K., Wang, H., Zhao, W.X., Zhu, Y., Wang, S., Zhang, F., Wang, Z., Wen, J.R.: S3-rec: Self-supervised learning for sequential recommendation with mutual information maximization. In: Proceedings of the 29th ACM international conference on information & knowledge management. pp. 1893–1902 (2020)

Attention-Based Causal Graph Convolutional Collaborative Filtering

Youhan Qi[1], Xinglin Liu[1], Chenyu Li[1], and Ying Wang[1,2](✉)

[1] College of Computer Science and Technology, Jilin University, Changchun 130012, Jilin, China
wangying2010@jiu.edu.cn
[2] Key Laboratory of Symbol Computation and Knowledge Engineering (Jilin University), Ministry of Education, Changchun 130012, Jilin, China

Abstract. Graph-structured data can naturally represent complex relationships between entities, hence graph neural networks have been widely employed in collaborative filtering methods to facilitate data mining on user-item graphs. However, the use of traditional methods may be affected by confounding factors in the prediction process, resulting in unfairness or suboptimality in the recommendation, and reducing the performance of the recommendation system. Causal Inference, as a powerful method, facilitates understand the dependency relationships between variables and identify potential data biases, thereby providing more accurate and fair recommendations for recommendation systems. Therefore, we propose a fair collaborative filtering model called Attention-based Causal Graph Convolutional Collaborative Filtering (ACGCF), which can address bias issues in recommendation systems. Firstly, ACGCF preprocess the dataset of the recommendation system and then the graph convolution method and attention mechanism are used to model the higher-order relationship on the user-item graph. Secondly, the backdoor adjustment in causal learning is used to eliminate the influence of confounding factors, and the do-operations are adjusted according to the user's feedback and behavior. Finally, integrate the user-user graph, item-item graph, and the causal debiasing user-item graph, and utilize graph matching to derive the final recommendation results. Furthermore, we have numerically verified the superiority of our method in terms of recommendation performance with other comparison methods.

Keywords: Recommender System · Graph Neural Networks aggregation · Causal inference · Back-door adjustment

1 Introduction

With the emergence of the information age, recommendation systems have become an indispensable part of major online platforms [1]. Through the collection and processing of user behavior data by data mining algorithms, we can discover hidden patterns and relationships in the data, thereby improving the accuracy and effectiveness of the recommendation system and greatly improving the user experience. Traditional recommender systems such as collaborative filtering (CF) [14] are one of the most important recommendation methods, which rely on a basic assumption that people with similar purchase experiences in the past will make similar decisions in the future. However, the efficacy of this approach is often impaired by challenges such as data sparsity, the cold start problem, and biases inherent in user-item interactions.

With the rise of Graph Neural Networks (GNNs) [2,3], the research of recommender system has made new progress. The history of user-item interactions is modeled as a user-item graph and applied to several domains such as knowledge graphs, heterogeneous graphs, and social recommendation. Examples include molecular property prediction [4,5] and object relation learning physical systems [6]. Some works consider user-item interactions as a bipartite graph, where edges between users and items represent interactions (e.g., clicks or rates) [7,15], and these models only use user-item interaction information in GNNs for learning, ignoring implicit complex interactions. Other works utilize GNNs to model knowledge graphs for reasoning [8,9]. These models treat edges as predefined relationships between attributes and items, users, rather than relationships between attributes and attributes. Previous research on GNN-enhanced recommendation has focused on developing effective messaging strategies to describe collaborative relationships between users and items [1,10]. Subsequent studies have deeply investigated simplified message passing methods, reducing the complexity of GNN models [10,11], and improving the quality of sampling methods [13]. Recently, progress has been made in this field with the integration of self-supervised learning (SSL) into GNN recommendation frameworks. The GDCCDR model [28], for instance, leverages graph disentangled contrastive learning to address the cross-domain recommendation challenge, while the F2PGNN [29] model introduces fairness in federated learning for personalized recommendations. The LGMRec [30] model further extends the capability to multimodal recommendations by jointly modeling local and global user interests. Although these models [2–4] improve the performance of recommender systems to some extent, they still analyze the interactions between all attributes equally, so they cannot capture the structural information of useful attribute interactions for joint decision making.

Although GNN has achieved remarkable results in recommender systems, there are still some challenges and problems. Firstly, the convolution operation of traditional GCN needs to aggregate the neighbor information of all nodes in the graph. For graphs with a large number of nodes and edges, the computational complexity of this operation is very high and consumes a large amount of computing resources. Second, these approaches assume that all interaction items can be used to model user preferences and that each item contributes equally in generating user preferences. However, the differences between items can affect the user's purchase decision. Finally, the bias problem in the user-item graph is still prominent. In the operation process of online shopping recommendation system, user interest is a core driving force, which directly affects the user's purchase decision. At the same time, the visibility or popularity of a product is an external factor that does not belong to the user's own information. Although it is independent of the user's interests, it also significantly affects the user's purchase decision. As shown in Fig. 1, in this example, User A's daily shopping preference is the electronics category. However, due to the influence of social events or external factors, a large number of users, including User A, have simultaneously purchased musical instruments. Such purchasing behavior can confuse the recommendation system's judgment, leading to recommendations outside of the user's preferences. Such external factors that affect the recommendation results are called confounding factors. Traditional GNN-based recommendation methods mainly focus on the learning of node representation and the generation of recommendation results, and lack of in-depth exploration of the causal relationship between nodes [12], so they are susceptible to the interference of confounding factors, resulting in the performance degradation of the recommendation system. In summary, this paper firstly performs data preprocessing to solve the problem of excessive graph data, then uses the attention mechanism to retain user preferences, and finally introduces causal reasoning into the recommendation system to reduce the impact of confounding factors on the system. In this paper, a model named Attention-based Causal Graph Convolutional Collaborative Filtering (ACGCF) is proposed to solve the bias problem in recommendation systems based on Attention mechanism and causal learning. The main work is summarized as follows:

- We preprocess datasets, considering the importance of attribute interactions for recommendation predictions, and categorize them into intrinsic and cross interactions based on their roles in the recommendation process, thus reducing computational complexity.
- We propose an adaptive preference retention mechanism to identify the differences between items, which are retained during the training process through the attention mechanism to capture user preferences and improve the recommendation performance of the model.
- Based on the idea of causal learning, we use causality diagram as a tool, and use a distribution intervention technique in causal learning to increase the diversity of causal features in the data distribution, reduce the false statistical correlation between user characteristics and goods, thereby weakening the impact of false correlation on the prediction process of the recommendation system.
- We verify the effectiveness of the proposed model ACGCF compared with other recommender system models on empirically acquired datasets, and verify the effectiveness of each module of the model through ablation experiments.

The paper is organized as follows: In Sect. 2, related work on causal inference and graph neural networks is reviewed. Section 3 introduces the proposed model, and Sect. 4 verifies the effectiveness of this model on real datasets. Finally, we conclude our work here in Sect. 5.

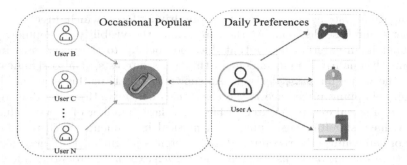

Fig. 1. The rise in the popularity of a musical instrument is an intervention for users who love electronics.

2 Preliminary

2.1 Graph Construction in Recommendation

In this section, we will introduce the notation used in this paper. We typically modeled the history of user-item interactions as a user-item graph $G = \{U, V, R, E\}$. Let $U = \{u_1, u_2, \ldots, u_n\}$ and $V = \{v_1, v_2, \ldots, v_m\}$ denote the sets of users and items, respectively, where n is the number of users and m is the number of items. We assume that $R \in R^{n*m}$ was the user-item graph and may contain several ordinal rating levels $\{1, \ldots, R\}$. Each edge $e = (u_i, v_j, r_{ij}) \in E$ show that user u_i assigned rating r_{ij} to item v_j. If u_i assigned a rating to v_j, r_{ij} is the rating; otherwise, $r_{ij} = 0$. Generally, we let R_{u_i} be the set of items that have interacted with the user u_i, R_{v_j} be the set of users who v_j have interacted with. We use an embedding vector $s^{u_i} \in R^d$ and an embedding vector $h^{v_j} \in R^d$ to denote a user u_i and an item v_j, respectively, where d was the dimension of the embedding vector.

2.2 Causal Graph

We construct a causal graph to explicitly analyze causality in traditional RS, as shown in Fig. 2(left) which contains five variables P, U, V, B, and Y. We now describe the plausibility of causal graph at a higher level as follows [17,18].

In Fig. 2, U denotes the user representation. $u_k \in R^p$ signifies the representation for k-th user, where $u_k[p]$ represents the value of the p-th attribute's value for k-th user. V denotes the item representation. $v_j \in R^q$ denotes the representation for j-th item, where $v_j[q]$ represents the value of the q-th attribute's value for j-th item. P denotes hidden confounders arising from user historical distributions, which can directly affect U. B is the inherent preference representation that can be computed from U and P, which describes the preference of k-th user under a historical distribution p_u. $Y \in \{0, 1\}$ denotes the prediction score for each user-item pair. $v_j[q]$ is q-th attribute's value for k-th item.

The edges in the graph represent the causal relations between each variable, in particular, the rationality of edges can be explained as follows:

- $P \to U$: Confounding factors can directly affect user representations, as users are influenced by their item popularity or their surrounding friends and interact with item groups at a higher frequency. The objective of causal learning is to eliminate the spurious correlation between P and U, thereby reducing the influence of confounders P on recommendation results.
- $(P, U) \to B$: The intrinsic preference representation is directly determined by P and U.
- $V \to Y$: Item representation V directly affects the result Y.
- $(U, B) \to Y$: This indirect path indicates that the outcome Y can be indirectly influenced by U and directly influenced by B.

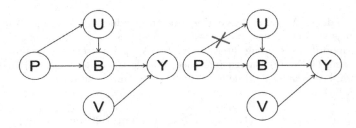

Fig. 2. The established causal graph(left) and the causal graph with confounding factors eliminated(right).

3 Methodology

In this section, we describe our model in detail. The overall framework of ACGCF is shown in Fig. 3. Then We uncover the details and ideas of each part of the model.

3.1 User and Item Graph Construction

We first utilize the message passing method to model the intrinsic interactions between nodes for feature learning. Inspired by the related literature [3,11], we chose MLP (Multilayer Perceptron) to model these internal interactions. Subsequently, we integrate the modeling results of these interactions into information about message passing. Specifically, we set up an MLP: $R^d \times R^d \to R^e$, which takes as input the embeddings of two nodes and outputs an interaction modeling result:

$$M_{ab} = g_{message}(v_a, v_b) \tag{1}$$

where, M_{ab} represents the interaction modeling result of node pair (a, b). We then summarize all interaction modeling results related to a particular node into message passing information. $g_{message}$ explicitly captures the interaction details

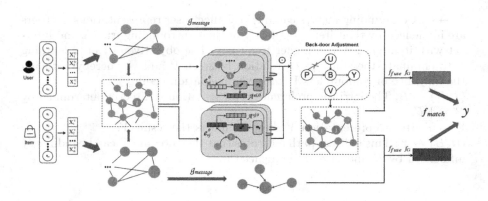

Fig. 3. The model of ACGCF. ACGCF is a recommender system framework designed to mitigate bias through causal inference, comprising four key components: (a) Graph construction module for parsing interactions. (b) Adaptive preference retention mechanism leveraging GNN. (c) Causal inference module to reduce bias. (d) Information fusion and graph matching module for enhanced recommendations.

between two attributes, effectively bridging the gap between modeling interactions for recommendation systems and message propagation in graph learning. Furthermore, inner interactions are employed to delineate the unique characteristics of users (items) and often exhibit intricate patterns. A high score for inner interaction does not necessarily indicate similarity between the two attributes in defining the specific traits of a user (item), contrasting with cross-interaction models that highlight similarity. Consequently, a neural approach that nonlinearly models the relationship between the two attributes is highly desirable.

3.2 Adaptive Preference Retention Mechanism

In this paper, we aimed to learn the fine-grained preference A^{u_i} of user u_i and representation A^{v_j} of item v_j from user-item graph. It was claimed that user u_i's preference relied on his/her characteristics s^{u_i} and neighborhood $h^{a,k}$, where $a \in R_{u_i}$. Where $e_{u_i}^{k+1}$ denotes the user u_i's embedding of GNN at layer k that is generated by user aggregation function. $k = \{1, 2, \ldots, K\}$ indicated the number of layers in GNN. R_{u_i} was the set of items user u_i has interacted with. R_{v_j} represents the set of users that have interacted with the item v_j. $s^{u_i,k}$ denoted the vector of user u_i at k-th layer. $h^{v_j,k}$ was the vector of item v_j at k-th layer. We adopted aggregation functions that is different from the traditional graph convolutional network on the user-item graph to capture the user's/item's representations. To be specific, the attention mechanism was applied to aggregate the neighborhood of user/item, as below:

$$e_{u_i}^{k+1} = \text{ReLU}(W_{user} \cdot (s^{u_i,k} \oplus \sum_{a \in R_{u_i}} \alpha_{i,a} \cdot h^{a,k}) + b_{user}) \quad (2)$$

$$e_{v_j}^{k+1} = \text{ReLU}(W_{item} \cdot (h^{v_j,k} \oplus \sum_{b \in R_{v_j}} \alpha_{j,b} \cdot s^{b,k}) + b_{item}) \quad (3)$$

where $\alpha_{i,a}$, user's neighborhood when characterizing user denoted the attention coefficient of items in the u_i's representation from the user-item interaction history A^{u_i}, $\alpha_{j,b}$ denoted the attention weight of users in the item v_j's neighborhood when characterizing item v_j's representation from the item-user interaction history R_{v_j}. W_{users}/W_{items} and b_{users}/b_{items} were weight and bias of aggregation functions, Take the aggregation function of user u_i as an example, we took user u_i's representation $s^{u_i,k}$ and u_i features of item interacted with user u_i as input, then use *Softmax* function to produce the attention coefficient $\alpha_{i,a}$ as follows:

$$\alpha_{i,a} = \frac{exp(\text{ReLU}(W_{i,j} \cdot (s^{u_i,k} \oplus h^{a,k})))}{\sum_{a \in R_{u_i}} exp(\text{ReLU}(W_{i,j} \cdot (s^{u_i,k} \oplus h^{a,k})))} \quad (4)$$

$$\alpha_{j,b} = \frac{exp(\text{ReLU}(W_{j,i} \cdot (s^{u_i,k} \oplus h^{a,k})))}{\sum_{b \in R_{v_j}} exp(\text{ReLU}(W_{j,i} \cdot (s^{u_i,k} \oplus h^{a,k})))} \quad (5)$$

where $W_{i,j}$ and $W_{j,i}$ are the attention weight of $\alpha_{i,a}$ and $\alpha_{j,b}$. The GNN can capture information on the neighborhood of a given node through the feature propagation mechanism. The nodes were assumed to contribute similarly in the neighborhood aggregation process, which led to the inevitable loss of fine-grained user preference-related information. In practice, the purchasing behavior of the user was usually determined by differences among items. Therefore, different items should make different contributions to generate the user embedding. This motivated us to design an adaptive preference retention mechanism to adaptively identify different kinds of information provided by different items.

By taking original user embedding generated by the GNN as input, we took advantage of a trainable mapping vector shared between layers of the GNN to learn the differences among items, and to output their preference retention scores, such as $a^{u_i,k}$ and $a^{v_j,k}$. These preference retention scores were used for representations that carried information on the differences among items in neighborhoods spanning a variety of ranges, and also for measuring the feature-related information of the items to identify their contributions as follows:

$$a^{u_i,k} = \text{ReLU}(m_1 \cdot e_{u_i}^k + n_1) \quad (6)$$

$$a^{v_j,k} = \text{ReLU}(m_2 \cdot e_{v_j}^k + n_2) \quad (7)$$

where $a^{u_i,k}$ was the preference retention scores of user u_i at k-th layer, and $a^{u_i,k}$ was the preference retention scores of v_j same as u_i. m_1 and m_2 indicated trainable mapping vectors shared by the processes of user and item preference modeling, respectively, and n_1 and n_2 were biases. $e_{u_i}^k$ and $e_{v_j}^k$ are user's/item's embeddings generated by GNN at the k-th layer, we model the user and item preferences through two trainable mapping vectors m_1 and m_2 and We use the adaptively adjusted weight m_1/m_2 shared between GNN layers was used to generate retention scores for different neighborhoods.

Finally, we confirm differences among the items by using the preference retention scores and preserving the differences in user embeddings automatically to model finegrained user preferences. The generated user/item preference embeddings $A^{u_i,k} = a^{u_i,k} / A^{v_j,k} = a^{v_j,k}$ were as follows:

$$A^{u_i,k} = a^{u_i,k} \odot e_{u_i}^k \qquad (8)$$

$$A^{v_j,k} = a^{v_j,k} \odot e_{v_j}^k \qquad (9)$$

3.3 Causal Intervention to Solve the Bias Problem

Due to the presence of confounders, existing recommender systems that estimate the conditional likelihood $P(Y|U,V)$ encounter spurious correlations, thereby contributing to the phenomenon of bias amplification [27]. Mathematically speaking, when considering a specific $U = u$ and $V = v$, the conditional probability $P(Y|U,V)$ can be calculated as follows:

$$P(Y|U=u, V=v)$$

$$= \frac{\sum_{d \in D} \sum_{m \in M} P(d)P(u|d)P(m|d,u)P(v)P(Y|u,v,m)}{P(u)P(v)} \qquad (10a)$$

$$= \sum_{d \in D} \sum_{m \in M} P(d|u)P(m|d,u)P(Y|u,v,m) \qquad (10b)$$

$$= \sum_{d \in D} P(d|u)P(Y|u,v,M(d,u)) \qquad (10c)$$

$$= P(d_u|u)P(Y|u,v,M(d_u,u)) \qquad (10d)$$

Where D and M are the sample spaces of D and M, Eq. (10a) adheres to the law of total probability, while Eq. (10b) is derived from Bayes rule. Since M can only assume the value $M(d,u)$ when $U = u$ and $D = d$, the conditional probability $P(M(d,u)|d,u)$ is set to 1, thereby eliminating the summation over M in Eq. (10b). Moreover, once $U = u$ is known, D is also known. Consequently, the probability of u having the distribution d is 1 only if d is identical to d_u, the historical distribution of user u over item groups. Otherwise, $P(d|u)$ is 0. From Eq. (10d), we observe that $P(d|u)$ not only shapes the user representation u but also influences Y through $M(d_u, u)$, introducing a spurious correlation.

To mitigate the impact of confounding factors, ACGCF assesses the causal influence of user representations on prediction scores. In practical applications, one approach is to gather intervention data where user representations are deliberately manipulated to isolate the effects of confounding factors. However, such interventions are often expensive to carry out on a large scale and may pose risks to user experience. Consequently, ACGCF utilizes a causal methodology known as backdoor adjustment [12,17,27], which allows estimating causal effects from existing data. According to the backdoor adjustment principle [12], the objective of ACGCF can be framed as: $P(Y|do(U=u), V=v)$, where $do(U=u)$ conceptually represents the removal of the causal links from P to U in the causal graph, thus preventing the influence of P on U (refer to Fig. 2, right panel).

We then develop a specific formula for implementing backdoor adjustment as follows:

$$P(Y|do(U=u), V=v)$$

$$= \sum_{d \in D} P(d|do(U=u))P(Y|do(U=u), v, M(d, do(U=u))) \quad (11a)$$

$$= \sum_{d \in D} P(d)P(Y|do(U=u), v, M(d, do(U=u))) \quad (11b)$$

$$= \sum_{d \in D} P(d)P(Y|u, i, M(d, u)) \quad (11c)$$

3.4 Information Fusion and Graph Matching

In the process of information fusion, in addition to the results of message passing, ACGCF further considers the node matching results to capture the matching information at the node level when generating the fused node representation. Distinctively, the $f_{f_{use}}$ fusion mechanism assimilates the outcomes of message propagation z_i and the consequences of causal manipulation s_i to create an amalgamated node representation. Mathematically, this is formulated as $u'_i = f_{f_{use}}(z_i, s_i)$, where u'_i signifies the consolidated representation of node i. After evaluating various approaches, we discovered that recurrent neural networks exhibited superior performance. Therefore, we selected the GRU (Global Recurrent Unit), a proficient recurrent neural network architecture, to fulfill the role of the $f_{f_{use}}$ function. Within this context, $f_{f_{use}}$ receives $f_{f_{use}}$ as a sequential input, and the ultimate hidden state of the GRU serves as the amalgamated node representation. Subsequently, the consolidated representations of individual nodes in a graph are accumulated to yield a holistic representation of the graph. We achieve this accumulation through the application of element-wise summation. In short, the f_G function in ACGCF is:

$$f_G(G, \hat{V}) = \sum_{i \in V} u'_i \quad (12)$$

During graph matching, we obtain the vector representation of the user attribute graph and the item attribute graph by f_G, respectively:

$$m_G^u = f_G(G^u, V^v) \quad (13)$$

$$m_G^v = f_G(G^v, V^u) \quad (14)$$

We use the dot product as the matching function f_{match} to match these two graphs to get the predicted output.

4 Evaluation

In this section, we perform experiments to evaluate the performance of ACGCF.

4.1 Dataset

We run ACGCF and baseline methods on two datasets containing user and item attributes. Here's a description of the dataset:

- **MovieLens1M** [20] is a dataset commonly used in the field of movie recommendation systems that contains user ratings for movies. Each data sample contains a user and a movie with their corresponding attributes. We further collected additional attributes of movies, containing data from 6,040 users who rated movies on a scale of 1 to 5, each rating at least 20 different movies.
- **Book-Crossing** [21] is a dataset for Book recommendation that contains both implicit and explicit ratings of books by users. Each data sample contains a user and a book with their corresponding attributes. Ratings range from 1 to 10 and clearly indicate how much a user likes a book. It also includes an implicit rating (represented by 0), which might indicate that the user viewed the book but didn't give it an explicit rating.

MovieLens1M and Book-crossing contain explicit ratings. We transform explicit ratings into implicit feedback. We treat scores greater than 3 as positive scores for MovieLens1M, and treat all scores with clear ratings as positive scores for Book-crossing because of its sparsity. We then randomly select a number of negative samples equal to the number of positive samples for each user. To ensure the quality of the dataset, we selected users who have more than 10 positive reviews for MovieLens1M and more than 20 positive reviews for Book-crossing.

4.2 Baselines

We compare our model with competitive baselines that can take user attributes and item attributes into account:

FM [22] simulates the interaction between each feature through dot product operation and then aggregates these simulation results to obtain the final prediction. **AFM** [26] assigns additional attention weights to each feature interaction during the aggregation process. Based on FM, **NFM** [23] introduces a MLP to perform nonlinear analysis of feature interactions, further improving the accuracy of the model. **W&D** [31] combines a linear model with a deep neural network, leveraging the strong learning capabilities of the deep neural network for recommendations. **DFM** [24] combines interaction analysis results from using MLP and FM for prediction. **LGCN** [25] captures high-order connectivity in user-item interactions through graph structures, generating high-quality recommendations. To fairly evaluate these baseline models, we ensure that their MLP structures, if used, remain consistent for direct comparison.

Experimental Set-Up. We use the following hyper-parameter settings: the node representation dimension is 64 and the number of units for the hidden layer is 128; We randomly initialized the model parameters using a Gaussian distribution N(0, 0.1). The batch used for training was 256 and the learning rate was 1×10^{-3} When evaluating the model, we used the area under the curve

(AUC), Logloss, and N@k as evaluation metrics. AUC and Logloss are often used for implicit feedback tasks, while N@k is a metric often used to evaluate top-k recommendations, higher AUC and N@k scores with lower Logloss indicate a good model. In this experiment, we set k to 5 and 10. To obtain more reliable results, all experiments were repeated five times and the average results were recorded.

Table 1. Summary of the performance in comparison with baselines.

model	MovieLens1M				Book-Crossing			
	AUC	Logloss	N@5	N@10	AUC	Logloss	N@5	N@10
FM	0.8824	0.4346	0.8464	0.8464	0.7364	0.5671	0.7597	0.8167
AFM	0.8649	0.4345	0.8289	0.8697	0.7341	0.5686	0.7971	0.8467
NFM	0.8863	0.3546	0.8276	0.8492	0.7823	0.5264	0.8023	0.8471
W&D	0.9124	0.3764	0.8405	0.8712	0.8813	0.5314	0.8131	0.8756
DFM	0.9017	0.4056	0.8206	0.8310	0.8181	0.5130	0.8317	0.8480
LGCN	0.9051	0.3518	0.8918	0.8816	0.8181	0.5024	0.8513	0.8801
ACGCF	**0.9387**	**0.3218**	**0.9486**	**0.9483**	**0.9034**	**0.4058**	**0.9347**	**0.9438**
Improv	0.76%	1.6%	6.21%	4.89%	0.89%	0.85%	7.38%	4.13%

4.3 Baseline Comparison

We have undertaken a comparative analysis of ACGCF's performance against established baselines. As depicted in Table 1, the predictive capabilities of each model are presented. The 'Improv' rows indicate the enhancement achieved by ACGCF over the optimal baseline results. These findings underscore the proficiency of ACGCF in harnessing the structural insights of user and item attributes for precise forecasting. Our model, the ACGCF, generally outperformed all the other baselines. The reason for the unsatisfactory performance of the FM and AFM models is that they do not take a neural network approach but rely solely on dot products to extract information from attribute interactions. To improve this, our model models the interactions inside the user attribute graph and the item attribute graph in detail, which provides powerful interaction modeling capabilities for accurate prediction. Compared with DFM and LGCN, the ACGCF model has further improved the recommendation performance. This fact proves that eliminating the interference of confounding factors in the modeling process can indeed significantly improve the accuracy and effect of recommendation.

4.4 Ablation Study

To verify the contribution of different modules to the performance of ACGCF, We designed three variants of the proposed model: ACGCF-all, ACGCF-A and

ACGCF-C. ACGCF-C and ACGCF-A include methods without causal intervention and methods with adaptive preference retention mechanism and ACGCF-all consisted of the proposed method without both. We conducted experiments were carried out on two actually acquired datasets and compared with the results of the proposed model. The experimental results are shown in Fig. 4. Figure 4 shows that ACGCF-all, ACGCF-C and ACGCF-A perform worse than the proposed method to varying degrees on the two datasets. There are three reasons. First, when we remove adaptive and causal interventions, our model degenerates to a normal GNN, yielding poor performance in terms of collaborative filtering. Secondly, since GNN does not have an adaptive preference retention mechanism, the contributions of all items are treated as the same, thus the items cannot be distinguished.

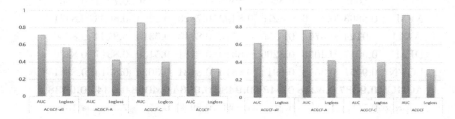

Fig. 4. Effect of the number of GCN layers on MovieLens1M dataset(left) and Book-Crossing dataset(right).

4.5 Effect of Hyperparameters

Since the number of GNN layers is critical to the model, we investigate its impact on the performance. The influence of embedding dimension on ACGCF is also analyzed. Effect of the number of GNN layers: To investigate the effect of the number of layers on the model, we vary the number of layers k of the GNN in the range of $\{1, 2, 3, 4, 5\}$ while keeping the other parameters constant. Figure 5 shows the performance obtained using different numbers of graph convolutional layers on the two datasets. When $k=3$, the model achieved the best performance on all two datasets. To investigate the influence of embedding dimensions on the model, we set the dimension of embedding d to $\{16, 32, 64, 128, 256\}$. Figure 6 shows the performance obtained using different embedding dimensions on the two datasets. Showed that the proposed method attained the best performance on MovieLens1M at $d = 64$, and on Book-Crossing at $d = 128$. This difference might have been acquired owing to the different sparsities of the datasets.

5 Conclusion

In this study, we propose the Attention-based Causal Graph Convolutional Collaborative Filtering (ACGCF) model to address confounding factor issues in recommender systems, thereby enabling more efficient data mining from user-item

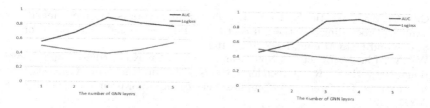

Fig. 5. Effect of the number of GCN layers on MovieLens1M dataset(left) and Book-Crossing dataset(right).

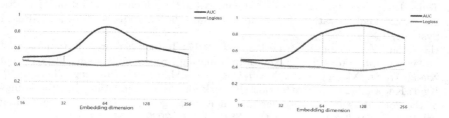

Fig. 6. Effect of embedding dimensions on MovieLens1M dataset(left) and Book-Crossing dataset(right).

data. Our approach includes preprocessing steps to manage computational complexity, introduces an attention mechanism to refine preference prediction, and incorporates causal reasoning to mitigate spurious correlations. Experimental results on real-world datasets demonstrate the ACGCF model's superior performance compared to traditional methods. Ablation studies confirm the individual contributions of each model component. Future research will focus on enhancing real-time capabilities, dynamic user preference tracking, multi-modal data integration, and privacy preservation to further enhance system performance and user experience.

Acknowledgements. This work is supported by the National Natural Science Foundation of China (No. 62272191, No. 62372211), the Science and Technology Development Program of Jilin Province (No. 20220201153GX), the International Science and Technology Cooperation Program of Jilin Province (No. 20230402076GH, No. 20240402067GH), the Open Project of the Fifth Electronics Research Institute of the Ministry of Industry and Information Technology (No. HK202303528).

References

1. Abdollahpouri, H., Burke, R.: Multistakeholder recommender systems. In: Recommender systems handbook, pp. 647–677. Springer (2022)
2. Datta G, Beerel P A. Can deep neural networks be converted to ultra lowlatency spiking neural networks?[C]//2022 Design, Automation & Test in Europe Conference & Exhibition (DATE). IEEE, 2022: 718–723

3. Gao C, Wang X, He X, Li Y. Graph neural networks for recommender system[C]//Proceedings of the Fifteenth ACM International Conference on Web Search and Data Mining. 2022: 1623–1625
4. Wei X , Liu J . Effects of Nonlinear Functions on Knowledge Graph Convolutional Networks for Recommender Systems with Yelp Knowledge Graph[C]// 11th International Conference on Computer Science and Inforation Technology (CCSIT 2021). 2021
5. Zhang, L., Kang, Z., Sun, X., et al.: KCRec: Knowledge-aware representation Graph Convolutional Network for Recommendation[J]. KnowledgeBased Systems **8**, 107399 (2021)
6. Tang, H. , Zhao, G. , Bu, X. , & Qian, X.(2021). Dynamic evolution of multigraph based collaborative filtering for recommendation systems. Knowledge-Based Systems, 228, 107251–
7. Hamilton, W. L., Ying, R., & Leskovec, J. (2017). Inductive representation learning on large graphs. Advances in Neural Information Processing Systems, 2017-Decem
8. Xiang Wang, Xiangnan He, Yixin Cao, Meng Liu, and Tat-Seng Chua. 2019. Kgat: Knowledge Graph Attention Network for Recommendation. In Proceedings of the 25th International Conference on Knowledge Discovery and Data Mining (SIGKDD).950–958.Processing (ICASSP). IEEE, 2020
9. Fuzheng Zhang, Nicholas Jing Yuan, Defu Lian, Xing Xie, and Wei-Ying Ma.2016. Collaborative Knowledge base Embedding for Recommender Systems. In Proceedings of the 22nd international conference on knowledge discovery and data mining (SIGKDD). 353–362
10. Zekun Li, Zeyu Cui, Shu Wu, Xiaoyu Zhang, and Liang Wang. 2019. Fi-GNN:Modeling Feature Interactions via Graph Neural Networks for CTR Prediction. In Proceedings of the 28th International Conference on Information and Knowledge Management (CIKM). 539–548
11. Yixin Su, Rui Zhang, Sarah Erfani, and Zhenghua Xu. 2021. Detecting BeneficialFeature Interactions for Recommender Systems. In Proceedings of the Conference on Artificial Intelligence (AAAI)
12. Pearl, J.: Causality. Cambridge University Press (2009)
13. He X,Deng K,Wang X, et al. Lightgcn: Simplifying and powering graph convolution network for recommendation[C]//Proceedings of the 43rd International ACM SIGIR conference on research and development in Information Retrieval. 2020: 639–648
14. Su, X., Khoshgoftaar, T.M.: A survey of collaborative filtering techniques[J]. Advances in artificial intelligence **2009**(1), 421425 (2009)
15. Jin, B., Gao, C., He, X., Jin, D., & Li, Y. (2020). Multi-behavior Recommendation with Graph Convolutional Networks. SIGIR 2020 - Proceedings of the 43rd International ACM SIGIR Conference on Research and Development in Information Retrieval, 659–668
16. Jin, B., Gao, C., He, X., Jin, D., & Li, Y. (2020). Multi-behavior Rec-ommendation with Graph Convolutional Networks. SIGIR 2020-Proceed-ings of the 43rd International ACM SIGIR Conference on Research and De-velopment in Information Retrieval, 659–668
17. Pearl J. Causal inference in statistics: An overview[J]. 2009.48(3): 1–40
18. Yu D, Li Q, Wang X, et al. Deconfounded recommendation via causal intervention[J]. Neurocomputing, 2023, 529: 128–139. recom-mendation
19. Su Y, Zhang R, M. Erfani S, et al. Neural graph matching based collaborative filtering[C]//Proceedings of the 44th international ACM SIGIR conference on research and development in information retrieval. 2021: 849–858

20. Maxwell Harper, F., Konstan, J.A.: The Movielens Datasets: History and Context. Transactions on Interactive Intelligent Systems (TIIS) **2015**, 1–19 (2015)
21. Cai-Nicolas Ziegler, Sean M McNee, Joseph A Konstan, and Georg Lausen. 2005. Improving Recommendation Lists Through Topic Diversification. In Proceedings of the 14th International Conference on World Wide Web (WWW). 22–32
22. Steffen Rendle. 2010. Factorization Machines. In Proceedings of the 10th International IEEE Conference on Data Mining (ICDM). 995–1000
23. Xiangnan He and Tat-Seng Chua. 2017. Neural Factorization Machines for Sparse Predictive Analytics. In Proceedings of the 40th International ACM conference on Research and Development in Information Retrieval (SIGIR). 355–364
24. Huifeng Guo, Ruiming Tang, Yunming Ye, Zhenguo Li, and Xiuqiang He. 2017.DeepFM: a Factorization-Machine based Neural Network for CTR prediction. In Proceedings of the 26th International Joint Conference on Artificial Intelligence(IJCAI). 1725–1731
25. He X, Deng K, Wang X, et al. Lightgcn: Simplifying and powering graph convolution network for recommendation[C]//Proceedings of the 43rd International ACM SIGIR conference on research and development in Information Retrieval. 2020: 639–648
26. Jun Xiao, Hao Ye, Xiangnan He, Hanwang Zhang, Fei Wu, and Tat-Seng Chua. 2017. Attentional Factorization Machines: Learning the Weight of Feature Interactions via Attention Networks. In Proceedings of the 26th International Joint Conference on Artificial Intelligence (IJCAI). 3119–3125
27. Wang W, Feng F, He X, et al. Deconfounded recommendation for alleviating bias amplification[C]//Proceedings of the 27th ACM SIGKDD Conference on Knowledge Discovery & Data Mining. 2021: 1717–1725
28. Liu J, Sun L, Nie W, et al. Graph Disentangled Contrastive Learning with Personalized Transfer for Cross-Domain Recommendation[C]//Proceedings of the AAAI Conference on Artificial Intelligence. 2024, 38(8): 8769–8777
29. Agrawal N, Sirohi A K, Kumar S. No Prejudice! Fair Federated Graph Neural Networks for Personalized Recommendation[C]//Proceedings of the AAAI Conference on Artificial Intelligence. 2024, 38(10): 10775–10783
30. Guo Z, Li J, Li G, et al. LGMRec: Local and Global Graph Learning for Multimodal Recommendation[C]//Proceedings of the AAAI Conference on Artificial Intelligence. 2024, 38(8): 8454–8462
31. Cheng H T, Koc L, Harmsen J, et al. Wide & deep learning for recommender systems[C]//Proceedings of the 1st workshop on deep learning for recommender systems. 2016: 7–10

Sequential Recommendation with Diverse Supervised Contrastive Views

Zitong Zhu, Meixiu Long, Junfa Lin, and Jiahai Wang[✉]

School of Computer Science and Engineering, Sun Yat-sen University,
Guangzhou, People's Republic of China
{zhuzt5,longmx7,linjf26}@mail2.sysu.edu.cn, wangjiah@mail.sysu.edu.cn

Abstract. Recently, contrastive learning has been used in sequential recommendation to address data sparsity issue. Among various contrastive learning methods, supervised contrastive learning has shown excellent performance by utilizing supervised contrastive views. However, current supervised contrastive views in sequential recommendation are sampled from existing sequences in a dataset, resulting in insufficient supervised contrastive information and low-quality contrastive views. To address the problem, this paper proposes sequential Recommendation with Diverse supervised contrastive views (DivRec). DivRec has two main components, contrastive views generation and contrastive views integration. Contrastive views generation constructs diverse supervised contrastive views from a generation perspective, where insertion augmentation and non-autoregressive generation are utilized to generate locally-diverse and globally-diverse supervised contrastive views, respectively. Contrastive views integration integrates multiple contrastive views for their complementary effects, where a leader-follower strategy in the integration component helps construct attribute-compatible contrastive pairs to mitigate conflict issues. Additionally, diffusion mechanism is introduced as a noise augmentation method to further improve the diversity of contrastive views. Extensive experiments demonstrate the effectiveness and the state-of-the-art performance of DivRec. The code and appendix are available at https://github.com/zzzzzdev/DivRec.

Keywords: Sequential Recommendation · Contrastive Learning

1 Introduction

Sequential recommendation (SR) predicts the next item that a user will interact with based on the user's historical interaction behaviors. However, user interaction records with items in the real world are limited, leading to data sparsity problems [27]. To cope with this, contrastive learning (CL) has been introduced into SR to enhance representations of items without additional interaction data [28].

Among various CL methods, supervised contrastive learning (SCL) [17] has shown excellent performance by leveraging supervised contrastive views. Current SCL mainly constructs supervised contrastive views through a heuristic strategy: in a dataset, if two behavior sequences have the same target items, they are considered supervised contrastive views of each other.

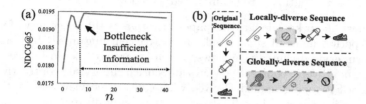

Fig. 1. (a) The relationship between the number of candidate supervised contrastive views n used in DuoRec and corresponding performance on Sports dataset. (b) DivRec modifies existing supervised contrastive views to construct locally-diverse sequences, and generates completely new sequences to construct globally-diverse sequences.

However, we believe that existing supervised contrastive views lack sufficient supervised contrastive information. Following DuoRec [17], we randomly sample one supervised contrastive view from Sports dataset (the number of candidate supervised contrastive views $n = 40$ on average) for each user in each epoch. We explore the relationship between n and the performance of DuoRec by modifying n from $[0, 40]$. As n increases, the performance rapidly reaches a bottleneck, with approximately $n = 8$. However, according to statistics, 70% of user behavior sequences contain more than 8 candidate supervised contrastive views. This shows that most supervised contrastive views do not provide sufficient supervised information to improve performance, resulting in low-quality contrastive views. This motivates us to enrich the supervised information by constructing diverse supervised contrastive views, which is more likely to yield high-quality contrastive views and break through performance bottlenecks [8].

To construct diverse contrastive views and enrich the supervised information, two aspects need to be considered: How to construct diverse contrastive views and how to integrate diverse contrastive views with existing supervised contrastive views. Therefore, this paper proposes sequential <u>R</u>ecommendation with <u>Div</u>erse supervised contrastive views (DivRec). DivRec contains two main components, contrastive views generation and contrastive views integration. Contrastive views generation constructs diverse supervised contrastive views using insertion augmentation and non-autoregressive generation. Existing supervised contrastive views are inserted by several correlated items to generate locally-diverse contrastive views, while completely new sequences are generated as globally-diverse contrastive views, as shown in Fig. 1(b). Contrastive views integration integrates diverse contrastive views with existing supervised contrastive views for their complementary effects, where a leader-follower strategy in the integration component helps construct attribute-compatible contrastive pairs to avoid conflict problems. In this strategy, locally-diverse and globally-diverse supervised contrastive views serve as follower while existing supervised contrastive views are considered leader. Additionally, a diffusion mechanism is introduced to vanilla Transformer encoder as a noise augmentation method to further boost the diversity of contrastive views.

Our main contributions are summarized as follows:

- To the best of our knowledge, this paper investigates the possibility of improving the performance of supervised contrastive learning in SR by generating diverse supervised contrastive views, which is rarely explored before.
- To construct diverse contrastive views and enrich supervised contrastive information, a SR model named DivRec with contrastive views generation and integration is proposed. Locally-diverse and globally-diverse contrastive views are generated and then integrated with existing supervised contrastive views using conflict-free strategy.
- Extensive experiments on three benchmark datasets demonstrate that DivRec outperforms the state-of-the-art methods by 1.23%-9.01% in performance.

2 Related Work

2.1 Contrastive Learning for SR.

In recent years, graph neural networks [26] and Transformer models [9,19] have become the mainstream backbones of sequential recommendation due to their ability to handle sequence modeling. Nevertheless, both of them are still struggling with the data sparsity caused by the limited user-item interactions [3].

Contrastive learning has been introduced into sequential recommendation over the past two years to address data sparsity issue [30]. CL4SRec [28] constructs contrastive views through three random operations: cropping, reordering, and masking. CoSeRec [12] retains these three random augmentations and proposes two additional informative augmentations: insertion and substitution, which consider correlations between items. DuoRec [17] introduces a novel supervised retrieval contrastive view. However, all of them ignore the insufficiency of supervised contrastive information in SR. MCLRec [15] utilizes learnable model augmentation and meta-learning strategy to adaptively construct diverse self-supervised contrastive views, which fails to take the diversity of supervised contrastive views into account. ICSRec [16] enhances the modeling of user intent by partitioning user behavior sequences into different subsequences and considering them as positive pairs in contrastive learning. Different from them, DivRec intends to generate diverse supervised contrastive views to enrich the supervised contrastive information. Moreover, the conflict-free integration of multiple contrastive views is investigated, which has been ignored by previous work.

2.2 Diffusion Model for SR

Diffusion model has achieved great success in various fields due to its ability to generate high-quality results, including computer vision [7], audio [10], natural language processing [4], and SR. In the field of SR, DiffRec [24] obtains final distributions of items by performing forward and reverse processes on a user's interaction matrix. DreamRec [29] directly utilizes diffusion model to generate user's preference embeddings based on their behavior sequences. PDRec [13] employs diffusion model as a flexible plugin to take advantage of the diffusion-generating user preferences on all items. Different from them, DivRec further explores the combination of diffusion model and contrastive learning, rather than solely using diffusion model as a generative model for the SR task.

3 Methodology

3.1 Task Formulation and Framework Overview

User and item sets are denoted as $\mathcal{U}$ and $\mathcal{V}$, respectively. $\mathcal{S}$ represents all users' interaction sequences arranged in chronological order. $|\mathcal{U}|$, $|\mathcal{V}|$, and $|\mathcal{S}|$ indicate the number of samples in each set. Given a user $u_i \in \mathcal{U}, 1 \leq i \leq |\mathcal{U}|$, the user's interaction sequence is $s_i = \{v_1, v_2, v_3, \ldots, v_n\} \in \mathcal{S}$, where $v_j \in \mathcal{V}$ represents the item interacted with at timestamp j. Sequential recommendation aims to predict the next item, denoted as v_{n+1}, that a user is most likely to interact with at timestamp $n + 1$, based on the user's historical interaction sequence s_i. It can be formulated as:

$$\arg\max_{v_k \in \mathcal{V}} P(v_{n+1} = v_k \mid s_i). \qquad (1)$$

The overview of DivRec is presented in Fig. 2. Locally-diverse and globally-diverse supervised contrastive views are constructed based on item correlations in the contrastive views generation component (Sect. 3.2). A diffusion mechanism is introduced to a vanilla Transformer encoder to further improve the diversity of contrastive views and the robustness of the sequence encoder (Sect. 3.3). Diverse contrastive views will be integrated with existing supervised contrastive views in the contrastive views integration component for complementary effects (Sect. 3.4). Finally, DivRec is trained jointly with recommendation loss, contrastive loss, and regularization loss (Sect. 3.5).

3.2 Contrastive Views Generation

Locally-Diverse Supervised Contrastive Views. Various augmentations are widely used to create contrastive views in SR, e.g., cropping, masking, and reordering [28], etc. These augmentations modify the original views, leading to the potential to improve the diversity of contrastive views locally. However, they may cause semantic inconsistency that damages the performance of contrastive

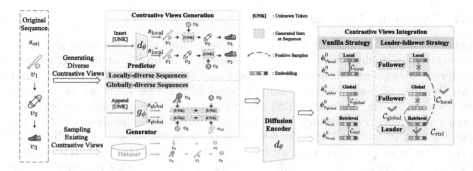

Fig. 2. The overall framework of DivRec. Contrastive views generation component generates locally-diverse and globally-diverse contrastive views. Corresponding sequence representations are obtained after being encoded by the diffusion encoder d_θ. Contrastive views integration integrates diverse contrastive views with existing supervised contrastive views via a leader-follower strategy to mitigate conflicts.

learning [15]. In order to overcome semantic inconsistency, locally-diverse supervised contrastive views are constructed from a generation perspective. Inspired by the use of insertion augmentations in CoSeRec [12], DivRec proposes a cloze task to insert more correlated items into original views. Given a user's original sequence $s_{\text{ori}} = \{v_1, v_2, v_3\} \in \mathcal{S}_{\text{ori}}$, some "[UNK]"[1] tokens are randomly inserted into it with a specific ratio $\alpha \in [0, 0.9]$. The resulting sequence is the locally-diverse supervised contrastive view: $s_{\text{local}} = \{v_1, [\text{UNK}], v_2, v_3\} \in \mathcal{S}_{\text{local}}$. The index of "[UNK]" in s_{local} is in $\mathcal{K} = \{k_1, k_2, \ldots\}$.

Next, s_{local} will be sent to a predictor to determine items to be inserted based on existing interaction records. We set the inference stage of the diffusion encoder d_θ (see Sect. 3.3 for details) as the predictor. During this process, in order to capture correlations of items in s_{local} more effectively, both the diffusion step and sampling noise are set as 0. Following this, a rounding operation is performed on the last layer of the Transformer's output to convert continuous embeddings to discrete items. This process can be formulated as follows:

$$\begin{aligned} \mathbf{h}^l_{\text{local}} &= d_\theta(s_{\text{local}}), \\ s_{\text{local},k} &= \text{Rounding}(\mathbf{h}^l_{\text{local},k}) \\ &= \arg\max_{v_i, v_j \in \mathcal{V}} \frac{\exp(\mathbf{h}^l_{\text{local},k} \odot \mathbf{e}_{v_i})}{\sum_{v_j} \exp(\mathbf{h}^l_{\text{local},k} \odot \mathbf{e}_{v_j})}, \end{aligned} \quad (2)$$

where $k \in \mathcal{K}$, $\mathbf{h}^l_{\text{local}}$ represents the final layer of the diffusion encoder's output, $\mathbf{h}^l_{\text{local},k}$ is the embedding of $\mathbf{h}^l_{\text{local}}$ at the index k, $d_\theta(\cdot)$ denotes diffusion encoder.

Different from CoSeRec [12] that employs heuristic and simple similarity measurements to determine inserted items, DivRec utilizes Transformer with

---

[1] "[UNK]" acts as an unknown item that needs to be predicted.

cloze task to more effectively capture item correlations. Furthermore, sharing the diffusion encoder helps reduce the overall number of parameters.

Finally, a locally-diverse supervised contrastive views s_{local} is generated as: $\{v_1, [P_1], v_2, v_3\}$, where $[P_1]$ is the inserted item. The same applies to another $\tilde{s}_{local}$: $\{v_1, v_2, [P_2], v_3\}$.

Globally-Diverse Supervised Contrastive Views. It is insufficient to construct diverse contrastive views by merely modifying the original sequence. To uncover additional supervised contrastive information, we also intuitively create globally-diverse supervised contrastive views by generating entirely new sequences through a generation task. For this generation task, the non-autoregressive generation task [4] conducts generation in parallel to achieve acceleration, which is suitable for SR. Therefore, DivRec takes the original sequence as the condition to parallelly generate completely new sequences. Specifically, given a user's original sequence $s_{ori} = \{v_1, v_2, v_3\} \in \mathcal{S}_{ori}$, a number of m tokens termed "[UNK]" are appended at the end of s_{ori} to obtain $s_{generate} = \{v_1, v_2, v_3, [UNK_1], \ldots [UNK_m]\}$.

Next, $s_{generate}$ will be fed into a generator to determine the items of the new sequence. Unlike generating locally-diverse supervised contrastive views that shares the diffusion encoder as the predictor, another Transformer encoder-only model $g_\phi(\cdot)$ serves as the non-autoregressive sequence generator to improve the generation quality. After the generation through $g_\phi(\cdot)$, last m embeddings in the last layer are truncated for the next rounding operation. Finally, globally-diverse supervised contrastive views $s_{global} = \{[P_1], \ldots [P_m]\}$ (m=2 in Fig. 2) can be obtained, where $[P_i]$ is the prediction of "[UNK]". The process can be formulated as follows:

$$\mathbf{h}^l_{global} = g_\phi(s_{generate})[-m:],$$
$$s_{global,k} = \text{Rounding}(\mathbf{h}^l_{global,k}), \quad (3)$$

where $[-m:]$ mimics the Python style of truncating last m embeddings. To further improve the semantic consistency between s_{global} and s_{ori}, a regularization term is proposed (see Sect. 3.5 for details).

3.3 Diffusion Encoder

Afterwards, in addition to s_{local} and s_{global}, two existing supervised contrastive views (i.e., s_{rtrl1} and s_{rtrl2}) are also incorporated for contrastive learning to guide the generation of diverse contrastive views and enrich the supervised contrastive information adaptively.

Regarding the sequence encoder, the application of the diffusion model in recommendation [2,29] has demonstrated potential generation ability, which may boost the generation of contrastive views. Motivated by this, DivRec extends a vanilla Transformer with a diffusion mechanism as the sequence encoder, referred as diffusion encoder $d_\theta(\cdot)$. Unlike the encoding process in the vanilla Transformer, the diffusion encoder incorporates several additional operations to adapt the diffusion mechanism.

Define a Diffusion Noise. The embedding of the target item $\mathbf{e}_{v_{\text{target}}}$ can be considered as the input data of diffusion. A diffusion step t_1 is uniformly sampled in $[0, T]$ to create a diffusion noise $\mathbf{e}^t_{v_{\text{target}}}$:

$$\mathbf{e}^{t_1}_{v_{\text{target}}} = \sqrt{\bar{\alpha}_{t_1}} \mathbf{e}^0_{v_{\text{target}}} + \sqrt{1 - \bar{\alpha}_{t_1}} \epsilon, \tag{4}$$

where T is the total steps of diffusion, $\bar{\alpha}_t$ can be calculated using different noise schedule strategies, and $\epsilon \sim \mathcal{N}(0, 1)$. During the inference stage, the diffusion noise is set to 0 in order to capture more accurate correlations of items.

Inject the Diffusion Noise Into a Sequence. Since the diffusion noise is a continuous embedding, it can be injected into an embedded sequence as follows:

$$\mathbf{h}_* = s_{\mathbf{e}_*} + \mathbf{P}(s_*) + \lambda \cdot (\mathbf{e}^{t_1}_{v_{\text{target}}} + \mathbf{T}(t_1)), \tag{5}$$

where $*$ represents *rtrl1, rtrl2, local,* and *global* for brevity. $s_{\mathbf{e}_*}$ is the embedded sequence, $\mathbf{P} \in \mathbb{R}^{n \times d}$ and $\mathbf{T} \in \mathbb{R}^{T \times d}$ are position and timestep embedding matrices. λ controls the scale of diffusion noise injected into the sequence.

Estimate the Input Data of Diffusion. The $\mathbf{h}_*$ is encoded by a vanilla Transformer. After that, the last embedding from the final layer's output of the Transformer is considered as the sequence representation $\hat{\mathbf{e}}^0_{v_{\text{target}}}$ as well as the estimation of diffusion input data. Specifically, six types of views, denoted as $\hat{\mathbf{e}}^0_{v_{\text{rtrl1}}}, \hat{\mathbf{e}}^0_{v_{\text{rtrl2}}}, \hat{\mathbf{e}}^0_{v_{\text{local}}}, \hat{\mathbf{e}}^0_{v_{\tilde{\text{local}}}}, \hat{\mathbf{e}}^0_{v_{\text{global}}}$, and $\hat{\mathbf{e}}^0_{v_{\tilde{\text{global}}}}$, can be used to estimate the same diffusion input data due to the sharing of the same target item. This can be formulated as:

$$\hat{\mathbf{e}}^0_{v_*} = \text{Transformer}(\mathbf{h}_*)[-1], \tag{6}$$

where $*$ represents *rtrl1, rtrl2, local,* and *global* for brevity. $[-1]$ mimics the Python style of indexing the last element. During inference, the estimation of the input data needs to iterate T steps [6].

As for the advantages of diffusion encoder, it is noteworthy that the diffusion mechanism can also be considered as a contrastive augmentation operation, in addition to its application as a generative model. During the forward process, the input data x_0 is broken to generate $x_1, x_2, \ldots, x_t$ by adding noise at different scales. When small-scale noise is applied, these variables exhibit certain semantic consistency, which can be considered as positive pairs in contrastive learning. Therefore, this noise augmentation operation combines diffusion mechanism with contrastive learning naturally and seamlessly, further improving the diversity of contrastive views.

3.4 Contrastive Views Integration

Different types of contrastive views have complementary effects, which is beneficial for improving the generalization of contrastive learning in the downstream recommendation task [8]. Therefore, we strive to explore the integration of diverse contrastive views (i.e., $\hat{\mathbf{e}}^0_{v_{\text{local}}}$ and $\hat{\mathbf{e}}^0_{v_{\text{global}}}$) with existing supervised contrastive views (i.e., $\hat{\mathbf{e}}^0_{v_{\text{rtrl1}}}$ and $\hat{\mathbf{e}}^0_{v_{\text{rtrl2}}}$) to enrich supervised contrastive information.

Vanilla Strategy. As shown in Fig. 2, the intuitive way to integrate multiple contrastive views is to create contrastive pairs among each type of contrastive views, respectively. Then the corresponding losses are jointly optimized, which is similar to integration methods in previous work [22].

However, the vanilla strategy used in previous work may suffer from conflict problems due to ignoring incompatible attributes of the contrastive pairs [17]. To mitigate conflict problems, a new strategy named leader-follower strategy is devised, which rearranges multiple contrastive views to construct attribute-compatible contrastive pairs.

Leader-Follower Strategy. Different from the vanilla strategy that constructs contrastive pairs among each type of contrastive views separately, the leader-follower strategy tends to rearrange multiple contrastive views, as shown in Fig. 2. Locally-diverse and globally-diverse contrastive views are completely new sequences for datasets, which need the guidance of existing supervised contrastive views to discover adaptive and richer supervised contrastive information. Therefore, locally-diverse and globally-diverse contrastive views are treated as follower, whereas existing supervised contrastive views are considered as leader. To construct attribute-compatible contrastive pairs, each contrastive pair contains at least one leader. As a result, there are three attribute-compatible contrastive pairs: $\hat{\mathbf{e}}^0_{v_{\text{rtrl1}}} \leftrightarrow \hat{\mathbf{e}}^0_{v_{\text{rtrl2}}}$, $\hat{\mathbf{e}}^0_{v_{\text{rtrl1}}} \leftrightarrow \hat{\mathbf{e}}^0_{v_{\text{local}}}$, and $\hat{\mathbf{e}}^0_{v_{\text{rtrl2}}} \leftrightarrow \hat{\mathbf{e}}^0_{v_{\text{global}}}$.

In this strategy, the attribute of each contrastive pair will largely depend on the leader view. The follower view serves as a supplement to the leader view to achieve attribute compatibility and diversity of contrastive views.

3.5 Training Loss

Recommendation task and contrastive learning task are closely related because they both rely on item correlations. Consequently, DivRec is trained jointly with three losses: recommendation loss, contrastive loss, and regularization loss.

Recommendation Loss. Similar to [17], Cross-Entropy loss is used as the recommendation loss. It can be calculated as follows:

$$\mathcal{L}_{\text{rec}} = \mathop{\mathbb{E}}_{i \in [0,b]} \left[-\log \frac{e^{d_\theta(s_i) \odot \mathbf{e}_{v_i}}}{\sum_{j=1}^{|\mathcal{V}|} e^{d_\theta(s_i) \odot \mathbf{e}_{v_j}}} \right], \tag{7}$$

where b is the training batch size, $d_\theta(\cdot)$ is the diffusion encoder, s_i is a user's historical interaction sequence, and $\mathbf{e}_{v_{\text{target}}}$ represents the embedding of target item.

Contrastive Loss. The InfoNCE loss [14] serves as the contrastive loss to optimize three contrastive pairs constructed by the leader-follower strategy. Taking $\hat{\mathbf{e}}^0_{v_{\text{rtrl1}}} \leftrightarrow \hat{\mathbf{e}}^0_{v_{\text{local}}}$ as an example, it can be optimized by the following formula:

$$\mathcal{C}_{\text{local}} = \mathop{\mathbb{E}}_{i,k \in [0,b]} \left[-\log \frac{\exp(\hat{\mathbf{e}}^0_{\text{local},i}{}^\top \hat{\mathbf{e}}^0_{\text{rtrl1},i}/\tau)}{\sum_k \exp(\hat{\mathbf{e}}^0_{\text{local},i}{}^\top \hat{\mathbf{e}}^0_{\text{rtrl1},k}/\tau)} \right], \tag{8}$$

where τ is the temperature coefficient, $\hat{\mathbf{e}}^0_{\text{local},i}$ and $\hat{\mathbf{e}}^0_{\text{rtrl1},i}$ represent a positive pair. The negative pairs consist of the remaining samples in the same batch [1]. The overall contrastive loss can be defined as:

$$\mathcal{C} = \omega \cdot (\mathcal{C}_{\text{local}} + \mathcal{C}_{\text{global}} + \mathcal{C}_{\text{rtrl}}), \qquad (9)$$

where ω controls the weight of the contrastive loss.

Regularization Loss. If the contrastive loss is exclusively optimized for locally-diverse and globally-diverse contrastive views, correlations between items may not be effectively captured, resulting in poor semantic meanings. Since both generated diverse contrastive views are derived from the original view, it is crucial for them to have similar semantic meanings to the original sequence. To ensure this, the Cross-Entropy loss is employed as a regularization loss to guide two types of diverse contrastive views to become as similar to the target item as the original view. Take the locally-diverse contrastive view $\hat{\mathbf{e}}^0_{\text{local}}$ as an example:

$$\mathcal{R}_{\text{local}} = \mathbb{E}\left[-\log \frac{\exp(\hat{\mathbf{e}}^0_{\text{local}} \odot \mathbf{e}_{v_{\text{target}}})}{\sum_{j=1}^{|\mathcal{V}|} \exp(\hat{\mathbf{e}}^0_{\text{local}} \odot \mathbf{e}_{v_j})}\right], \qquad (10)$$

where $\mathbf{e}_{v_{\text{target}}}$ is the embedding of the target item. The overall regularization loss can be calculated as:

$$\mathcal{R} = \beta \cdot \mathcal{R}_{\text{local}} + \gamma \cdot \mathcal{R}_{\text{global}}, \qquad (11)$$

where β and γ are the weights of the regularization loss for locally-diverse and globally-diverse contrastive views.

Overall Training Loss. In summary, the overall training loss is as follows:

$$\mathcal{L} = \mathcal{L}_{\text{rec}} + \mathcal{C} + \mathcal{R}. \qquad (12)$$

3.6 Discussion

Characteristics of DivRec. The characteristics of DivRec include:

- The supervised contrastive views in current studies(e.g.,DuoRec [17],ICSRec [16], EC4SRec [22], and DSCBR [25]) are all sampled from datasets or artificially defined using heuristic methods, resulting in lack of adaptability and insufficient supervised contrastive information. In DivRec, contrastive views generation constructs diverse contrastive views at local-level and global-level, which promotes the discovery of more supervised contrastive information.
- The appropriate integration of generated diverse contrastive views with existing supervised contrastive views is significantly crucial; otherwise, it may result in conflict issues. The leader-follower strategy is a smart design to mitigate conflicts caused by incompatible attributes of different types of contrastive views, the effectiveness of which is verified in Sect. 4.4.

– Different from previous work (e.g., DiffRec [2], DreamRec [11], and PDRec [13]) that merely takes the diffusion model as a generative model for SR task, DivRec further combines diffusion model with contrastive learning in a natural and seamless way. Diffusion model serves not only as a generative model, but also as a noise augmentation method that enhances the diversity of contrastive views and the robustness of the sequence encoder.

Time Complexity Analysis of DivRec. d is the embedding size and $|\mathcal{V}|$ is the number of items set. The time complexity of our DivRec is $\mathcal{O}(|\mathcal{V}| * d)$, which is consistent with baselines [15, 17]. The details of time complexity analysis are available in Appendix B.

Table 1. The statistics of experimented datasets.

Dataset	#Users	#Items	Avg.Length	#Actions	Sparsity
Sports	36598	18357	8.3	296337	99.95%
Toys	19412	11924	8.6	167597	99.93%
Yelp	30449	20038	10.4	316541	99.95%

4 Experiment

4.1 Experimental Settings

Datasets. We collect three publicly released datasets from two platforms: Amazon and Yelp. *Sports and Toys* are the two subcategories of Amazon Dataset[2]. *Yelp*[3] is a business recommendation dataset, and we only include records after January 1st, 2019. We keep "5-core" interactions for users and items. The historical interactions of users are sorted chronologically with a length set as 50. The processed dataset statistics are presented in Table 1.

Evaluation Metrics. We split our datasets into train, validation, and test sets using a "leave one out" strategy. We evaluate all items in the dataset without negative sampling. The evaluation metrics used in our experiments are top-k Hit Ratio (HR@k) and top-k Normalized Discounted Cumulative Gain (NDCG@k), where k is set as $\{5, 10, 20\}$.

---

[2] https://cseweb.ucsd.edu/~jmcauley/datasets.html.
[3] https://www.Yelp.com/dataset.

Baselines. DivRec is compared to various baselines grouped into four categories as follows. 1) **Non-sequential model**: BPR-MF [18]. 2) **Vanilla sequential models**: GRU4Rec [5], Caser [20], and SASRec [9]. 3) **Diffusion model for SR**: T-DiffRec [2]. 4) **Sequential models with contrastive learning**: CL4SRec [28], CoSeRec [12], DuoRec [17], and MCLRec [15]. The details of baselines are available in Appendix C.

Implementation Details. BPR-MF and CL4SRec are implemented based on public resources, while the remaining baselines are reproduced using the provided implementations by the authors. DivRec is implemented based on Pytorch. We follow [17] to set Transformer's hyperparameters and sample existing supervised contrastive views from datasets. We set the learning rate as 0.001, and the batch size as 256. We tune weights of losses: ω, β, and γ within the range of [0,1].

4.2 Overall Performance

From results presented in Table 2, several observations can be made as follows. Firstly, Transformer-based models with contrastive learning, e.g., CL4SRec, CoSeRec, DuoRec, and MCLRec, outperform other models, including BPR-MF, GRU4Rec, SASRec, and Caser. Additionally, T-DiffRec, which applies diffusion mechanism, performs similarly to SASRec on Sports and Toys, even better on Yelp. Therefore, these findings inspire the use of Transformer with CL to model the SR task, as well as the effective application of diffusion mechanism.

Secondly, DuoRec, which utilizes SCL, significantly enhances recommendation performance compared to other CL methods, e.g., CL4SRec and CoSeRec. Hence, supervised contrastive learning shows promise in achieving further superior performance. MCLRec demonstrates outstanding outcomes by introducing learnable augmented views, which serves as the motivation for constructing contrastive views from an adaptive or generation perspective.

Finally, benefiting from contrastive views generation and contrastive views integration components, DivRec significantly outperforms all other baselines on all datasets. This can be attributed to two main factors: 1) Diverse contrastive views are generated at the local-level and global-level, which facilitates the discovery of more supervised contrastive information. 2) Diverse contrastive views are integrated with existing supervised contrastive views to achieve complementary effects without any conflicts.

4.3 Ablation Study

To verify the effectiveness of each design in DivRec, five variants are conducted as follows: 1) **w/o local**: delete locally-diverse supervised contrastive views. 2) **w/o global**: delete globally-diverse supervised contrastive views. 3) **w/o diffusion**: delete diffusion mechanism of diffusion encoder to degenerate into a vanilla Transformer encoder. 4) **w/o $\mathcal{R}$**: delete regularization term in the overall

Table 2. Performance comparisons of different methods. The best and second best results are bolded and underlined. Improv. represents the relative improvements compared to the best baseline. ⋆: significant improvement of DivRec over the best baseline with p-value ≤ 0.01.

Dataset	Metric	BPR-MF	GRU4Rec	Caser	SASRec	T-DiffRec	CL4SRec	CoSeRec	DuoRec	MCLRec	DivRec	Improv.
Sports	HR@5	0.0123	0.0162	0.0154	0.0214	0.0198	0.0231	0.0290	<u>0.0307</u>	0.0304	**0.0328**⋆	6.84%
	HR@10	0.0215	0.0258	0.0261	0.0333	0.0296	0.0369	0.0439	<u>0.0478</u>	0.0471	**0.0515**⋆	7.74%
	HR@20	0.0369	0.0421	0.0399	0.0500	0.0427	0.0557	0.0636	<u>0.0710</u>	0.0710	**0.0774**⋆	9.01%
	NDCG@5	0.0076	0.0103	0.0114	0.0144	0.0130	0.0146	0.0190	0.0190	<u>0.0191</u>	**0.0205**⋆	7.33%
	NDCG@10	0.0105	0.0142	0.0135	0.0177	0.0161	0.0191	0.0244	0.0245	<u>0.0245</u>	**0.0264**⋆	7.76%
	NDCG@20	0.0144	0.0186	0.0178	0.0224	0.0194	0.0238	0.0293	0.0303	<u>0.0304</u>	**0.0329**⋆	8.22%
Toys	HR@5	0.0122	0.0121	0.0205	0.0429	0.0313	0.0503	0.0533	<u>0.0634</u>	0.0621	**0.0649**	2.37%
	HR@10	0.0197	0.0184	0.0333	0.0652	0.0453	0.0736	0.0755	<u>0.0924</u>	0.0911	**0.0956**⋆	3.46%
	HR@20	0.0327	0.0290	0.0542	0.0957	0.0627	0.0990	0.1037	0.1263	<u>0.1273</u>	**0.1319**⋆	3.61%
	NDCG@5	0.0076	0.0077	0.0125	0.0245	0.0213	0.0264	0.0370	<u>0.0379</u>	0.0369	**0.0386**	1.85%
	NDCG@10	0.0100	0.0097	0.0168	0.0320	0.0259	0.0339	0.0442	<u>0.0472</u>	0.0462	**0.0485**⋆	2.75%
	NDCG@20	0.0132	0.0123	0.0221	0.0397	0.0303	0.0404	0.0513	<u>0.0557</u>	0.0553	**0.0577**⋆	3.59%
Yelp	HR@5	0.0127	0.0152	0.0142	0.0160	0.0228	0.0227	0.0241	<u>0.0441</u>	0.0436	**0.0455**⋆	3.17%
	HR@10	0.0216	0.0248	0.0254	0.0260	0.0350	0.0384	0.0395	<u>0.0637</u>	0.0623	**0.0670**⋆	5.18%
	HR@20	0.0346	0.0371	0.0406	0.0443	0.0551	0.0623	0.0649	<u>0.0917</u>	0.0912	**0.0994**⋆	8.40%
	NDCG@5	0.0082	0.0091	0.0080	0.0101	0.0144	0.0143	0.0151	<u>0.0325</u>	0.0323	**0.0329**	1.23%
	NDCG@10	0.0111	0.0124	0.0113	0.0133	0.0183	0.0194	0.0205	<u>0.0388</u>	0.0383	**0.0398**⋆	2.58%
	NDCG@20	0.0143	0.0145	0.0156	0.0179	0.0233	0.0254	0.0263	<u>0.0458</u>	0.0456	**0.0478**⋆	4.37%

Table 3. Ablation study with important designs (HR@20 and NDCG@20).

Model	Sports		Toys		Yelp	
	HR	NDCG	HR	NDCG	HR	NDCG
w/o local	0.0722	0.0306	0.1272	0.0558	0.0983	0.0462
w/o global	0.0705	0.0303	0.1278	0.0567	0.0914	0.0460
w/o diffusion	0.0752	0.0323	0.1302	0.0574	0.0986	0.0475
w/o $\mathcal{R}$	0.0741	0.0322	0.1294	0.0567	0.0956	0.0467
w/ vanilla	0.0755	0.0323	0.1310	0.0564	0.0972	0.0475
DivRec	**0.0774**	**0.0329**	**0.1319**	**0.0577**	**0.0994**	**0.0478**

training loss. 5) **w/ vanilla**: replace leader-follower strategy with vanilla strategy in the contrastive views integration component.

The results in Table 3 show the significant effectiveness of locally-diverse and globally-diverse supervised contrastive views. When comparing **w/o diffusion** with DivRec, the diffusion mechanism improves performance due to its remarkable generation ability. It can also be considered as a noise augmentation operation to enhance contrastive learning. The performance of **w/o $\mathcal{R}$** and DivRec demonstrates that the regularization term boosts performance, aligning with our design motivation as it improves the semantic consistency between the generated diverse contrastive views and the original views. By comparing **w/ vanilla** with

DivRec, it can be seen that leader-follower strategy achieves better results than vanilla strategy.

4.4 Effectiveness Analysis

Effectiveness of Contrastive Views Generation. Different contrastive views have varying effects on learned embedding matrices in SR. A more effective contrastive view promotes more representative embeddings [21]. To further analyze the effectiveness of locally-diverse and globally-diverse supervised contrastive views, we conduct the visualization of the learned item embedding matrices and their singular values. This visualization projects the embedding matrix into 2D and outputs the normalized singular values based on SVD decomposition. To ensure fairness, the comparison is based on DuoRec [17], which only uses the existing supervised contrastive views.

From Fig. 3(a) and (c), it is clear that DivRec produces a larger and more uniform distribution of embeddings for both low- and high-frequency items, particularly for low-frequency items. This suggests that DivRec generates more representative embeddings. Additionally, as illustrated in Fig. 3(b) and (d), the singular values of DivRec decrease slower than those of DuoRec and MCLRec, indicating that the embedding matrix of DivRec has a higher rank and is more representative. This is due to diverse contrastive views, which enrich the supervised contrastive information [21] and enable more items to participate in CL.

Effectiveness of Contrastive Views Integration. Apart from insertion augmentation used in DivRec, the original sequence can also be modified by random augmentations (e.g., cropping and reordering, etc.) to construct locally-diverse supervised contrastive views. Therefore, to further analyze the effectiveness of the contrastive views integration component, we consider two situations. The locally-diverse supervised contrastive views are constructed by random augmentations in the first situation (Fig. 4(a)(b)), while they are constructed by insertion augmentation in the second situation (Fig. 4(c)(d)). The relationships of alignment and uniformity (defined in Appendix A) are presented to investigate the impact of different integration strategies.

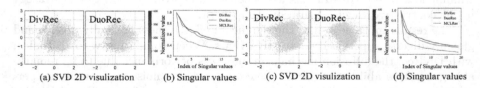

(a) SVD 2D visulization   (b) Singular values   (c) SVD 2D visulization   (d) Singular values

Fig. 3. The SVD 2D visualization and singular values of item embedding matrix on Toys ((a)(b)) and Sports ((c)(d)) datasets. The colorbar in the SVD 2D visualization corresponds to the frequency of items.

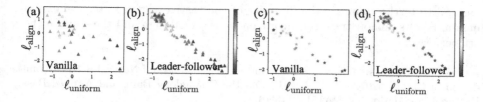

Fig. 4. The relationship between alignment and uniformity when integrating multiple contrastive views via vanilla strategy (a)(c) and leader-follower strategy (b)(d) on Toys dataset. Locally-diverse supervised contrastive views are constructed by random augmentations in the first situation (a)(b), while they are constructed by insertion augmentation in the second situation (c)(d).

From Fig. 4(a)(c), it can be observed that integrating multiple contrastive views via vanilla strategy leads to an unstable training process, especially when locally-diverse contrastive views are constructed by random augmentation. This is because multiple contrastive views have different attributes [23], and the random augmentation may cause semantic inconsistencies, resulting in more serious conflicts. However, as shown in Fig. 4(b)(d), leader-follower strategy could contribute to a more stable training process in both situations, indicating that conflicts are effectively mitigated. This is due to the leader-follower strategy taking the attribute differences between different views into account and constructing attribute-compatible contrastive pairs.

4.5 Hyper-Parameter Analysis

Hyper-parameter analyses are conducted on Sports and Yelp datasets. The diffusion noise scale λ, the length of globally-diverse supervised contrastive view m, and the insertion ratio α are tuned in the range of $[0, 0.9]$, $\{1, 4, 8, 16, 32\}$, and $\{0, 1e-4, 1e-3, 1e-2\}$, respectively. From Fig. 5 in appendix, a relatively small noise scale λ is appropriate, bringing in a slight improvement in performance and robustness (see Appendix D for robustness analysis). Inserting a specific ratio of items into the original sequence yields superior results compared to not inserting any items, which further demonstrates the effectiveness of locally-diverse contrastive views. However, the effects of m are not significant.

5 Conclusion

This paper investigates the possibility of improving the performance of supervised contrastive learning in sequential recommendation by generating diverse contrastive views. To achieve this, we propose a new SR model named DivRec, which incorporates contrastive views generation and contrastive views integration. Additionally, diffusion mechanism is introduced as a noise augmentation method to further improve the diversity of contrastive views. Extensive experimental results demonstrate the effectiveness of each design and the state-of-the-art performance of DivRec.

Acknowledgments. This work is supported by the National Natural Science Foundation of China (62072483, 62472461), and the Guangdong Basic and Applied Basic Research Foundation (2022A1515011690).

Appendix

A Alignment and Uniformity

To validate the quality of contrastive learning guided by the InfoNCE loss, alignment and uniformity [23] are defined as follows:

$$\ell_{\text{align}} \triangleq \mathop{\mathbb{E}}_{(x,x^+)\sim p_{\text{pos}}} \left\| \varphi\left(x^+\right) - \varphi(x) \right\|^2,$$
$$\ell_{\text{uniform}} \triangleq \log \mathop{\mathbb{E}}_{x^- \sim p_{\text{data}}} e^{-2\|\varphi(x^-)-\varphi(x)\|^2}, \tag{13}$$

where p_{pos} denotes the distribution of samples of the positive pair and p_{data} is the distribution of independent samples. $\varphi(x)$ is the original contrastive view, and $\varphi(x^+)$, $\varphi(x^-)$ are corresponding positive and negative contrastive views, respectively. Intuitively, a smaller alignment loss indicates that $\varphi(x^+)$ and $\varphi(x)$ are more similar, while a smaller uniformity loss represents that $\varphi(x^-)$ and $\varphi(x)$ are pushed further apart. Therefore, a smaller alignment and uniformity loss represent better performance of contrastive learning. The relationships of alignment and uniformity represent the attributes of contrastive learning.

B Time Complexity Analysis of DivRec

The time complexity of our DivRec mainly comes from training and inference phase. d is the embedding size, L is the length of truncated sequences, and $|\mathcal{V}|$ is the number of items set.

In the training phase, the time complexity mainly lies in three parts. Their time complexities are as follows. 1) non-autoregressive generation of diverse sequences: $\mathcal{O}(|\mathcal{V}| * d)$, 2) sequence encoding: $\mathcal{O}(L^2 * d)$, 3) calculation of training loss: $\mathcal{O}(|\mathcal{V}| * d)$. As $|\mathcal{V}|$ is much larger than L^2, the overall time complexity of DivRec is dominated by $\mathcal{O}(|\mathcal{V}| * d)$, which is consistent with baselines MCLRec [15] and DuoRec [17].

In the inference phase, the overall time complexity of our DivRec depends on the number of reverse steps of diffusion, denoted as k, which ranges from 1 to 32. When $k = 1$, DivRec exhibits the same time complexity as baselines, however, DivRec could also outperform baselines (see performance of "w/o diffusion" in Table 3).

Based on the above analysis, our DivRec adaptively generates contrastive views with rich supervised information without additional time complexity compared to baselines [15,17].

C Baselines

We compare DivRec to various baselines organized into four groups as follows:

- **Non-sequential model.** BPR-MF [18] directly optimizes the personal rank objective based on SGD and training triples.
- **Vanilla sequential models.** Different from non-sequential models, sequential models intend to capture sequential patterns of a user's historical interactions. GRU4Rec [5] and Caser [20] are the first to use an RNN-based and CNN-based model for SR, respectively. SASRec [9] is the first to use an attention-based model for SR, which serves as the backbone of most recent models.
- **Diffusion model for SR.** DiffRec [2] is the first to apply diffusion model to SR. It leverages the diffusion process to destruct and reconstruct the interaction matrix for predicting the next item. T-DiffRec, an extension of DiffRec specially designed for sequential recommendation, serves as a baseline for comparison.
- **Sequential models with contrastive learning.** CL4SRec [28] is the first to introduce contrastive learning into SR. CoSeRec [12] enhances CL4SRec by proposing two additional informative augmentations: insertion and substitution. DuoRec [17] introduces a new supervised augmentation strategy, where two sequences that share the same target item are considered as a positive pair. MCLRec [15] develops a learnable model augmentation with meta-learning optimization to enhance the quality of contrastive pairs. DuoRec and MCLRec are two of the most powerful baselines.

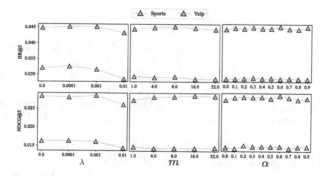

Fig. 5. Hyperparameter analysis on Sports and Yelp datasets. Diffusion noise scale λ, the length of globally-diverse supervised contrastive view m, and insertion ratio α are tuned in the range of $[0, 0.9]$, $\{1, 4, 8, 16, 32\}$, and $\{0, 1e\text{-}4, 1e\text{-}3, 1e\text{-}2\}$, respectively.

D Robustness Analysis

The utilization of the diffusion encoder with noise injection theoretically makes DivRec robust against noisy sequences. To verify this claim, noisy items are inserted into the sequence at varying ratios during the testing stage. Figure 6 shows that as more noisy items are injected into sequences, the performance of all three models decreases. However, the drop ratio of DivRec is smaller than DuoRec and MCLRec, especially in the Yelp dataset. This suggests that DivRec has better noise robustness than baselines. We speculate that the robustness of DivRec be attributed to two factors: the diffusion mechanism and locally-diverse contrastive views. On the one hand, the diffusion mechanism injects noise into embedded sequences during training, which improves the robustness at the embedding level. On the other hand, DivRec is trained using locally-diverse contrastive views which contain inserted noisy items in sequences, contributing to the robustness at the sequence level.

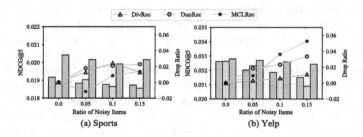

Fig. 6. The robustness analysis on Sports and Yelp datasets. The left Y-axis (bar) represents the performance of NDCG@5, while the right Y-axis (line) represents the corresponding drop ratio of performance.

References

1. Chen, T., Kornblith, S., Norouzi, M., Hinton, G.: A simple framework for contrastive learning of visual representations. In: ICML. pp. 1597–1607 (2020)
2. Du, H., Yuan, H., Huang, Z., Zhao, P., Zhou, X.: Sequential recommendation with diffusion models. arXiv preprint arXiv:2304.04541 (2023)
3. Fang, H., Zhang, D., Shu, Y., Guo, G.: Deep learning for sequential recommendation: Algorithms, influential factors, and evaluations. ACM TOIS **39**(1), 1–42 (2020)
4. Gong, S., Li, M., Feng, J., Wu, Z., Kong, L.: Diffuseq: Sequence to sequence text generation with diffusion models. In: ICLR (2023)
5. Hidasi, B., Karatzoglou, A.: Recurrent neural networks with top-k gains for session-based recommendations. In: CIKM. pp. 843–852 (2018)
6. Ho, J., Jain, A., Abbeel, P.: Denoising diffusion probabilistic models. In: NeurIPS. pp. 6840–6851 (2020)
7. Ho, J., Salimans, T.: Classifier-free diffusion guidance. arXiv preprint arXiv:2207.12598 (2022)
8. Huang, W., Yi, M., Zhao, X., Jiang, Z.: Towards the generalization of contrastive self-supervised learning. In: ICLR (2023)
9. Kang, W., McAuley, J.J.: Self-attentive sequential recommendation. In: ICDM. pp. 197–206 (2018)

10. Kong, Z., Ping, W., Huang, J., Zhao, K., Catanzaro, B.: Diffwave: A versatile diffusion model for audio synthesis. In: ICLR (2021)
11. Li, Z., Sun, A., Li, C.: Diffurec: A diffusion model for sequential recommendation. ACM TOIS **42**(3), 1–28 (2023)
12. Liu, Z., Chen, Y., Li, J., Yu, P.S., McAuley, J., Xiong, C.: Contrastive self-supervised sequential recommendation with robust augmentation. arXiv preprint arXiv:2108.06479 (2021)
13. Ma, H., Xie, R., Meng, L., Chen, X., Zhang, X., Lin, L., Kang, Z.: Plug-in diffusion model for sequential recommendation. In: AAAI. pp. 8886–8894 (2024)
14. Oord, A.v.d., Li, Y., Vinyals, O.: Representation learning with contrastive predictive coding. arXiv preprint arXiv:1807.03748 (2018)
15. Qin, X., Yuan, H., Zhao, P., Fang, J., Zhuang, F., Liu, G., Liu, Y., Sheng, V.S.: Meta-optimized contrastive learning for sequential recommendation. In: SIGIR. pp. 89–98 (2023)
16. Qin, X., Yuan, H., Zhao, P., Liu, G., Zhuang, F., Sheng, V.S.: Intent contrastive learning with cross subsequences for sequential recommendation. In: WSDM. pp. 548–556 (2024)
17. Qiu, R., Huang, Z., Yin, H., Wang, Z.: Contrastive learning for representation degeneration problem in sequential recommendation. In: WSDM. pp. 813–823 (2022)
18. Rendle, S., Freudenthaler, C., Gantner, Z., Schmidt-Thieme, L.: BPR: bayesian personalized ranking from implicit feedback. In: UAI. pp. 452–461 (2009)
19. Sun, F., Liu, J., Wu, J., Pei, C., Lin, X., Ou, W., Jiang, P.: Bert4rec: Sequential recommendation with bidirectional encoder representations from transformer. In: CIKM. pp. 1441–1450 (2019)
20. Tang, J., Wang, K.: Personalized top-n sequential recommendation via convolutional sequence embedding. In: WSDM. pp. 565–573 (2018)
21. Tian, Y., Sun, C., Poole, B., Krishnan, D., Schmid, C., Isola, P.: What makes for good views for contrastive learning? In: NeurIPS. pp. 6827–6839 (2020)
22. Wang, L., Lim, E.P., Liu, Z., Zhao, T.: Explanation guided contrastive learning for sequential recommendation. In: CIKM. pp. 2017–2027 (2022)
23. Wang, T., Isola, P.: Understanding contrastive representation learning through alignment and uniformity on the hypersphere. In: ICML. pp. 9929–9939 (2020)
24. Wang, W., Xu, Y., Feng, F., Lin, X., He, X., Chua, T.: Diffusion recommender model. In: SIGIR. pp. 832–841 (2023)
25. Wu, C., Yuan, H., Zhao, P., Qu, J., Sheng, V.S., Liu, G.: Dual-supervised contrastive learning for bundle recommendation. IEEE TCSS **11**(3), 3955–3965 (2024)
26. Wu, S., Tang, Y., Zhu, Y., Wang, L., Xie, X., Tan, T.: Session-based recommendation with graph neural networks. In: AAAI. pp. 346–353 (2019)
27. Xia, L., Huang, C., Xu, Y., Dai, P., Zhang, X., Yang, H., Pei, J., Bo, L.: Knowledge-enhanced hierarchical graph transformer network for multi-behavior recommendation. In: AAAI. pp. 4486–4493 (2021)
28. Xie, X., Sun, F., Liu, Z., Wu, S., Gao, J., Zhang, J., Ding, B., Cui, B.: Contrastive learning for sequential recommendation. In: ICDE. pp. 1259–1273 (2022)
29. Yang, Z., Wu, J., Wang, Z., Wang, X., Yuan, Y., He, X.: Generate what you prefer: Reshaping sequential recommendation via guided diffusion. In: NeurIPS (2024)
30. Yu, J., Yin, H., Xia, X., Chen, T., Li, J., Huang, Z.: Self-supervised learning for recommender systems: A survey. IEEE TKDE **36**(1), 335–355 (2024)

Disentangled Causal Embedding with Unbiased Knowledge Distillation for Recommendation

Nan Liu, Shunmei Meng[✉], Xiao Liu, Qianmu Li, and Yu Jiang

Nanjing University of Science and Technology, Nanjing, China
{liunan,mengshunmei,liuxiao,qianmu,jiangyu}@njust.edu.cn

Abstract. Knowledge Distillation, a model compression method that transfers knowledge from a complex teacher model to a simpler student model, has been widely applied in recommendation systems. Existing KD recommendation models primarily focus on improving the performance and inference speed of the student model during the inference stage, while overlook the significant issue of popularity bias. It is evident that the issue of bias exists, arising from the soft labels provided by the teacher model, is further propagated and intensified throughout the distillation process. Using causal embeddings or propensity-based unbiased learning to eliminate the bias effect is proved to be effective. Nevertheless, by solely amalgamating various causes of popular bias and interactions into a unified representation, the robustness and interpretability of the model cannot be assured. In view of these challenges, we present DCUKD, a new unbiased knowledge distillation approach with disentangling causal embedding. It addresses the root causes of deviation, aiming to produce unbiased results and significantly enhance the model's recommendation performance. In particular, we introduce a causal mechanism with popularity bias to verify that popularity bias is not entirely detrimental. Then DCUKD decomposes popularity bias based on time effects from causal graph, which eliminates detrimental bias and retains benign bias. Through causal intervention, the recommendation result is adjusted with the desired bias to guide the student model. Finally, we explore the effectiveness of DCUKD on two real datasets, showing that our method outperforms all considered baselines.

Keywords: Recommender systems · Popularity bias · Knowledge distillation · Causal embedding

1 Introduction

With the information overload on websites, there is a growing need to strike a better balance between recommendation accuracy and efficiency. The extraction of information from large and highly redundant datasets is essential. However, the most effective models frequently entail large-scale or even combinations of multiple models. To address the real-time requirements of recommendation systems (RS), knowledge distillation (KD) has been employed. KD is a model compression method [4] designed to compress and transfer knowledge from intricate models to more lightweight ones. The performance is preserved and may even be improved by utilizing soft labels learned from teacher models that contain rich information.

Despite the improvement in performance, we observe that the distillation process disproportionately favors popular items. Combined with the existing research methods, although it becomes apparent that KDs have general improvement in the popular category, the performance in the unpopular category remains less optimistic.

Addressing the significant bias problem in knowledge distillation is crucial. Failure to overcome this issue not only hampers the learning of the student model but also substantially diminishes the fairness and diversity of the recommendation system, thereby impacting the overall user experience. In regard to this phenomenon [2,14,19], we pinpoint the source of the bias - the biased soft labels produced by the teacher model, which are further amplified as the student model learns these soft labels. To eliminate the bias problem of soft labels, the easiest way is to change the teacher model so that it produces unbiased soft labels. However, achieving this in real-life scenarios is challenging due to the large scale of teacher models and the complexity of model adjustments. To address these challenges, we propose incorporating a debiasing process into the student model.

Obviously, part of the bias comes from the popularity bias, where recommendation systems tend to frequently recommend popular items, leading to a decline in the popularity of long-tail items. This bias may be caused by the user's conformity, rather than real interest [7,8]. As a result, the recommendation model may produce unexpected results in such biased data, capturing distorted user preferences and amplifying the long-tail impact. Given the widespread popularity bias and its negative effects on recommendations, many studies have focused on eliminating the popularity bias completely to restore genuine user preferences, for example, the most pertinent piece of work is the recently proposed UnKD [3], which mainly aims at addressing a single popularity bias. Nevertheless, we discover that not all popularity bias is detrimental [22,28]. Along with the conformity effect, the popularity of items is often closely related to the quality of items. The complete elimination of popularity bias leads to the loss of this crucial signal, posing a challenge for the model to distinguish and recommend appropriate items. Secondly, regarding the sources of bias, we posit that another significant portion stems from time-aware factors, among which we believe factors with a greater impact include user dynamic selection bias and item change bias. For example, the emergence of new items, seasonal changes, regional differences and other factors give rise to distinct shopping patterns that impact the entire user group. These phenomena align with the domain of traditional concept drift research [11,16]. Our work model user preferences and item changes to eliminate bias in the RS, using time as the primary factor to capture time drift patterns in user behavior and item changes. In reality, user behavior and item changes encompass various concept drifts, all of these work at different times. We eliminate time-aware factor bias by learning from successful time modeling cases [5,11].

Regarding the recommendation bias, our DCUKD selectively eliminates both the popularity bias and the time-aware factor bias with significant effects. We propose specific approaches to learn disentangled causal embeddings as opposed to the previous strategy of employing basic scalar popularity values.

To summarize, we make the following contributions:

– As far as we are aware, this is the new work to reveal the popularity bias and time-aware factor bias of knowledge distillation in RS by causal graph.
– We propose a disentangled causal embedding with unbiased knowledge distillation method, which preserves only the quality attribute of items through disentangling and considers the influence of time-aware factor bias to guide student model.
– Extensive experiments on two real-world datasets show that DCUKD outperforms various state-of-the-art recommendation methods. In addition, the detailed model ablation study aids in justifying the effectiveness of the model and understanding the component-wise impact on performance improvement.

2 Related Work

In this section, we review relevant works from the two perspectives listed below.

Knowledge Distillation. KD is a model compression method based on "teacher-student" network thought. It initially trains a large-scale teacher model and utilizes soft labels for knowledge transfer [24]. Now, more and more researches are introducing KD into recommendation systems to reduce delay in order to get real-time recommendation. For instance, RD [20] ranks soft labels from teachers, considering the top K items as positive samples, which is used to train student model; UnKD [3] groups soft labels by popularity and draws positive and negative samples from each group to train the student model. CD [13] uses soft labels for both positive and negative knowledge distillation samples; DERRD [9] also considers soft labels to generate list distillation loss functions. In addition to the research on soft labels, there is a growing body of research on teacher models. For example, HTD [10] extracts topological knowledge based on relationships in teacher embedding spaces. Despite these advancements, it's noteworthy that existing distillation methods still face serious long-tail problems.

Causal Recommendation. The fields of machine learning has been increasingly focused on causal inference. Causal inference is currently used to solve bias problems, make interpretable recommendations, and improve model generalization ability. For example, PD [27] uses causal diagram to remove the influence of popularity pointing to items, and extends to PDA [27], which can artificially add bias and simulate popularity; DICE [29] utilizes causal embedding to decompose user interest and conformity in recommendation; TIDE [28] uses causal diagram to distinguish between beneficial and harmful biases in popularity. Additionally, causal inference is applied as a counterfactual learning strategy [1,15,23]. MACR [23] uses model-independent counterfactual reasoning to eliminate popularity bias in RS. MMGCN [21] uses causal graphs combined with counterfactual to eliminate the click bait problem. Mehrotra et al. [17] employ a quasi-experimental Bayesian framework to quantify the impact of interventions on outcomes, thereby providing counterfactual data. Yuan et al. [26] utilize a simple to learn paradigm where unbiased information is used to create labels for unseen data, addressing the Missing-Not-At-Random problem. Despite these studies focus on the issue of causal recommendations, they do not expressly take into account popularity bias and time-aware factors that boost recommendation performance.

3 Problem Definition

Suppose a recommender system that has a user set $\mathcal{U} = \{u_1, ..., u_n\}$ and an item set $\mathcal{I} = \{i_0, i_1, ..., i_m\}$. Let u (or i) represents a user (or an item) in set $\mathcal{U}$(or $\mathcal{I}$) and let t for time. A rating $r_{ui}(t)$ reflects the user's preference for item i on day t, with higher ratings reflects greater preferences. Let G be the set of triples that represents the previous user behavior data which was systematically gathered before the time T, i.e., $G = \{(u_j, i_j, t_j) \mid r_{ui}(t) is known\}$, where the triple (u, i, t) indicates that the user u clicked the item i at time t, and the rating is $r_{ui}(t)$. For ease of use, prior to time t, we gather users' feedback on the interactive item i as $G_i^t = \{(u_j, i_j, t_j) \in G | i_j = i, t_j < t, r_{ui}(t)\}$. Additionally, we define the item's popularity P_i as the quantity of interactions on i that have been detected, i.e., $P_i = |G_i^T|$. We define b is the average rating of all user interactions with the item, b_u represents user's dynamic preference bias, and b_i represents item change bias. The recommendation system's task description is as follows:

Input: Observational interaction data G.

Output: A prediction model that predicts user click behavior taking into account two different types of bias.

4 Methodology

In this section, we propose a general framework named DCUKD to disentangle causal embedding for unbiased knowledge distillation. Figure 2(b) illustrates the causal graph of DCUKD and the unbiased knowledge distillation framework is shown in Fig. 1.

4.1 Causal View of Recommendation

Figure 2(a) shows a causal diagram of distillation bias in existing methods. Next, we interpret this causal diagram:

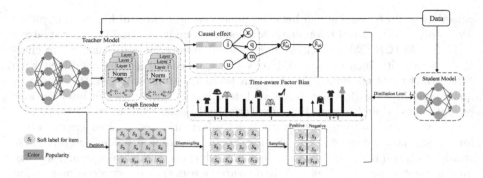

Fig. 1. Overview of the general DCUKD framework.

- U represents the user node, e.g., user ID or feature that is used for representing a user.
- I represents an item node.
- M represents the matching score between U and I.
- P represents the popularity of item I.
- Y represents the soft labels that teacher model predicted.
- S represents the learned student model.
- Edges $\{U, I\} \to M$ represents the match between user and item.
- Edge $I \to P$ represents the impact of item popularity.
- Edges $\{M, P\} \to Y$ represents that the final prediction score of the teacher model affected by the degree of match and the popularity of the item.
- Edge $Y \to S$ represents that the student model relies on soft labels for training.

According to the analysis, users' motivations for clicking on items are mainly due to two aspects: (1) an interest in the item itself, often associated with its quality. (2) the influence of the item's popularity, reflecting users' conformity. We employ an additive model to elucidate the factors influencing the occurrence of a click event. The predicted scores y_{ui} for a user u and an item i are determined by the following formula:

$$y_{ui} = y_{ui}^{quality} + y_{ui}^{conformity} \tag{1}$$

where $y_{ui}^{quality}$ and $y_{ui}^{conformity}$ represent the prediction scores for specific causes.

Next, we use counterfactual inference to quantify popularity bias effect (namely $y_{ui}^{quality}$ and $y_{ui}^{conformity}$). In accordance with the causal graph, the trained soft labels would deviate from accurately reflecting the real preferences of the user due to an additional path ($I \to P \to Y$) connecting I and Y. For instance, an item may receive a higher score merely because it belongs to a popular group rather than it actually satisfies the user's preferences. Such biases would be reinforced and further propagated into the student model, significantly diminishing the quality of its recommendations. Therefore, addressing the problem of bias in knowledge distillation is crucial.

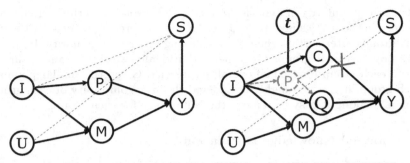

(a) Causal graph of traditional KD methods. (b) Causal graph of our DCUKD.

Fig. 2. The causal graph that shows the knowledge distillation. U: user, I: item, M: match score, P: item popularity, C: conformity, Q: quality, Y: soft labels, S: student model. The causal impact of P on Y is where the bias originates. Our DCUKD utilizes time-aware factors to disentangle P. Obviously, other causal paths from time-aware factor might exist, but here we only focus on the U and I.

4.2 Disentangled Representation Learning

First of all, $C \to Y$ represents the effect of user behavior on time-aware conformity. The impact of conformity depends not only on the current time when user u interacts with item i, but also on the time and number of the last interactions. Therefore, we utilize the parameterized function h_g to assess the intensity of the conformity impact of item i at time t:

$$c_i^t = h_g(t, G_i^t) = w_i \sum_{(u_j, i_j, t_j) \in G_i^t} \exp\left(-\frac{|t - t_j|}{\tau}\right) \quad (2)$$

where introduces the parameter w_i for each item i to readjust the impact, recognizing that conformity can have a significant effect on certain items. Here, the effect of past conformity is simply accumulated because the effect of current popular items is generally more substantial than that of past popular items. We introduce the temperature parameter t to control the sensitivity of c_i^t to time, with a smaller t directing the model to focus more on current interactions and diminishing the influence in the past popular items.

We know that users are prone to favor items of higher quality. Here, we simply capture the intrinsic quality of each item i using the time-independent specific variable q_i.

The matching score corresponding to user u and item i is $m_{ui} = f_\theta(u, i)$, $f_\theta(u, i)$ can be realized by numerous recommendation models, including BPRMF [4] and LightGCN [6].

Combine three parts yields the preliminary prediction:

$$\hat{y}_{ui}^* = \tanh(q_i + c_i^t) \times \text{Softplus}(m_{ui}) \quad (3)$$

Next, the popularity of items is grouped into K groups and the current work is referenced in the partitioning procedure. To be more precise, we classify the items into K groups after sorting them in descending order by popularity. Items with similar popularity are grouped together. Additionally, we observe and ensure that items in each group have equal overall popularity. Given the similar popularity of each group, it can be roughly considered that the conformity of each group is the same, making it easy to remove the conformity for each group.

4.3 Unbiased Knowledge Distillation

In the proposed framework, we aim to reduce the latency during the inference stage by incorporating unbiased knowledge distillation in the final stage. To be specific, our DCUKD with time-aware factors is debiased in the learning process of student model. The subtlety of DCUKD is that it breaks down the popularity bias while taking into account the time-aware factor bias.

Time-Aware Factor. A baseline predictor for a rating y_{ui} takes into account the user and item major impacts, the following is a suitable method to construct a static baseline predictor:

$$\hat{y}_{ui} = b_{ui} + \hat{y}_{ui}^* \tag{4}$$

where b_{ui} indicates the observed bias from the average of user u and item i, respectively.

Much of the temporal variability is included within the baseline predictors. Firstly, the rating of an item changes over time. For example, a movie starring a well-known actor becomes more popular. In our model, the item deviation b_i is not a constant but a function of time. Secondly, user bias also influences their baseline ratings over time, such as users who typically rated average movies as "3 stars" now rate them as "4 stars." Hence, in our model, we consider the parameter b_u as a function of time. This modification renders the baseline predictor a time-dependent function:

$$b_{ui} = b + \mu b_u(t) + \gamma b_i(t) \tag{5}$$

where μ and γ are set to control the impact of the different time-aware factor.

Estimating Clicks. The detailed training process of DCUKD is shown in Fig. 1, including grouping, separation, sampling and training.

Firstly, group and separate items through the disentangled learning process, then we extract the positive and negative item pairs (i^+, i^-) from non-conformity sequence, whose probability distribution is $p_i \propto e^{-rank(i)/\mu}$, where $rank(i)$ represents the rank position i in the group, μ is a hyperparameter.

Finally, we use group distillation losses to train the student model:

$$L_s = -\sum_u \frac{1}{|U|} \sum_{g \in G} \sum_{(i^+, i^-) \in S_{ug}} \log \sigma\left(\hat{y}_{ui^+}(t) - \hat{y}_{ui^-}(t)\right) \tag{6}$$

The item pairs sampled for the loss calculation are taken from $((i^+, i^-) \in Sug)$. Note that the sampled pairs of items come from the same group, which remove the influence of item conformity and retains only the quality of items. This approach is consistent with the user's real preferences, ensuring the accuracy of distillation losses and providing correct and useful knowledge for training the student model. $\hat{y}_{ui+}(t)$ and $\hat{y}_{ui-}(t)$ represent the prediction scores for positive and negative samples, respectively. σ represents the sigmoid function.

$$L_r = \sum_{(u,i) \in G} -\log \sigma (y_{ui+} - y_{ui-}) \tag{7}$$

where L_r represents the loss of supervision from the raw training data, σ represents the sigmoid function, i^+ and i^- represent the positive and negative sample pairs of the item from the training data.

The final loss function used to train student model is:

$$L = \alpha L_s + \beta L_r \tag{8}$$

where the hyperparameters α and β are used to balance their contributions.

5 Experiments

In this section, we do experiments to demonstrate the usefulness of the proposed framework. In particular, our goal is to respond to the following research questions:

RQ1: Does DCUKD achieve the goal of removing the bias of soft labels? How does our proposed DCUKD framework perform compared with state-of-the-art knowledge distillation methods?

RQ2: Does the disentangling popularity bias model improve performance?

RQ3: Is it beneficial to model the time-aware factor? How does the effect change when different factors are introduced separately?

5.1 Experimental Settings

Datasets. The Movielens-1M dataset and the CiteULike dataset are two commonly used datasets collected from real-world applications. Table 1 lists the statistics for both datasets.

Table 1. Statistics of the datasets.

Dataset	User	Item	Interaction	Sparsity
MovieLens	6,040	3,706	1,000,209	95.54%
CiteULike	5,219	25,181	125,580	99.91%

Data Preprocessing. To evaluate the performance of DCUKD, we discard users and items with fewer than 20 interaction records. For training and testing,

90% of users are regarded as the training set, and the remaining 10% are allocated to the testing set. The last step involves validation, where we randomly partition 10% of the interactions in the training set.

Recommendation Models. We employ Matrix Factorization (MF) [12], the most widely used recommendation model. Additionally, include the cutting-edge Graph Convolutional Networks (GCN) [25] collaborative filtering mechanism. Specifically, we use the latest recommendation models BPR-MF [18] and Light-GCN [6].

Hyper-parameters and Metrics. For the quality c of items, we set it as 0 for practical and experimental convenience. Next, for the number of groups K, for better visualization, let's set $K = 2$, dividing it into a popular group and an unpopular group. To compare the effectiveness of the proposed model and the baseline method, we adopt two regular Top-N metrics as follows, and we set $N = 10$ and $N = 20$ respectively in the experiment to evaluate models more comprehensively.

- **NDCG@N (Normalized Discounted Cumulative Gain):** NDCG is a position-aware metric which assigns larger weights to higher positions.
- **Recall@N:** Recall is the proportion of positive samples recommended by the system.

Compared Methods. We contrast our DCUKD with the following baselines: RD [20], CD [13], DERRD [9], HTD [10] and UnKD [3].

5.2 Performance Comparison (RQ1)

The results of the performance comparison between DCUKD and the baseline methods are shown in Table 2, our DCUKD consistently outperforms other methods in backbone model and two real-world dataset. For instance, in the MovieLens dataset, using the LightGCN backbone model, our method improves over UnKD by 30.12% and 24.09% in Recall@20 and NDCG@20, respectively. With fewer training data, our method successfully completes the scenario, from which the following observations and analysis may be drawn:

- Across the board for both datasets, DCUKD regularly outperforms the baseline methods, which verifies the superiority of DCUKD in removing the popularity bias and time-aware factor bias in the distillation recommendation process.
- In comparison to MovieLens, all the approaches exhibit inferior performance on CiteULike, which demonstrates the severe sparsity of CiteULike adversely impacts the learning of the student model. Despite the limited training data, our method performs well in this challenging scenario. With insufficient data, similar conclusions can be made for various combinations. When comparing our DCUKD to other baselines, it produces the best overall performance.

Table 2. Performance comparisons on two real-world datasets (all $p-$**value** < 0.01). The top 10 results provide the basis for all metrics and the best results are in bold.

Dataset	Backbone Model	LightGCN		BPRMF	
	Method	Recall	NDCG	Recall	NDCG
MovieLens	Teacher	0.181	0.2951	0.185	0.3012
	Student	0.1456	0.2581	0.1435	0.2511
	RD	0.1473	0.2559	0.1471	0.2583
	CD	0.1445	0.2534	0.1477	0.2602
	DERRD	0.1436	0.2532	0.1487	0.2606
	HTD	0.1441	0.2539	0.1472	0.2592
	UnKD	0.1547	0.2615	0.1569	0.2672
	DCUKD	**0.2013**	**0.3245**	**0.1822**	**0.2989**
	impv-e%	30.12%	24.09%	16.12%	11.86%
CiteULike	Teacher	0.1518	0.1016	0.1657	0.1139
	Student	0.076	0.0477	0.0783	0.051
	RD	0.0808	0.0514	0.0833	0.0538
	CD	0.0801	0.0518	0.0936	0.0616
	DERRD	0.0793	0.0511	0.0809	0.0527
	HTD	0.0788	0.0485	0.0958	0.0628
	UnKD	0.0863	0.055	0.1006	0.0654
	DCUKD	**0.1086**	**0.0905**	**0.1088**	**0.0695**
	impv-e%	25.84%	64.54%	8.15%	6.27%

- Surprisingly, DCUKD delivers more than 8.15% improvement even on the sparser dataset CiteULike. This outcome shows DCUKD is more flexible to sparse training datasets. For a training item, DCUKD disentangles popularity bias, retains only benign bias, and introduces the influence of time-aware factors, which is effectively solve the deviation problem.
- Basically, the debiasing knowledge distillation method based on causal reference perform better than the other approaches on both datasets, because the causal reference offers more bias information to assist learn more accurate soft labels.

5.3 Ablation Study

Performance of Disentangling Popularity Bias (RQ2)
Now we investigate the effectiveness of the disentangling popularity bias. For this purpose, we compare DCUKD with its two variants as follows:

DCUKD-DP: where item quality and conformity are disentangled effect in both training and inference stage.

Table 3. Characteristics of DCUKD and its variants in LightGCN backbone model, we also present their results on the challenge of predicting preferences. The best results are indicated in bold.

Methods	Teacher Model				Student Model				Performance			
									MovieLens		CiteULike	
	User?	Items?	Quality?	Conformity?	Users?	Items?	Quality?	Conformity?	Recall	NDCG	Recall	NDCG
DCUKD-U	✗	✗	✓	✓	✓	✗	✓	✗	0.2013	0.3244	0.105	0.086
DCUKD-I	✗	✗	✓	✓	✗	✓	✓	✗	0.1874	0.3062	0.0833	0.072
DCUKD-DP	✗	✗	✓	✓	✗	✗	✓	✗	0.1954	0.3169	0.1051	**0.093**
DCUKD-UnDP	✗	✗	✓	✓	✗	✗	✓	✓	0.1935	0.3149	0.0983	0.082
DCUKD	✗	✗	✓	✓	✓	✓	✓	✗	**0.2013**	**0.3245**	**0.1086**	0.0905

Table 4. Characteristics of DCUKD and its variants in BPRMF backbone model, we also present their results on the challenge of predicting preferences. The boldest performance is displayed.

Methods	Teacher Model				Student Model				Performance			
									MovieLens		CiteULike	
	User?	Items?	Quality?	Conformity?	Users?	Items?	Quality?	Conformity?	Recall	NDCG	Recall	NDCG
DCUKD-U	✗	✗	✓	✓	✓	✗	✓	✗	0.1819	0.2988	0.1052	0.0635
DCUKD-I	✗	✗	✓	✓	✗	✓	✓	✗	0.1536	0.2425	0.0964	0.0565
DCUKD-DP	✗	✗	✓	✓	✗	✗	✓	✗	0.1816	**0.3000**	0.1023	0.0603
DCUKD-UnDP	✗	✗	✓	✓	✗	✗	✓	✓	0.1811	0.2988	0.1021	0.0617
DCUKD	✗	✗	✓	✓	✓	✓	✓	✗	**0.1822**	0.2989	**0.1088**	**0.0695**

DCUKD-UnDP: there are no disentangled effect.

The results are visually presented in Fig. 3, on both datasets, we observe that DCUKD consistently outperforms the two variations. In particular, DCUKD-DP is significantly superior to DCUKD-UnDP, especially on CiteULike. The results of DCUKD, DCUKD-DP, and DCUKD-UnDP are clearly displayed in Table 3 and Table 4. Using the LightGCN backbone model, on the MovieLens dataset, DCUKD-DP compared Recall@20 and NDCG@20 in DCUKD-UnDP rose by 0.982% and 0.635%, respectively. On the CiteULike dataset, DCUKD-DP compared Recall@20 and NDCG@20 in DCUKD-UnDP grew by 6.918% and 13.415%, respectively. The improvement in disentangling popularity bias on the CiteULike dataset is notably significant, even though both datasets show improvement. This highlights the substantial variability in popularity bias across different datasets and underscores the significant impact of conformity in the CiteULike dataset. In our DCUKD method, it has marked improvement compared to the improved DCUKD-DP. Recall@20 and NDCG@20 have increased on the MovieLens dataset by 3.019% and 2.398%, respectively. On the CiteULike dataset, Recall@20 has increased by 3.330% and NDCG@20 has decreased by 2.688%. In conclusion, through the combination of disentangling popularity bias and time-aware factors, we draw three key conclusions: (1) popularity bias contains both positive and negative signals, and a model that indiscriminately retains or eliminates all bias lead to undesirable performance; (2) our method's disentanglement of popularity bias is effective in extracting useful bias while removing harmful bias; (3) the greater the dataset influenced by conformity, the more significant the improvement in results.

Performance of Introducing Time-aware Factor (RQ3)

Next, we explore whether it is essential to introduce time-aware factors. We compare our DCUKD-U and DCUKD-I with the following special cases: DCUKD-U

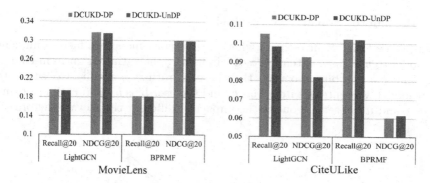

Fig. 3. Performance comparison of Recall@20 and NDCG@20 for disentangling popular bias.

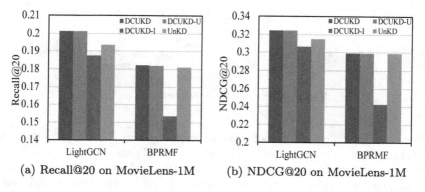

Fig. 4. Performance comparison of Recall and NDCG for introducing time-aware factor on the MovieLens dataset.

indicates the introduction of user bias and DCUKD-I indicates the introduction of item bias. The experimental results are shown in Fig. 4. The results of DCUKD and DCUKD-U basically coincide with each other, with a small difference. However, the performance of DCUKD-I is inferior to them. This shows that considering the time-aware factors is crucial. But the influence of each time-aware factor on the model prediction varies for different datasets, prompting us to set hyperparameters γ and μ.

In Table 3 and Table 4, it is evident that the performance of DCUKD surpasses various previous research methods. Additionally, the effect of DCUKD-U is significantly superior to DCUKD-I, indicating that the influence of user bias is greater than that of item bias in the time-aware factors. Finally, compared with the two variants, the performance of our DCUKD is optimal.

5.4 Parameter Sensitivity Analysis

In this subsection, we investigate whether different time-aware factors influence the model to the same extent. We conclude that the rating of an item changes over time, and user bias also alters their baseline ratings over time. Consequently, in our model, we represent item bias b_i and the user bias b_u as functions of time, where γ and μ are set to control the impact of different time-aware factors.

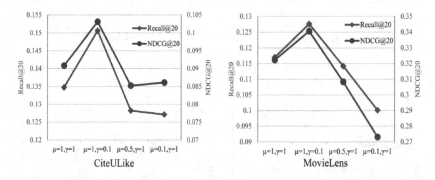

Fig. 5. Effect of hyperparameters γ and μ

Hence we set different values for the hyperparameters γ and μ, and the experimental results are shown in Fig. 5. The results reveal that, under the condition of unchanged $\mu(\mu = 1, \gamma = 0.1 \, or \, 0.5)$, the larger the γ, the worse the performance. This suggests that the model is adversely affected by the bias introduced by item modifications. However, in the case of unchanged γ ($\gamma = 1, \mu = 0.1 \, or \, 0.5$), the smaller the μ, the worse the model performance, indicating that the bias caused by user changes has a positive effect on the model. We conduct experiments on the dataset CiteULike to determine the values of hyperparameters γ and μ where the model performed best. Therefore, we conclude that different time-aware factors have different degrees of influence on the model.

6 Conclusion

In this work, we propose a new debiased distillation method DCUKD with time-aware factors. Initially, we determined that the bias in the distillation process originated from the soft labels of the teacher model, leading to further propagation and reinforcement during the distillation process. Subsequently, we establish a causal diagram to explore and identify the source of the bias in soft labels, and then disentangling popularity bias, which eliminates detrimental bias and retains benign bias. We also introduce time-aware factor bias to guide student model learning. Our experiments on two real-world datasets validate that our DCUKD is largely superior to state-of-the-art methods, especially for long-tail items.

Acknowledgments. This work is supported in part by the project of Frontier Technologies R&D Program of Jiangsu (No. BF2024071), the Science and Technology Project of State Grid Jiangsu Electric Power Company Ltd (No. J2023179), the Open Research Project of State Key Laboratory of Novel Software Technology (No. KFKT2022B28), National Natural Science Foundation of China (No. 61702264).

References

1. Agarwal, A., Takatsu, K., Zaitsev, I., Joachims, T.: A general framework for counterfactual learning-to-rank. In: Proceedings of the 42nd International ACM SIGIR Conference on Research and Development in Information Retrieval. pp. 5–14 (2019)
2. Ben-Porat, O., Torkan, R.: Learning with exposure constraints in recommendation systems. In: Proceedings of the ACM Web Conference 2023. pp. 3456–3466 (2023)
3. Chen, G., Chen, J., Feng, F., Zhou, S., He, X.: Unbiased knowledge distillation for recommendation. In: Proceedings of the Sixteenth ACM International Conference on Web Search and Data Mining. pp. 976–984 (2023)
4. Cheng, Y., Wang, D., Zhou, P., Zhang, T.: A survey of model compression and acceleration for deep neural networks. arXiv preprint arXiv:1710.09282 (2017)
5. Hao, D., Guangwei, H., Ting, W., Wei, S.: Recommendation method for potential factor model based on time series drift. Data Analysis and Knowledge Discovery **6**(10), 1–8 (2022)
6. He, X., Deng, K., Wang, X., Li, Y., Zhang, Y., Wang, M.: Lightgcn: Simplifying and powering graph convolution network for recommendation. In: Proceedings of the 43rd International ACM SIGIR conference on research and development in Information Retrieval. pp. 639–648 (2020)
7. Huang, J., Oosterhuis, H., De Rijke, M.: It is different when items are older: Debiasing recommendations when selection bias and user preferences are dynamic. In: Proceedings of the fifteenth ACM international conference on web search and data mining. pp. 381–389 (2022)
8. Jia, J., Shang, T., Li, L., Chen, S.: De-biasing user conformity bias and item popularity bias in group recommendation. In: 2023 IEEE 6th Information Technology, Networking, Electronic and Automation Control Conference (ITNEC). vol. 6, pp. 1559–1563. IEEE (2023)
9. Kang, S., Hwang, J., Kweon, W., Yu, H.: De-rrd: A knowledge distillation framework for recommender system. In: Proceedings of the 29th ACM International Conference on Information & Knowledge Management. pp. 605–614 (2020)
10. Kang, S., Hwang, J., Kweon, W., Yu, H.: Topology distillation for recommender system. In: Proceedings of the 27th ACM SIGKDD Conference on Knowledge Discovery & Data Mining. pp. 829–839 (2021)
11. Koren, Y.: Collaborative filtering with temporal dynamics. In: Proceedings of the 15th ACM SIGKDD international conference on Knowledge discovery and data mining. pp. 447–456 (2009)
12. Koren, Y., Bell, R., Volinsky, C.: Matrix factorization techniques for recommender systems. Computer **42**(8), 30–37 (2009)
13. Lee, J.w., Choi, M., Lee, J., Shim, H.: Collaborative distillation for top-n recommendation. In: 2019 IEEE International Conference on Data Mining (ICDM). pp. 369–378. IEEE (2019)
14. Liang, D., Charlin, L., McInerney, J., Blei, D.M.: Modeling user exposure in recommendation. In: Proceedings of the 25th international conference on World Wide Web. pp. 951–961 (2016)

15. Liu, D., Cheng, P., Dong, Z., He, X., Pan, W., Ming, Z.: A general knowledge distillation framework for counterfactual recommendation via uniform data. In: Proceedings of the 43rd International ACM SIGIR Conference on Research and Development in Information Retrieval. pp. 831–840 (2020)
16. Lu, J., Liu, A., Dong, F., Gu, F., Gama, J., Zhang, G.: Learning under concept drift: A review. IEEE Trans. Knowl. Data Eng. **31**(12), 2346–2363 (2018)
17. Mehrotra, R., Bhattacharya, P., Lalmas, M.: Inferring the causal impact of new track releases on music recommendation platforms through counterfactual predictions. In: Proceedings of the 14th ACM Conference on Recommender Systems. pp. 687–691 (2020)
18. Rendle, S., Freudenthaler, C., Gantner, Z., Schmidt-Thieme, L.: Bpr: Bayesian personalized ranking from implicit feedback. arXiv preprint arXiv:1205.2618 (2012)
19. Shanmugam, K.: Self selection bias in the estimates of compensating differentials for job risks in india. J. Risk Uncertain. **22**, 263–275 (2001)
20. Tang, J., Wang, K.: Ranking distillation: Learning compact ranking models with high performance for recommender system. In: Proceedings of the 24th ACM SIGKDD international conference on knowledge discovery & data mining. pp. 2289–2298 (2018)
21. Wang, W., Feng, F., He, X., Zhang, H., Chua, T.S.: Clicks can be cheating: Counterfactual recommendation for mitigating clickbait issue. In: Proceedings of the 44th International ACM SIGIR Conference on Research and Development in Information Retrieval. pp. 1288–1297 (2021)
22. Wang, X., Jin, H., Zhang, A., He, X., Xu, T., Chua, T.S.: Disentangled graph collaborative filtering. In: Proceedings of the 43rd international ACM SIGIR conference on research and development in information retrieval. pp. 1001–1010 (2020)
23. Wei, T., Feng, F., Chen, J., Wu, Z., Yi, J., He, X.: Model-agnostic counterfactual reasoning for eliminating popularity bias in recommender system. In: Proceedings of the 27th ACM SIGKDD Conference on Knowledge Discovery & Data Mining. pp. 1791–1800 (2021)
24. Yim, J., Joo, D., Bae, J., Kim, J.: A gift from knowledge distillation: Fast optimization, network minimization and transfer learning. In: Proceedings of the IEEE conference on computer vision and pattern recognition. pp. 4133–4141 (2017)
25. Ying, R., He, R., Chen, K., Eksombatchai, P., Hamilton, W.L., Leskovec, J.: Graph convolutional neural networks for web-scale recommender systems. In: Proceedings of the 24th ACM SIGKDD international conference on knowledge discovery & data mining. pp. 974–983 (2018)
26. Yuan, B., Hsia, J.Y., Yang, M.Y., Zhu, H., Chang, C.Y., Dong, Z., Lin, C.J.: Improving ad click prediction by considering non-displayed events. In: Proceedings of the 28th ACM International Conference on Information and Knowledge Management. pp. 329–338 (2019)
27. Zhang, Y., Feng, F., He, X., Wei, T., Song, C., Ling, G., Zhang, Y.: Causal intervention for leveraging popularity bias in recommendation. In: Proceedings of the 44th International ACM SIGIR Conference on Research and Development in Information Retrieval. pp. 11–20 (2021)
28. Zhao, Z., Chen, J., Zhou, S., He, X., Cao, X., Zhang, F., Wu, W.: Popularity bias is not always evil: Disentangling benign and harmful bias for recommendation. IEEE Transactions on Knowledge and Data Engineering (2022)
29. Zheng, Y., Gao, C., Li, X., He, X., Li, Y., Jin, D.: Disentangling user interest and conformity for recommendation with causal embedding. In: Proceedings of the Web Conference 2021. pp. 2980–2991 (2021)

Multi-Attribute Sequential Recommendation

Shuhan Qiu, Shanming Wei, and Qianmu Li[✉]

Nanjing University of Science and Technology, Nanjing, China
qianmu@njust.edu.cn

Abstract. Sequential recommendation has gained significant attention in the realm of recommender systems, especially with the recent advancements in artificial intelligence. Many researchers have been integrating deep learning techniques to enhance the effectiveness of sequential recommendation. However, many existing methods rely on modeling with specific Item ID, which inevitably overlooks the inherent attribute features of the items and resulting in challenges such as the long-tail effect and cold start problems. In this paper, we propose a novel sequential recommendation model named **MASRec**, which integrates multiple attribute information from items. Our approach divides an interaction sequence into multiple attribute sequences, enabling independent learning of user interests on each attribute. Furthermore, we introduce a weight parameter for each attribute to capture user preferences across different attributes. The prediction for the next item is derived by aggregating and weighting the prediction results from each attribute. Finally, we have conducted experiments on four real-world datasets with MASRec, and the results demonstrate its ability to significantly enhance recommendation effectiveness, better than mainstream baseline models currently available.

Keywords: Sequential Recommendation · Self-Attention · Item Attribute

1 Introduction

With the development of the internet and information technology, people face information overload problem in their daily lives. Recommender systems can efficiently filter out resources that meet the user's needs from a huge amount of information. Sequential recommendation has recevied wide attention in the research of recommender systems, which captures the interaction information between users and items to recommend the next item to user.

Many model have been proposed to enhance the effectiveness of sequential recommendation and help it to be applied to different scenarios, such as product recommendation, advertisement recommendation, news recommendation and so on. Existing models include: Collaborative Filtering based models [21], Markov Chain based models [6,19], CNN/RNN based models [7,24,25] and Attention

Mechanism based models [12,23,28]. Despite the progress in all of the above methods, most of them are use the specific Item ID to develop the sequential recommender. The main issue associated with this approach are that models relying on Item ID tend to overlook the wealth of item attribute information. Additionally, user may encounter difficulties interacting with new items in the domain, leading to potential item cold start problems.

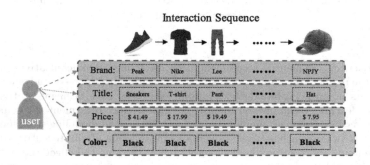

Fig. 1. Diagram illustrating the varying levels of user attention towards different attributes within an interaction sequence, where the arrows indicate the degree of attention allocated by users. In this case, the user is more interested in the color attribute.

Nearly everything in our daily life carry some attributes that characterize them, and inspiration from Natural Language Processing (NLP) techniques prompting us to explore whether items could be characterized using text as an alternative to traditional ID representation methods. In our survey and research, it has been proposed to use item content to generate item embedding that can be used as generic item representation and replacing the traditional ID characterisation approach in recommendation task. [3] After that, several researchers have embraced and enhanced text-based item representation by integrating various techniques. These approaches have demonstrated promising capabilities in addressing the cold-start problem and facilitating cross-domain recommendations. [11,14] However, most of the existing methods have simply stitched the item text and input it into recommender for processing, which does not learn importance of different item attributes. For example, a user from a rich family might prefer the brand value and popularity of an item, while an ordinary family might be more concerned with item's price. We have drawn a Fig. 1 for this phenomenon to help readers understand it more directly.

Therefore, this idea inspires us to develop a new sequential recommender, which we name it **M**ulti-**A**ttribute **S**equential **Rec**ommendation model, or **MASRec** for short in this paper. MASRec can recommend the next item to every user based on multiple item attribute information and the whole recommendation process can be briefly expressed as follows: we assign a learnable weight parameter to each attribute and independently predicting the features of the next potential interaction item for the user on each attribute. After that, we aggregate the prediction results from each attribute using weighting based on the learned weight parameters to generate the prediction for the next item. In MASRec, a new or long-tail item has the same probability of being recommended as popular items during the candidate stage if they share similar attribute information. This is because our embedding of items does not involve any interaction information, and warm item (interacted item) would not have advantage of being preferred for recommendation.

The main contributions of this paper can be summarized as follows:

- This paper proposes MASRec, a multi-attribute sequential recommendation model that use attention network and Transformer architecture, which is able to use the item attribute information to help improve the effectiveness of sequential recommendations.
- Our method can independently learn user preferences on each item attribute. Additionally, it learns the user's attention towards different attribute categories through weight parameters. This advantage enables a deeper understanding of user interests.
- We conducted comprehensive experiments on MASRec to evaluate how well MASRec compares to baseline models in terms of recommendation effectiveness on four real-world datasets. And we also investigated how the preferences of users at the attribute level affect recommendation results.

2 Related Work

2.1 Sequential Recommendation

Sequential recommendation occupies an important position in the field of recommender system, which is mainly based on the user's interaction history to predict the items that user is likely to interact with next. Early sequential recommendation algorithms mainly include Collaborative Filtering (CF) [9] based methods and Markov Chain based methods [19]. CF based methods tend to suffer from data sparsity problems and severe Cold-Start problems, while Markov Chain based methods are often significantly drop effecitve when the interaction sequence is too long. Inspired by deep learning techniques in recent years, there are many works applying deep learning techniques to sequential recommendation, and the existing works can be classified into Convolutional Neural Network (CNN) [25] based methods, Recurrent Neural Network (RNN) [7] based methods, Self-Attention mechanisms based methods [12], etc. Most of these techniques are to encode user's interaction sequences and learn the user's interests through various neural networks to make predictions about the next possible interaction items. Although there is improvements in recommendation effectiveness compared to old methods, but most approaches use explicit *Item ID* when embedding them into the feature space, and this make recommendation algorithms difficult to generalize to different domains. Recently, many works have fused textual information about items into sequential recommendation tasks for learning item representation [3,10,11,14], which is an opportunity to combine Natural Language Processing (NLP) techniques with recommandation system.

2.2 Attribute-Based Recommendation

Existing research has extensively explored the utilization of item attribute information to enhance recommendation tasks. For example, in the field of session-based recommendation, there are researchers who use a parallel computing RNN architecture to process rich attribute information [8]. Recommendation algorithm based on RNN for processing textual information in attributes are also available in the "Latent Cross" technique [2], it can embed textual information into the feature space and then perform element-wise product with the hidden states of the model, similar work has been done by [15]. However, with the advancement of the Transformer architecture, the attention mechanism has attracted growing interest among researchers. Meanwhile, some scholars have also noticed the diverse user interactions and proposed a new collaborative filtering model, CDMF [20], which can sample multiple user interaction types pre-learned knowledge. In the realm of sequential recommendation, some researchers have also chosen the self-attention network to learn the information at the attribute level of the item [28], the proposed model called FDSA. This model aggregates all item attributes into a single sequence, but MASRec leverages multiple self-attention modules to concurrently process diverse attributes. Other researchers have introduced a multi-factor generative adversarial network (MFGAN) [18] to explore the utility of various attribute factors for sequence recommendation tasks.

3 Methodology

In this section, we introduce a Sequential Recommendation method based on Multiple Attributes of items, named MASRec.

3.1 Approach Overview and Problem Setup

Problem Setup. We denote the set of users and items as U and I respectively. And the sequence of interactions for each user is notated as $S_u = (i_1^u, i_2^u, ..., i_{|S_u|}^u)$ where $i \in I$ and $u \in U$. In our method, we added a set of attributes denoted as A. This set contains the attribute categories, each of which is denoted as a. a is also a set comprising the elements within each attribute category. Then, the interaction sequence is segmented into multiple interaction attribute sequence, also known as sub-sequence of original sequence, with each attribute sequence denoted as $(\alpha_1^u, \alpha_2^u, ..., \alpha_{|S_u|}^u)$. The goal of our model MASRec is to input a interaction sequence $(i_1^u, i_2^u, ..., i_{|S_u|-1}^u)$ and output the prediction of next item.

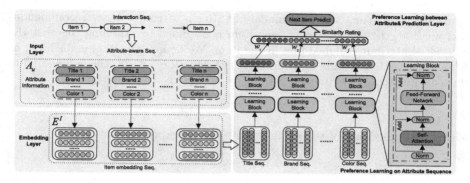

Fig. 2. The Model Architecture of MASRec, where "Seq." means "Sequence".

Approach Overview. In this paper, we treat sequential recommendation task as making predictions about next item which likely to be interact with. Different from existing methods, we refine the granularity of prediction by decompose user interaction sequence into several attribute sequences. For each attribute, we predict the next possible interaction attribute for the user. After that, we aggregate these prediction results with weighting to forecast the next item. Inspired by [12], we use attention mechanism on each attribute interaction sequence for prediction. The structure of MASRec is shown schematically in Fig. 2.

3.2 The Network Architecture of Multi-Attribute Sequential Recommendation

Input Layer. First we need to preprocess the interaction sequence for subsequent batch processing. We adjust all sequences to a fixed length n. If sequence length exceeds n, our model considers only the n nearest neighbor interaction items, and if the length is less than n, we add padding items to the left side of the sequence until the length reaches n. Similarly, we set a length limit on the length of the text within the attribute to prevent too much extraneous information from interfering with user preference learning.

Since we need to make predictions at the attribute level, the sequence of user interactions can then be extended into a matrix of user interaction attributes consisting of attribute information for each item:

$$A_u = \begin{bmatrix} \alpha_{1,1}^u & \alpha_{1,2}^u & \cdots & \alpha_{1,n}^u \\ \alpha_{2,1}^u & \alpha_{2,2}^u & \cdots & \alpha_{2,n}^u \\ \cdots & \cdots & \cdots & \cdots \\ \alpha_{j,1}^u & \alpha_{j,2}^u & \cdots & \alpha_{j,n}^u \end{bmatrix} \quad (1)$$

where $A_u \in \mathbb{N}^{n \times j}$, j is the number of attributes considered in each item. Obviously, each row in A_u is a attribute interaction sequence and each column represesnts an interacted item. It is important to note that there may be duplication of elements in attribute interaction sequence because user may favour items with the same characteristics on particular attribute.

Embedding Layer. Since our model MASRec is attribute-aware recommendation method, the embedding of each item is obtained by stitching together the individual attributes embedding in the item. We denote the item embedding as $E^I \in \mathbb{R}^{|I| \times (d \times j)}$, attribute embedding is notated as $E^A \in \mathbb{R}^d$, the embedding of the k-th item can be expressed as:

$$E_k^I = \left[E_{1k}^A, E_{2k}^A, ... E_{jk}^A\right] \quad (2)$$

where the embedding of the j-th attribute can be expressed as:

$$E_{jk}^A = \left[x_{j1}, x_{j2}, ..., x_{jd}\right]^T \quad (3)$$

In fact, since MASRec is trained independently on each attribute, the feature vector dimensions can be different between attributes, but due to the desire to simplify the complexity of subsequent operations, we set the feature vector dimensions on each attribute to be d in MASRec and the experimental part. The embedded representation of the j-th attribute sequence in a interaction sequence S_u can be written as:

$$Embed(\alpha_{j,1}^u, \alpha_{j,2}^u, ..., \alpha_{j,n}^u) = \left[E_{j1}^A, E_{j2}^A, ..., E_{jn}^A\right]^T \quad (4)$$

Next, we want to learn the user's preference among different attribute, so we multiple each attribute sequence by a learnable weight parameter. All weight parameters can be written as:

$$\boldsymbol{w} = (w_1, w_2, ..., w_j) \quad (5)$$

the corresponding attribute sequence should be modified to:

$$w_j \left[E_{j1}^A, E_{j2}^A, ..., E_{jn}^A\right]^T \quad (6)$$

In order to make the model be able to aware of the position of each element in the sequence, we also ad positional embedding to each element in attribute sequence:

$$E_{inp} = w_j \left[E_{j1}^A + P_1, E_{j2}^A + P_2, ..., E_{jn}^A + P_n\right]^T \quad (7)$$

Now, E_{inp} is prepared to be inputted into the model for attribute-level recommendation. It's worth noting that for any attribute of padding item, we utilize a zero vector of the corresponding dimension as its embedding vector.

Preference Learning on Attribute Sequence. In this subsection, we present method for learning user's perference on each attribute sequence. For the learning of single attribute sequence, MASRec also adopt Transformer architecture and attention mechanism. In Transformer, the formula for scaled dot-product attention score is defined as:

$$Attention(\mathbf{Q}, \mathbf{K}, \mathbf{V}) = softmax(\frac{\mathbf{Q}\mathbf{K}^T}{\sqrt{d}})\mathbf{V} \quad (8)$$

where **Q**, **K**, **V** stands for queries, keys and values respectively. $\sqrt{d}$ is a scaling operation to prevent the value from being too large due to the inner product of vectors. Self-attention differs from the attention mechanism in that the queries, keys and values in the model are derived from the same sequence, which has led to self-attention being widely used in the sequential recommandation field. Similarly in this paper, we use the attribute embedding E_{inp} obtained at the embedding layer to compute the attention scores between attribute elements by feeding them into the self-attention mechanism through a linear transformation:

$$SA(E_{inp}) = Attention(E_{inp}W^Q, E_{inp}W^K, E_{inp}W^V) \qquad (9)$$

where W^Q, W^K, W^V are three learnable parameter matrices. Processing through the self-attention makes it possible to take into account all previously interacted elements when predicting the next element that user is likely to interact with. Inspired by previous work [12,26], in order to learn the information between different embedding dimensions and enhance the effectiveness of self-attention, we connected a Feed-Forward Network (FFN) which consist of two fully-connected layer after self-attention block, using ReLU as activation function:

$$FFN(x) = \max(0, xW_1 + b_1)W_2 + b_2 \qquad (10)$$

where x represents the output tensor of the self-attention block, which is further processed and fed into the FFN, next we describe how to do some necessary operations on the input and output of self-attention as well as FFN.

In the deep learning field, stacking modules is often necessary to make the model able to extract sequential information as much as possible, a principle that holds particularly true for the attention mechanism. However, as the size of the model increases and the network layer goes deeper, training the model tends to become unstable. Issues such as overfitting and gradient vanishing may happen. To mitigate these problems, it is essential to fine-tune each core module accordingly. For the heart of MASRec, the self-attention module, we use layer normalization [1], dropout operation [22] and residual connection [4] on the input and output of each block:

$$x_i = x_{i-1} + Dropout(SA(LayerNorm(x_{i-1}))) \qquad (11)$$

Since many researchers face constraints on computational resources, we incorporate layer normalization into MASRec to normalize the input of each layer. This not only enhances the model's generalization ability but also stabilizes the training process. Furthermore, it enables us to utilize smaller batch size, thereby reducing the GPU memory required for training. The formula for layer normalization is as follows:

$$LayerNorm(x) = \alpha \odot \frac{x - \mu}{\sqrt{\sigma^2 + \epsilon}} + \beta \qquad (12)$$

where $\odot$ is the Hadamard product, σ and μ are the variance and mean of input tensor x, α and β are learned scaling factors and bias terms. To address the risk of overfitting, we also employ Dropout regularization, a widely adopted technique in the academic community. During training, dropout operation randomly drops out several neurons, preventing the network from relying too heavily on any single neuron and thereby preventing overfitting. The residual connection is also a very well-known technology in deep learning. By adding input information directly to the output, it alleviates the problems of gradient vanishing and gradient explosion that may exist in deep neural network training, and also speeds up the network training speed. We also operate each FFN block accordingly:

$$x_i = x_{i-1} + Dropout(FFN(LayerNorm(x_{i-1}))) \tag{13}$$

In Eq. 11 and Eq. 13, x_{i-1} represents the input when stacking the self-attention block and the FFN block, and x_i represents the corresponding output.

Preference Learning Between Attributes and Prediction Layer. After learning on each attribute sequence, we further delve into understanding the user's preferences among attributes. For instance, some users may prioritize the price of an item in their interaction history, while others may focus more on the brand.

In MASRec, the sequence of input model is $(i_1^u, i_2^u, ..., i_{t-1}^u)$ then predicts the interaction item at the t-th position, so in each attribute sequence the input is $(\alpha_1^u, \alpha_2^u, ..., \alpha_{t-1}^u)$. Before forecasting the next item, the output in each attribute like $w_j \left[\hat{E}_{j2}^A, \hat{E}_{j3}^A, ..., \hat{E}_{jt}^A \right]^T$, so weighted splicing of the outputs at the t-th position on different attribute gives a final output:

$$\hat{E}_t^I = [w_1 \hat{E}_{t1}^A, w_2 \hat{E}_{t2}^A, ..., w_j \hat{E}_{tj}^A] \tag{14}$$

we use embedding of the item at the t-th position E_t^I as ground truth, then the sum of the error can be expressed as:

$$\mathcal{L}_{err} = \sum_{t=2}^{n-1} \| \hat{E}_t^I - E_t^I \|^2 \tag{15}$$

where n is fixed length of the sequence as mentioned above. The training process adjusts the weighting parameters to minimize the sum of errors, thereby achieving the goal of learning the user's preferences among different attributes. We use $\hat{E}_t^I$ in MASRec model to make a prediction of the next item, and the score for each candidate item is first calculated using the vector dot product:

$$Score_{t,k} = \hat{E}_t^I E_k^I \tag{16}$$

A higher score item means that MASRec considers it more likely to interact with the user, $Score_{t,k}$ is the score of the k-th item at position t. After calculating the scores of all the candidate items, the recommendation is made according to the scores in descending order.

3.3 Model Training

In the previous section, our description of MASRec is mainly divided into learning attribute sequences and learning user preferences between attributes, so the training stage is also divided into two parts.

The first part is learning on attribute sequence. In Subsect. 3.2 we describe the inputs on each attribute sequence, since we have fixed the sequence to length n, the input during model training is $(\alpha_{j,1}^u, \alpha_{j,2}^u, ..., \alpha_{j,n-1}^u)$, and the ground truth sequence in learning process is $(\alpha_{j,2}^u, \alpha_{j,3}^u, ..., \alpha_{j,n}^u)$. For ease of representation we rewrite the sequence as:

$$g = (g_1, g_2, ..., g_{n-1}) = (\alpha_{j,2}^u, \alpha_{j,3}^u, ..., \alpha_{j,n}^u) \tag{17}$$

We use binary cross entropy as the loss function for this part:

$$\mathcal{L}_{ce} = - \sum_{a_j^u \in a_j} \sum_{t=1}^{n} [\log(\sigma(r_{t,g_t})) + \log(1 - \sigma(r_{t,g_t'}))] \tag{18}$$

where g_t' is a negative element which is not interacted with the user and r_{t,g_t} is dot product operation which is similar to prediction layer:

$$r_{t,g_t} = \hat{E}_{tj}^A E_{g_t}^A \tag{19}$$

where $\sigma(\cdot)$ is a sigmoid function, used to transform r_{t,g_t} into a probability value between 0 and 1:

$$\sigma(r_{t,g_t}) = \frac{1}{1 + e^{-r_{t,g_t}}} \tag{20}$$

Another part is learning user preferences among multi-attributes. We have actually given the corresponding loss function $\mathcal{L}_{err}$ in Subsect. 3.2, so the loss function for training MASRec can be defined as:

$$\mathcal{L} = \mathcal{L}_{err} + \sum \mathcal{L}_{ce} \tag{21}$$

The optimiser used in the training process is the Adam optimiser [13], which is a variant algorithm derived from the stochastic gradient descent algorithm with adaptive moment estimation.

4 Experiments

In this section, we will present our experiments around the following Research Questions (RQs) to demonstrate the effectiveness of our proposed model MASRec:

- **RQ1**: How does MASRec compare to existing state-of-the-art (SOTA) models in terms of performance on the sequential recommandation task?
- **RQ2**: Is the original Item ID still necessary in MASRec?
- **RQ3**: How do the parameters in MASRec affect model performance?

Table 1. Experimental results for each baseline model and MASRec. We have bolded the best performance and underlined the second best performance."Improv." means the degree of improvement of MASRec.

Datasets	Metrics	GRU4Rec	TransRec	NPE	BERT4Rec	SRGNN	FDSA	UniSRec	MASRec	Improv.
Office	Recall@10	0.1406	0.0825	0.1328	0.0734	0.1496	0.1115	0.1224	**0.1563**	4.48%
	NDCG@10	0.124	0.0659	0.1002	0.0448	0.1311	0.0854	0.0724	**0.1351**	3.05%
	Recall@20	0.1523	0.0951	0.1452	0.0913	0.1624	0.1327	0.1502	**0.1686**	3.82%
	NDCG@20	0.127	0.0691	0.1034	0.0493	0.1343	0.0907	0.0794	**0.1382**	2.90%
CDs	Recall@10	0.1065	0.0647	0.0846	0.0573	0.1078	0.0921	0.1066	**0.1209**	12.15%
	NDCG@10	0.0806	0.0425	0.0576	0.0282	0.0819	0.0574	0.0564	**0.0908**	10.87%
	Recall@20	0.127	0.0824	0.1062	0.0873	0.1283	0.1246	**0.1495**	0.1425	-
	NDCG@20	0.0858	0.047	0.0631	0.0358	0.087	0.0656	0.0674	**0.0962**	10.57%
Cell	Recall@10	0.1023	0.096	0.0981	0.0306	0.1186	0.0673	0.0775	**0.1227**	3.46%
	NDCG@10	0.0656	0.0541	0.0596	0.0172	0.0765	0.0433	0.0426	**0.0811**	6.01%
	Recall@20	0.1376	0.1329	0.1312	0.0464	0.1555	0.0912	0.1075	**0.1561**	0.39%
	NDCG@20	0.0749	0.0634	0.0679	0.0209	0.0857	0.0493	0.0501	**0.0895**	4.43%
Kindle	Recall@10	0.1124	0.0337	0.0684	0.0793	0.1027	0.1442	**0.1504**	0.1342	-
	NDCG@10	0.0734	0.0179	0.0359	0.0364	0.0732	0.0948	0.0731	**0.0972**	2.53%
	Recall@20	0.1359	0.0526	0.0939	0.1146	0.1241	0.182	**0.2046**	0.1597	-
	NDCG@20	0.0863	0.0226	0.0424	0.0454	0.0786	0.1044	0.1008	**0.1073**	2.78%

4.1 Experimental Setup

Datasets. We selected four real-world datasets comes from Amazon platform [17] for experimental evaluation of MASRec. They come from four different domain: *Office Products (Office), CDs and Vinyl (CDs), Cell Phones and Accessories (Cell)* and *Kindle Store (Kindle)*. Each of these four datasets varies in size, and each item contains rich attribute information such as: *Item ID, Title, Price, Sales_type, Sales_rank, Brand, Categories*. And these datasets are public datasets, which can be obtained via internet[1]. In each dataset, we consider the user's behavior of rating or reviewing a product as user interaction, and generate the interaction sequence by timestamp.

Baselines. In order to compare the effects of MASRec, the following methods were selected as baseline models:

- **GRU4Rec** [24]: A session-based recommendation model based on RNNs, which can learn the user's interaction history.
- **TransRec** [5]: A sequential recommendation model, which can model "third-order" relationships and user are modeled as translation vectors based on interaction sequence.
- **NPE** [16]: A collaborative filtering model, which can efficiently learn item representations for improving recommendations to cold-start users.

---
[1] https://cseweb.ucsd.edu/~jmcauley/datasets/amazon_v2.

- **BERT4Rec** [23]: A sequential recommendation model based on bi-directional Transformer encoder, which can model the user's behavior sequence and designed a Cloze task to train the model.
- **SRGNN** [27]: A session-based recommendation model based on graph neural networks, which can model user session sequence and learning the representation of items. They also employs attention networks to learn user interests in the sequence.
- **FDSA** [28]: A sequential recommendation model based on attention networks, which can fuse attribute information of item to recommender. Different from FDSA combines all attribute information into a single sequence, our method has the capability to process each attribute sequence independently.
- **UniSRec** [11]: A sequential recommendation model which can learning universal representation of item using textual information from item, replacing the specific Item ID used in traditional methods.

Evaluation Metrics. In our experiments, we use $Recall@K$ and $NDCG@K$ as the evaluation metrics, where $K = 10, 20$. During the training and evaluation stage, we use the last item in each sequence as test item, and the rest as validation items, which is also known as leave-one-out strategy. We finally calculate the average of the test score on all sequence as the MASRec evaluation score.

4.2 Performance on Sequential Recommendation (RQ1)

We have compared MASRec with seven baseline models on four datasets, and the experimental results indicate that our method generally enhances the effectiveness of sequential recommendation task compared to existing models. However, the extent of improvement varies across different datasets. Next we will analyse the experimental results in detail. The specific experimental data are presented in Table 1.

Among all baseline models, SRGNN demonstrates superior performance, surpassing other baseline models on the *Office, CDs* and *Cell* datasets. This highlights the efficacy of attention networks in sequential recommendation tasks. FDSA and UniSRec, which leverage item attribute information, outperform most models on the *Office, CDs* and *Kindle* datasets among the remaining baseline models. Furthermore, UniSRec utilizing pre-trained language model for text encoding surpassing our model in terms of the *Recall* metric on the *Kindle* dataset. Lastly, GRU4Rec also performs admirably among various baseline models, further cementing its status as a widely referenced baseline model in recommender field.

MASRec generally outperforms baseline models on all datasets we used, except for the *Recall@20* metric on the *CDs* dataset and the *Recall@10,20* metrics on the *Kindle* dataset, where UniSRec performs better than our model. UniSRec is the only algorithm in the baseline models that uses a pre-trained language model with strong text encoding capabilities. The output of the BERT model it uses is a 768-dimensional vector, prompting us to consider leveraging large language model for more feature extraction of text in our future research. Besides, it is worth noting that the enhancement effects on the *CDs* dataset exceed 10%, perhaps because of users' preference for music often exhibiting clear tendencies towards specific attributes, such as favorite artists or music styles, enabling our model to capitalize on such scenarios. The degree of improvement on the *Office* and *Cell* datasets mostly ranges between 3% and 6%, which proves that attribute information is helpful for the recommendation task. We also need to discuss the performance on the *Kindle* dataset: our model does not exactly outperform FDSA and UniSRec. The average length of the interaction sequences on the *Kindle* dataset is the longest among the four datasets, and the user's preference for book-like items is difficult to describe in terms of a few specific attributes, and the same user in different scenarios of his life also requires many different types of books, which limits the performance of MASRec. All the experimental results and our analyses illustrate that MASRec performs better in specific scenarios, highlighting the need to further improve its generalization ability in future research.

4.3 Analysis Between Original ID and Attributes (RQ2)

In previous works [11,14], both discuss whether to retain Item ID in their recommendation models. Thus, we conducted experiments on four datasets to investigate whether Item ID should be included in MASRec. The results of these experiments are shown in Fig. 3.

Our experimental approach involves comparing the performance of MASRec with and without Item ID. From the four histograms, it is evident that in most cases MASRec outperforms $MASRec_{Item\ ID}$, only the two NDCG metrics on the dataset *CDs* show slightly lower performance of MASRec, but it dose not significantly impact the overall performance. It is understandable that in previous Item ID-based recommendation algorithms, the recommendations were not based on the information provided by the Item ID itself, but rather on user interaction data to learn embedding vectors, which were utilized for recommendation tasks while distinguishing between items. Conversely, MASRec leverages information from item attributes to enhance the sequential reommmandation tasks. As a result, Item ID has become redundant, lacking the ability to provide meaningful semantic information. Therefore, we have excluded Item ID as an attribute in our experiments for RQ1.

4.4 Impact of Model Parameters (RQ3)

We also investigated the impact of key parameters in MASRec to optimize its performance. Specifically, we focus on tuning the embedding dimension d during the embedding layer, also known as the embedding size. We varied this parameter from 60 to 480 with a step size 60, and additionally included two values of 400 and 500, obtained by further reducing the step size at a later stage of the experiment. To evaluate the performance of different embeddding size d on the model, we

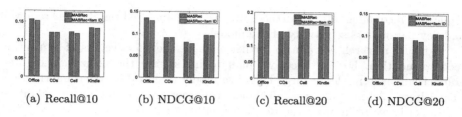

(a) Recall@10   (b) NDCG@10   (c) Recall@20   (d) NDCG@20

Fig. 3. Comparing the performance between MASRec+Item ID and original model on four datasets.

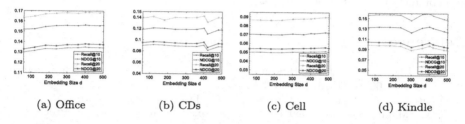

(a) Office   (b) CDs   (c) Cell   (d) Kindle

Fig. 4. Comparing the impact of different embedding size d on MASRec performance across four datasets.

ran full experiments on all dataset which introduced in Subsect. 4.1. The results of these experiments are illustrated in line graphs, Fig. 4.

We can intuitively observe that when d is less than 400, the evaluation metrics for the *Office* dataset exhibit an upward trend, while those for the *CDs* and *Cell* datasets tend to plateau, and those for the *Kindle* dataset show slight fluctuations. When d exceeds 400, the metrics for the *Office* and *Cell* datasets stabilize, those for the *CDs* dataset oscillate, and those for the *Kindle* dataset exhibit a downward trend. This behavior can be explained by the occurrence of overfitting phenomena during training when the vector dimension is excessively high, thereby affecting the overall model performance. Out of the 16 metrics across all four datasets, the maximum value for 7 metrics is observed at $d = 400$, and the remaining metrics also either achieve their second-highest value or perform better at this dimension. Considering the constraints of computational resources and the analysis of the experimental results, we set the embedding size d to 400 in our experiments related to RQ1.

5 Conclusion

In this paper, we have proposed a multi-attribute sequential recommendation method called MASRec. Unlike existing methods, our model independently learns user preferences for each attribute sequence and aggregates prediction results from each attribute using weighting to make predictions for the next item. We have conducted comprehensive experiments on MASRec, demonstrating that leveraging item attribute information can notably enhance the effectiveness of sequential recommendation. Furthermore, we have analysed the Item ID removal and key parameters sensitivity of the model.

In future research, we aim to further enhance the effectiveness of sequential recommendation. For instance, we plan to investigate the utilization of pre-trained language model or large language model to encode textual information in attributes. Additionally, we will explore the incorporation of multimodal information of items to improve the recommendation task.

Acknowledgments. This work was supported by Frontier Technologies R&D Program of Jiangsu, under Grant BF2024071.

References

1. Ba, J.L., Kiros, J.R., Hinton, G.: Layer normalization. arXiv preprint arXiv:1607.06450 (2016)
2. Beutel, A., Covington, P., Jain, S., Xu, C., Li, J., Gatto, V., Chi, E.H.: Latent cross: Making use of context in recurrent recommender systems. In: Proceedings of the eleventh ACM international conference on web search and data mining. pp. 46–54 (2018)
3. Ding, H., Ma, Y., Deoras, A., Wang, Y., Wang, H.: Zero-shot recommender systems. arXiv preprint arXiv:2105.08318 (2021)
4. He, K., Zhang, X., Ren, S., Sun, J.: Deep residual learning for image recognition. In: Proceedings of the IEEE conference on computer vision and pattern recognition. pp. 770–778 (2016)
5. He, R., Kang, W., McAuley, J.: Translation-based recommendation. In: Proceedings of the eleventh ACM conference on recommender systems. pp. 161–169 (2017)
6. He, R., McAuley, J.: Fusing similarity models with markov chains for sparse sequential recommendation. In: 2016 IEEE 16th international conference on data mining (ICDM). pp. 191–200. IEEE (2016)
7. Hidasi, B., Karatzoglou, A., Baltrunas, L., Tikk, D.: Session-based recommendations with recurrent neural networks. arXiv preprint arXiv:1511.06939 (2015)
8. Hidasi, B., Quadrana, M., Karatzoglou, A., Tikk, D.: Parallel recurrent neural network architectures for feature-rich session-based recommendations. In: Proceedings of the 10th ACM conference on recommender systems. pp. 241–248 (2016)
9. Hidasi, B., Tikk, D.: General factorization framework for context-aware recommendations. Data Min. Knowl. Disc. **30**(2), 342–371 (2016)
10. Hou, Y., He, Z., McAuley, J., Zhao, W.X.: Learning vector-quantized item representation for transferable sequential recommenders. In: Proceedings of the ACM Web Conference 2023. pp. 1162–1171 (2023)
11. Hou, Y., Mu, S., Zhao, W.X., Li, Y., Ding, B., Wen, J.: Towards universal sequence representation learning for recommender systems. In: Proceedings of the 28th ACM SIGKDD Conference on Knowledge Discovery and Data Mining. pp. 585–593 (2022)
12. Kang, W., McAuley, J.: Self-attentive sequential recommendation. In: 2018 IEEE international conference on data mining (ICDM). pp. 197–206. IEEE (2018)
13. Kingma, D.P., Ba, J.: Adam: A method for stochastic optimization. arXiv preprint arXiv:1412.6980 (2014)

14. Li, J., Wang, M., Li, J., Fu, J., Shen, X., Shang, J., McAuley, J.: Text is all you need: Learning language representations for sequential recommendation. arXiv preprint arXiv:2305.13731 (2023)
15. Mizrachi, S., Levin, P.: Combining context features in sequence-aware recommender systems. In: RecSys (Late-Breaking Results). pp. 11–15 (2019)
16. Nguyen, T., Takasu, A.: Npe: neural personalized embedding for collaborative filtering. arXiv preprint arXiv:1805.06563 (2018)
17. Ni, J., Li, J., McAuley, J.: Justifying recommendations using distantly-labeled reviews and fine-grained aspects. In: Proceedings of the 2019 conference on empirical methods in natural language processing and the 9th international joint conference on natural language processing (EMNLP-IJCNLP). pp. 188–197 (2019)
18. Ren, R., Liu, Z., Li, Y., Zhao, W.X., Wang, H., Ding, B., Wen, J.: Sequential recommendation with self-attentive multi-adversarial network. In: Proceedings of the 43rd international ACM SIGIR conference on research and development in information retrieval. pp. 89–98 (2020)
19. Rendle, S., Freudenthaler, C., Schmidt-Thieme, L.: Factorizing personalized markov chains for next-basket recommendation. In: Proceedings of the 19th international conference on World wide web. pp. 811–820 (2010)
20. Sar S., O., Roitman, H., Amir, A., Karatzoglou, A.: Collaborative filtering method for handling diverse and repetitive user-item interactions. In: Proceedings of the 29th on Hypertext and Social Media, pp. 43–51 (2018)
21. Schafer, J.B., Frankowski, D., Herlocker, J., Sen, S.: Collaborative filtering recommender systems. In: The adaptive web: methods and strategies of web personalization, pp. 291–324. Springer (2007)
22. Srivastava, N., Hinton, G., Krizhevsky, A., Sutskever, I., Salakhutdinov, R.: Dropout: a simple way to prevent neural networks from overfitting. The journal of machine learning research **15**(1), 1929–1958 (2014)
23. Sun, F., Liu, J., Wu, J., Pei, C., Lin, X., Ou, W., Jiang, P.: Bert4rec: Sequential recommendation with bidirectional encoder representations from transformer. In: Proceedings of the 28th ACM international conference on information and knowledge management. pp. 1441–1450 (2019)
24. Tan, Y.K., Xu, X., Liu, Y.: Improved recurrent neural networks for session-based recommendations. In: Proceedings of the 1st workshop on deep learning for recommender systems. pp. 17–22 (2016)
25. Tang, J., Wang, K.: Personalized top-n sequential recommendation via convolutional sequence embedding. In: Proceedings of the eleventh ACM international conference on web search and data mining. pp. 565–573 (2018)
26. Vaswani, A., Shazeer, N., Parmar, N., Uszkoreit, J., Jones, L., Gomez, A., Kaiser, L., Polosukhin, I.: Attention is all you need. In: NIPS (2017)
27. Wu, S., Tang, Y., Zhu, Y., Wang, L., Xie, X., Tan, T.: Session-based recommendation with graph neural networks. In: Proceedings of the AAAI conference on artificial intelligence. vol. 33, pp. 346–353 (2019)
28. Zhang, T., Zhao, P., Liu, Y., Sheng, V.S., Xu, J., Wang, D., Liu, G., Zhou, X., et al.: Feature-level deeper self-attention network for sequential recommendation. In: IJCAI. pp. 4320–4326 (2019)

Layer Transformer-Powered Graph Convolutional Networks for Enhanced Recommendation

Shicong Lin[1], Zhilong Shan[1(✉)], and Su Mu[2]

[1] School of Computer Science, South China Normal University, Guangzhou 510631, People's Republic of China
{20182232032,zlshan}@m.scnu.edu.cn
[2] Institute of Artificial Intelligence in Education, South China Normal University, Guangzhou 510631, People's Republic of China
musu@m.scnu.edu.cn

Abstract. Graph neural networks have been widely used in recommendation systems and have achieved outstanding performance. Graph convolutional neural networks (GCNs) are classic graph neural networks that can better extract features and representations from graph-structured data by stacking multiple layers of GCNs. However, stacking GCNs can lead to over-smoothing, making nodes difficult to distinguish and affecting model performance. Meanwhile, the natural noise in the user-item interaction graph also affects the model's ability to learn better embeddings. In this paper, we propose the Layer Transformer-Powered Graph Convolutional Networks (LayerTrans) for recommendation to address the over-smoothing problem. The model dynamically uses cross-attention to extract information during the information propagation process of GCNs and then updates the nodes to learn better embedding representations of users and items. In addition, we use degree centrality to measure the importance of each node and calculate the probability of edge deletion between nodes based on their importance. This paper conducts sufficient experimental evaluation on four real public datasets, and the comparison results with multiple existing methods demonstrate the superior performance of our method.

Keywords: Recommendation systems · Graph convolutional networks · Over-smoothing · Layer Transformer

1 Introduction

With the development of the Internet, society has transitioned from an era of information scarcity to one of information overload. To find information of interest from a vast amount of data or to achieve precise information delivery, Resnick et al. proposed recommendation systems [1]. These systems analyze users' historical behaviors to model their interests, thereby achieving accurate information recommendations. Earlier work like collaborative filtering [2,3], have been extensively studied and applied. However, in practice, user-item interaction data is often sparse, and traditional methods suffer from low prediction accuracy and an inability to capture complex relationships. Since user-item interactions can naturally be represented as graph structures, early works [4,5] introduced graph neural networks into recommendation systems, achieving notable performance. Graph convolutional neural networks (GCNs) [6] aggregate neighborhood information through convolution operations to update node representations. After multiple layers of iteration, nodes can learn high-order structural information from high-order neighbors, discover potential relationships between nodes, and make better recommendations.

In GCNs, as features propagate along the edges, after multiple iterations, each node will learn similar features from neighboring nodes and gradually become more similar, causing the nodes to lose their unique characteristics. This phenomenon is called over-smoothing. Figure 1 shows the over-smoothing phenomenon of GCNs on the Cora dataset. When using three layers of GCNs, the nodes can be classified normally. The classification effect becomes worse at the fifth layer. At the sixth layer, serious over-smoothing occurs, nodes aggregate together and can't be distinguished. Although methods based on GCNs [4,7–10] have achieved excellent performance in recommendation systems, they all suffer from over-smoothing. LayerGCN extracts the structural information of the initial layer by calculating the similarity between the initial embedding layer and the current embedding layer, thereby alleviating over-smoothing [11]. However, it only extracts the initial structural information by simply calculating the similarity, lacks the learning of high-order structures, and finally uses the sum aggregation function for simple information extraction, resulting in a suboptimal model. We propose the Layer Transformer-Powered Graph Convolutional Networks (LayerTrans), which uses attention to dynamically extract the structural information of the initial layer and the current layer for updating node representations. Moreover, LayerTrans uses the LSTM aggregation function to mine the fine-grained information between layers, making the model perform better. In addition to the over-smoothing problem, the noise in the user-item interaction data in real scenarios will also affect the model performance. This noise usually comes from accidental clicks of users and wrong product recommendations. Existing models either randomly drop interactions in the presence of noise in the graph [12] or learn how to drop the noise [13]. In this paper, we use degree centrality to measure the importance of nodes to achieve edge clipping and reduce noise interference.

In this paper, we propose LayerTrans. It uses Transformer to extract neighborhood information of the graph to update the node representation, avoid nodes tending to be similar, and alleviate the over-smoothing problem. We indirectly calculate the edge importance based on degree centrality to decide whether to prune the edge. Through pruning, we further sparsify the graph, which can speed up the training of the model and reduce the interference of noise. The main contributions of this paper are summarized as follows:

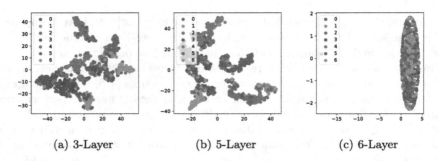

Fig. 1. Visualization of GCN on the Cora dataset.

In summary, this paper makes the following contributions:

- We propose a recommendation model called LayerTrans. It can dynamically extract the neighborhood information obtained by GCN aggregation and use this information to update the node representation to alleviate the over-smoothing.
- To alleviate the impact of natural noise in recommendation scenarios, we use node centrality to measure edge importance and prune edges according to edge importance to reduce noise interference.
- Finally, we conduct experiments on four models to demonstrate the superiority and effectiveness of LayerTrans.

2 Related Work

2.1 Classic Collaborative Filtering Models

Collaborative filtering (CF) [3] is a widely used technology in recommendation systems and it is mainly divided into user-based CF models and item-based CF models. User-Based Collaborative Filtering (UBCF) identifies users with similar preferences to the target user and makes recommendations based on the preferences of these similar users. This approach was effectively demonstrated by Resnick et al. [3] in the GroupLens system, which provided collaborative filtering for netnews. Herlocker et al. [14] further refined UBCF by introducing an algorithmic framework to enhance recommendation accuracy and scalability. Item-Based Collaborative Filtering (IBCF) emerged as an alternative to UBCF. Sarwar et al. [15] proposed item-based CF algorithms that compute item-item similarities, allowing recommendations based on the items' co-occurrence patterns. Deshpande and Karypis [16] expanded on this by developing top-N recommendation algorithms that efficiently generate high-quality recommendations through item similarity. Latent Semantic Models leverage matrix factorization techniques to uncover latent relationships between users and items. SVD [17] has been instrumental in enhancing collaborative filtering methods by reducing dimensionality and uncovering latent factors in user-item interaction matrices and NMF [18] ensures that the matrix after decomposition is non-negative and more interpretable. Graph-Based Collaborative Filtering represents users and items as nodes in a graph, utilizing graph connectivity and path information to make recommendations.

2.2 Graph Neural Networks for Recommendation

In recent years, graph neural networks have received widespread attention and application in the recommendation field [4,6]. Compared with traditional recommendation algorithms, GNN-based recommendation systems [7,9,12] can more effectively capture the complex relationships between users and items and mine high-order correlation characteristics in user behavior data. In the GCN-based model, PinSage [5] effectively captures the high-order relationships between users and items by sampling local subgraphs and performing convolution operations on the graph. NGCF [4] encodes collaborative signals by performing embedding propagation using GCN to capture high-order connections. UltraGCN [19] bypasses infinite message passing in graph convolution by imposing a constraint loss on user-item interactions. IMP-GCN [20] performs graph convolution on the generated subgraphs. LightGCN [9] simplifies GCN by removing feature transformation and non-linear layers, enhancing model performance. While LightGCN is computationally efficient, it is prone to suboptimal spaces with shallow layers due to the over-smoothing problem. [10,21] indicate that GCN achieves optimal performance after stacking two or three layers. [22,23] improve GCN through different methods to mitigate the effects of over-smoothing. Inspired by ResNet [24] and LightGCN [9], recent studies use the skip connection method to integrate the embeddings from the first layer with the embedding information from the $(l+1)th$ layer and effectively alleviate the impact of the over-smoothing [6,8]. In other works, Zhu et al. use generative models to generate additional contrastive views [13]. Yang et al. adaptively adjust the contrast loss scale of each node for optimization and compare it with other contrastive learning models to achieve advanced performance [25].

This paper designs a new recommendation framework (LayerTrans). It dynamically extracts information provided by neighboring nodes before node update to solve the over-smoothing problem.

3 Methodology

3.1 Preliminaries

To facilitate reading this article, we will first introduce some important symbols and briefly introduce GCN and LightGCN. Given a graph $\mathcal{G} = \{\mathcal{V}, \mathcal{E}\}$, where $\mathcal{V}$ is the node set of the given graph, $\mathcal{E}(|\mathcal{E}| = \mathcal{K})$ is the edge set of the given graph. $A \in \{0,1\}^{\mathcal{N} \times \mathcal{N}}$ represents the adjacency matrix of a given graph $\mathcal{G}$ and D denotes the diagonal matrix. In the recommendation systems, a user-item interaction graph can be viewed as a bipartite graph. $X^l \in \mathbb{R}^{\mathcal{N} \times \mathcal{T}}$ represents the embedding matrix of the user-item interaction graph, the dimension after node mapping is T, l represents the number of hidden layers, and X^0 represents the initial layer embedding of nodes.

Inspired by CNN, GCN [6,11] is a convolution-based model specifically designed to process graph-structured data. In GCN, the embedding X^{l+1} of the $(l+1)th$ hidden layer can be recursively written as follows:

$$X^{l+1} = \sigma(\hat{A}X^l W^l) \quad (1)$$

where $\sigma(\cdot)$ is a nonlinear activation function, $\hat{A} = \hat{D}^{-\frac{1}{2}}(A+I)\hat{D}^{-\frac{1}{2}}$ is the normalized adjacency matrix, and $\hat{D}$ is the diagonal degree matrix of $(A+I)$. LightGCN [9] simplifies GCN by removing the feature filter matrix W and the nonlinear activation function $\sigma(\cdot)$. The embedding X^{l+1} of the $(l+1)th$ hidden layer can be recursively defined as:

$$X^{l+1} = (\hat{D}^{-\frac{1}{2}}A\hat{D}^{-\frac{1}{2}})X^l \quad (2)$$

The $(l+1)th$ node embedding is aggregated by the lth embedding, where $\hat{D}^{-\frac{1}{2}}A\hat{D}^{-\frac{1}{2}}$ is normalized adjacency matrix in LightGCN.

3.2 Model Overview

The architecture of LayerTrans is shown in Fig. 2, which consists of three components: edge dropout based on node centrality, inter-layer graph Transformer embedding module, and prediction layer. To further sparse the user-item interaction graph and alleviate the impact of natural noise in the user-item interaction graph, we prune the edges based on the centrality of the nodes. The LayerTrans block focuses on both the initial information of users and items and the structural information of the graph. During each message passing process, it aggregates the embeddings from the initial layer with the information from the current layer to obtain refined node representations. The prediction module aims to predict users' preferences for items they have not interacted with or the likelihood of future interactions based on the learned latent representations of users and items.

3.3 Degree-Centrality Edge Dropout

To alleviate the impact of natural noise in the user-item interaction graph, we prune the edges of the graph based on the idea of model sparsification [12,26]. Research [8] has verified that high-influence nodes are more likely to experience over-smoothing problems. To better learn the potential representation of users or items, we trim the edges of the original graph according to the centrality of the nodes to reduce the signal interference from noisy neighbors. Research [12] shows that edge trimming can effectively alleviate the overfitting problem in GCN. In a directed graph, the degree centrality of a node i can be defined as:

$$d_i = d_{in} + d_{out} \quad (3)$$

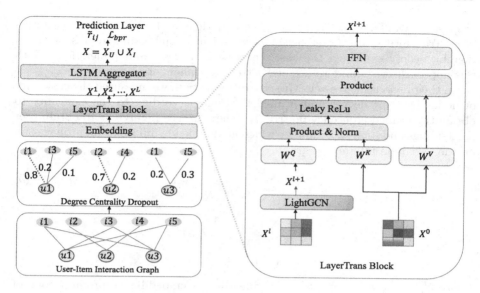

Fig. 2. Overview of the proposed LayerTrans.

where d_{in} and d_{out} represent the out-degree and in-degree of node i respectively. In particular, $d_{in} = d_{out}$ in an undirected graph. For the interaction $e_{ij} \in \mathcal{E}$ between node i and node j, the probability of it being deleted in the original graph $\mathcal{G}$ is:

$$p_{e_{ij}} = 1 - \frac{1}{\sqrt{d_i d_j}} \quad (4)$$

where d_i, d_j denote the degrees of nodes i and j, respectively. According to this method, we delete k edges ($k < \mathcal{K}$) from the original graph $\mathcal{G}$. Then, we construct a new sparse matrix $\tilde{A}_p$ with the obtained $\mathcal{K}$-k edges. $\tilde{A}_p$ is the adjacency matrix of the graph $\mathcal{G}$ after edge pruning. In our subsequent embedding process, we use $\tilde{A}_p$ to replace $D^{-\frac{1}{2}} A D^{-\frac{1}{2}}$ in the LightGCN information propagation process.

As shown in Fig. 2, we calculate the deletion probability of each edge based on the degree of centrality. The larger the value, the greater the probability of edge deletion. From Eq. 4, we can see that the edge pruning algorithm based on node centrality we use is more inclined to delete interactions with high-influence nodes. Research [11] shows that alternating between random pruning and nonrandom pruning methods can speed up the convergence of the model.

3.4 LayerTrans

The over-smoothing of nodes is because the GCN-based model obtains all the information of the neighborhood during the aggregation process, and does not distinguish whether this information is necessary, and directly uses it to update the node's representation. Therefore, in the forward propagation process of our proposed LayerTrans, we use residual connections and FFN to build an intra-layer Transformer to obtain better user and item embedding representations, thereby deeply exploring the user's potential interests and preferences. With this approach, we expect to achieve better recommendation results.

In the previous degree-sensitive centrality section, we measured the centrality of nodes by their degree and pruned the edges according to the centrality to obtain the pruned adjacency matrix $\tilde{A}_p$. Then, we use LightGCN to obtain user embeddings and item embeddings, which are learned by information propagation on the user-item interaction graph.

$$e_u^{l+1} = \sum_{i \in N_u} \frac{1}{\sqrt{|N_u|}\sqrt{|N_i|}} e_i^l \tag{5}$$

$$e_i^{l+1} = \sum_{u \in N_i} \frac{1}{\sqrt{|N_i|}\sqrt{|N_u|}} e_u^l \tag{6}$$

$$X^{l+1} = \tilde{A}_p X^l \tag{7}$$

where e_u^{l+1} and e_i^{l+1} represent the fine-grained embedding representations of users and items at the $(l+1)th$ layer, respectively. X^{l+1} represents the adjacency matrix of the $(l+1)th$ layer, and the symmetric normalization term $\frac{1}{\sqrt{|N_i|}\sqrt{|N_u|}}$ is to reduce the embedding scale in the convolution operation. After obtaining the temporary X^{l+1} through the forward propagation process of LightGCN, we input X^{l+1} and X^0 into the inter-layer Transformer to mine the potential interests of users and the potential information of projects:

$$Q = X^l W^Q, K = X^0 W^K, V = X^0 W^V \tag{8}$$

$$CrossAttention(Q, K, V) = LeakyReLU(\frac{QK^T}{\sqrt{d}})V \tag{9}$$

specifically $W^Q \in R^{d_q \times d_n}, W^K \in R^{d_k \times d_n}, W^V \in R^{d_v \times d_n}$ are the mapping matrices, and $d_k = d_v$. By applying these mapping matrices, we map X^0, X^{l+1} into a low-dimensional vector space. Subsequently, cross-attention [27] is employed to dynamically extract useful high-order information from X^{l+1}, and combine it with the fixed information of the initial embedding X^0 to explore the deep relationships between users and items. In particular, the use of cross-attention better captures subtle changes in the input data, and by analyzing these changes layer by layer, this approach can explore the dynamic changes of user interests and make more accurate recommendations. Here we use the LeakyReLU activation function instead of softmax, because in the recommendation system, the relationships between users and items are usually complex and nonlinear. we enhance the model's non-linear expressiveness and ensure a higher number of active neurons. Subsequently, the computed matrix is further processed using layer normalization and an FFN network. The final embedding representation of the $(l+1)th$ layer, X^{l+1} is defined as:

$$X^{l+1} = LayerNorm(FFN(CrossAttention(Q, K, V))) \tag{10}$$

$$FFN(\cdot) = Linear(GELU(Linear(f(\cdot)))) \tag{11}$$

where $FFN(\cdot)$ is a feedforward propagation network, which includes two linear layers, and $GELU(\cdot)$ is a nonlinear activation function. Since the design of GELU is based on Gaussian distribution, using it to process data can further reduce noise interference and enhance the robustness of the model. After the CrossAttention layer and the FFN layer, the model further aggregates the structural information of the initial embedding layer X^0 and the high-order neighborhood information in X^l to obtain the updated embedding representation X^{l+1}, reducing the influence of other hidden layers on the X^{l+1} layer. Research shows that the initial embedding layer X^0 has a greater impact on the learning of the model [11].

3.5 Prediction

Finally, we input the learned embeddings of users and items into an LSTM-based aggregation layer, using the aggregated results for prediction. After L layers of forward propagation, we discard the initial embedding layer X^0 and aggregate $\{X^1, X^2, \cdots, X^L\}$ using the LSTM aggregation function, rather than using the $(L+1)th$ layer's result as the input to the prediction layer. This approach is taken because the information contained in X^0 has already been refined and included in $\{X^1, X^2, \cdots, X^L\}$. In GCN, after multiple layers of forward propagation, the model will suffer from over-smoothing. However, by using an aggregation function to obtain the final embeddings helps extract semantic information at different levels, alleviates over-smoothing, and inherently includes the effect of residual connections. Particularly, the aggregation function based on the LSTM architecture has stronger expressive power and can extract more fine-grained semantic information between different levels. Therefore, we use the aggregation function to obtain the final embeddings from the forward propagation, rather than using the embeddings from the last layer. The final embeddings can be represented as:

$$X = A_{GGREGATOR}(LSTM(X^1, X^2, \cdots, X^L)) \tag{12}$$

The final embedding X obtained by Eq. 12 is used to predict the candidate items for each user, and these candidate items are ranked. The recommended items for the top-K scores are as follows:

$$\tilde{r}_{ij} = x_i x_j^T, x_i \in X_U, x_j \in X_I \tag{13}$$

where $X = X_I \cup X_U$, X_I is the final embedding representation of all items, and X_U is the final embedding representation of all users. After the forward propagation of the Layer Trans model, we get the refined user representation and item representation, where $x_i \in R^d, x_j \in R^d$ represent the final embedding representation of user u and item i respectively. Generally, we determine whether a user is interested in a certain item by calculating the similarity between the user and the item.

To improve the model performance, we modified the loss function in the case of supervised learning. Given a pair of positive samples (u, i) of user-item interactions, we sample a negative sample (u, j) that has never interacted with the user for comparison. Because the task focuses on users' ratings of items, we use Bayesian personalized ranking (BPR) [28] loss as the main loss function. BPR encourages the prediction score of positive samples to be higher than that of negative samples:

$$\mathcal{L}_{bpr} = \sum_{(u,i,j) \in S} -ln\sigma(\tilde{r}_{ui} - \tilde{r}_{uj}) \quad (14)$$

where S is training data set, $i \in N_u, j \notin N_u$, N_u represents the neighborhood of user u. To further optimize the embedding of users and items to alleviate the overfitting problem. The final loss function of the model is defined as follows:

$$\mathcal{L} = \mathcal{L}_{bpr} + \lambda ||X^0||_2^2 \quad (15)$$

It includes a main prediction loss function $\mathcal{L}_{bpr}$ and a regularization term, λ is an adjustable regularization coefficient, and $|| \cdot ||_2$ is an L_2 norm.

4 Experiments

4.1 Overview of Datasets

We conducted experiments on four public datasets to verify the effectiveness of our proposed model. Table 1 shows the statistics of all datasets. The datasets are described as follows:

- **Movielens**: It is a classic dataset which is widely used in recommendation tasks.
- **MOOC** [29]: It is an educational dataset that records users' course viewing status from October 1, 2016, to March 31, 2018.
- **Amazon Video Games**: A subset of the Amazon Review Dataset, which mainly contains data in the field of video games.
- **Amazon Grocery and Gourmet Food**: Also from the Amazon Review Dataset. This dataset has more samples and a larger interaction graph.

Table 1. Statistics of the datasets

Datasets	Users	Items	Interactions	Sparsity
Movielen	674	1573	73980	93.0221%
MOOC	82535	1302	458453	99.5734%
Games	50677	16897	454529	99.9469%
Food	115144	39688	1025169	99.9506%

4.2 Baseline Methods

- **BPR** [28]: This is a matrix factorization method optimized using Bayesian ranking loss.
- **MultiVAE** [30]: This is an item-based collaborative filtering (CF) method based on variational autoencoders. It optimizes the model by reconstructing an objective.
- **LightGCN** [9]: This is a simplified graph convolutional neural network that discards the redundant feature transformations and non-linear activations in GCN and uses an aggregation function to obtain the final embedding.
- **UltraGCN** [19]: This model introduces graph structure information into MF through two auxiliary losses instead of stacking GCN.
- **IMP-GCN** [20]: This is an improved graph convolutional network that performs convolutions on subgraphs to prevent introducing high-order noise.
- **Layer-GCN** [11]: The model optimizes the embedding of deep networks during forward propagation by calculating inter-layer similarity to alleviate over-smoothing.

4.3 Experiments Setting

To ensure the fairness of the experiments, we follow the methodology of [11] and use a unified framework for all models. The dataset is divided into training, validation, and test sets in a ratio of 7:2:1. In the experiments, the initial embedding dimension for users and items is set to 64, the embeddings of Q, K, and V are set to 256. The implementation is done using the Pytorch framework [31], with a learning rate of 0.001, Xavier initialization [32] for parameters, and Adam [33] as the optimizer. For LayerTrans, the number of layers is fixed at 4, and the regularization coefficient λ is chosen from $\{1e^{-2}, 1e^{-3}, 1e^{-4}, 1e^{-5}\}$, and the edge deletion probability is set to $\{0.0, 0.05, 0.1, 0.15, 0.2\}$. All models are trained on a single V100 with 32 GB memory.

The experiment uses Recall@K and NDCG@K as evaluation indicators of model performance. NDCG@K is a commonly used indicator in recommendation systems. It takes into account both ranking and recommendation quality. In the experiment, we set $K = 10$, $K = 20$ and $K = 50$. For simplicity and formatting considerations, we denote Recall@K as R@K and NDCG@K as N@K.

4.4 Performance Comparison

Table 2 summarizes the performance of different models. Overall, our proposed model performs excellently on all four datasets. With BRP as the baseline, our LayerTrans improves the N@50 metric by 7.15%, 6.17%, 1.89%, and 2.35% on the respective datasets.

It can be observed that UltraGCN outperforms LightGCN across all datasets. This is because UltraGCN introduces auxiliary loss to learn graph structural information, avoiding the stacking of graph convolution layers and alleviating the over-smoothing problem. MultiVAE also shows good performance on each

Table 2. Overall Performance Comparison. The best results are in bold and the suboptimal results are underlined.

Datasets	Metrics	BPR	MultiVAE	LightGCN	UltraGCN	IMP-GCN	Layer-GCN	Layer-Trans
Movielens	R@10	0.0769	0.1025	0.0851	0.0827	0.0784	0.1146	**0.1316**
	R@20	0.1645	0.1803	0.1485	0.1343	0.1506	0.2143	**0.1889**
	R@50	0.2599	0.3011	0.2485	0.2322	0.2620	0.3205	**0.3488**
	N@10	0.1132	0.1324	0.1124	0.1314	0.1166	0.1465	**0.1865**
	N@20	0.1358	0.1504	0.1294	0.1499	0.1343	0.1731	**0.1899**
	N@50	0.1630	0.1846	0.1582	0.1694	0.1671	0.2037	**0.2345**
MOOC	R@10	0.2395	0.2752	0.2480	0.2453	0.2411	0.2743	**0.2996**
	R@20	0.3101	0.3261	0.2947	0.2805	0.2967	0.3601	**0.3752**
	R@50	0.4851	0.5261	0.4732	0.4575	0.4873	0.5460	**0.5756**
	N@10	0.1624	0.1824	0.1636	0.1802	0.1672	0.1957	**0.2102**
	N@20	0.1830	0.1989	0.1855	0.1899	0.1824	0.2202	**0.2357**
	N@50	0.2228	0.2439	0.2170	0.2286	0.2259	0.2628	**0.2845**
Games	R@10	0.0192	0.0218	0.0257	0.0279	0.0260	0.0278	**0.0315**
	R@20	0.0702	0.0709	0.0791	**0.0837**	0.0810	0.0830	0.0832
	R@50	0.0952	0.0971	0.0795	0.1132	0.1106	0.1170	**0.1386**
	N@10	0.0105	0.0124	0.0143	0.0161	0.0154	0.0159	**0.0199**
	N@20	0.0265	0.0270	0.0311	0.0322	0.0317	0.0315	**0.0385**
	N@50	0.0308	0.0323	0.0369	0.0394	0.0382	0.0399	**0.0497**
Food	R@10	0.0341	0.0336	0.0388	0.0393	0.0387	0.0398	**0.0490**
	R@20	0.0556	0.0541	0.0621	0.0620	0.0622	0.0652	**0.0845**
	R@50	0.1024	0.1012	0.1131	0.1137	0.1143	0.1176	**0.1530**
	N@10	0.0196	0.0191	0.0227	0.0228	0.0230	0.0238	**0.0342**
	N@20	0.0260	0.0252	0.0293	0.0299	0.0294	0.0313	**0.0447**
	N@50	0.0369	0.0361	0.0413	0.0420	0.0419	0.0436	**0.0604**

dataset, as generative models have strong learning and reconstruction capabilities, making them less susceptible to over-smoothing. Particularly, graph-based VGAE also exhibits outstanding performance in many works [34]. Layer-GCN demonstrates strong capabilities by removing the ego layer and reducing neighbor signal interference through refined message propagation. Finally, Layer-Trans considers the effectiveness of different neighbor messages within layers, using attention to extract useful information and avoid aggregating redundant information. Additionally, it uses the LSTM aggregation function to extract fine-grained information across different layers for the final prediction. The results indicate that our proposed model is more effective in further uncovering users' potential interests.

4.5 Ablation Study and Effectiveness Analyses

Ablation Study. To evaluate the impact of different parts of the model on the performance, we remove different components of the model and perform ablation experiments.

- **LayerTrans-RD**: Use random edge dropout method to replace the degree centrality edge dropout method.
- **LayerTrans-Att**: Instead of using Cross Attention, simple similarity calculation is used.
- **LayerTrans-Sum**: Use a simple sum aggregation function instead of the LSTM aggregation function.

According to the results of Fig. 3, we can get the following conclusions:

- Instead of using the random edge dropout, there is a significant decrease in both metrics, which shows the effectiveness of the edge removal method based on node centrality.
- The performance degradation after removing Cross-Attention shows that the use of attention is conducive to mining the deep relationship between users and items for better embedding.
- The significant drop in LayerTrans-AggSum performance shows that the LSTM aggregation function can extract more fine-grained information between layers in the final information extraction process, enriching the embedding of users/items. Getting a better final embedding can help the model reach better performance.

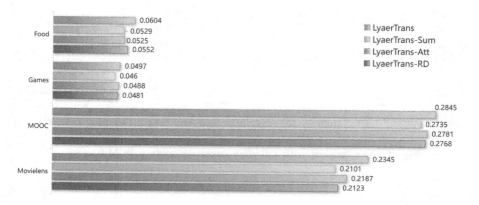

Fig. 3. Ablation study of LayerTrans. Performance on NDCG@50.

Table 3 shows the effectiveness of our model in alleviating the over-smoothing problem. Considering the space occupied, we only show the best performance on the Movielens dataset. From Table 3, we can observe that in terms of the

Table 3. Performance comparison between LayerTrans and LightGCN

Model	R@20	R@50	N@20	N@50
LayerTrans - 4 Layers	0.1889	0.3488	0.1899	0.2345
LightGCN - 4 Layers	0.1444	0.2587	0.1339	0.1583
LightGCN - 3 Layers	0.1473	0.2524	0.1295	0.1532
LightGCN - 2 Layers	0.1383	0.2307	0.1509	0.1654
LightGCN - 1 Layers	0.1359	0.2277	0.1484	0.1645

NDCG@K indicator, LightGCN achieves the best results when stacking 2 layers. Due to the over-smoothing issue, the performance of stacking 3 and 4 layers decreases. In contrast, our model can still achieve good results after stacking multiple layers, indicating that our model can alleviate the over-smoothing problem.

Analysis of Parameters. As shown in Fig. 4, our proposed LayerTrans does not exhibit a decline in performance as the number of layers increases to four or five, indicating that LayerTrans can alleviate the over-smoothing problem.

The regularization coefficient significantly impacts the model's performance, and selecting an appropriate value can optimize the results.

The learning rate determines the model's convergence speed and whether it can reach optimal performance. The results indicate that the model's performance decreases notably when the learning rate is too low. Therefore, in our experiments, we set the hyperparameters to $L = 4$, $\lambda = 1e^{-4}$, and $lr = 1e^{-4}$.

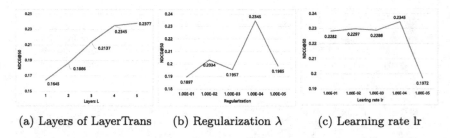

(a) Layers of LayerTrans (b) Regularization λ (c) Learning rate lr

Fig. 4. Comparisons with different hyperparameters on Movielens.

5 Conclusion

In this paper, we propose LayerTrans to alleviate the over-smoothing problem in GCN and better complete the recommendation task. We first use degree centrality to measure the importance of nodes, and calculate the deletion probability of each edge based on this importance to reduce the impact of noise on the model.

Next, we use inter-layer Transformer to extract embedding , and aggregate the inherent structural information and high-order structural information of X^l in the initial embedding X^0 during the forward propagation process to gain a new embedding, and learn the deep relationship between users and items. Finally, we use the LSTM aggregator to explore fine-grained information between layers and optimize the final representation.

Our method optimizes the aggregation of information from multiple perspectives, reduces the interference of redundant information, and conducts experiments on four widely used datasets. The results are better than some existing representative methods.

Acknowledgments. This research was supported by the National Key Research and Development Program of China under grant number 2023YFC3305704, the National Social Science Fund of China under grant number BCA240050 and the National Natural Science Foundation of China under grant number 62377015.

References

1. Resnick, P., Varian, H.R.: Recommender systems. Commun. ACM **40**(3), 56–58 (1997)
2. He, X., He, Z., Song, J., Liu, Z., Jiang, Y.G., Chua, T.S.: Nais: Neural attentive item similarity model for recommendation. IEEE Trans. Knowl. Data Eng. **30**(12), 2354–2366 (2018)
3. Resnick, P., Iacovou, N., Suchak, M., Bergstrom, P., Riedl, J.: Grouplens: An open architecture for collaborative filtering of netnews. In: Proceedings of the 1994 ACM conference on Computer supported cooperative work. pp. 175–186 (1994)
4. Wang, X., He, X., Wang, M., Feng, F., Chua, T.S.: Neural graph collaborative filtering. In: Proceedings of the 42nd international ACM SIGIR conference on Research and development in Information Retrieval. pp. 165–174 (2019)
5. Ying, R., He, R., Chen, K., Eksombatchai, P., Hamilton, W.L., Leskovec, J.: Graph convolutional neural networks for web-scale recommender systems. In: Proceedings of the 24th ACM SIGKDD international conference on knowledge discovery & data mining. pp. 974–983 (2018)
6. Kipf, T.N., Welling, M.: Semi-supervised classification with graph convolutional networks. arXiv preprint arXiv:1609.02907 (2016)
7. Berg, R.v.d., Kipf, T.N., Welling, M.: Graph convolutional matrix completion. arXiv preprint arXiv:1706.02263 (2017)
8. Chen, M., Wei, Z., Huang, Z., Ding, B., Li, Y.: Simple and deep graph convolutional networks. In: International conference on machine learning. pp. 1725–1735. PMLR (2020)
9. He, X., Deng, K., Wang, X., Li, Y., Zhang, Y., Wang, M.: Lightgcn: Simplifying and powering graph convolution network for recommendation. In: Proceedings of the 43rd International ACM SIGIR conference on research and development in Information Retrieval. pp. 639–648 (2020)
10. Li, Q., Han, Z., Wu, X.M.: Deeper insights into graph convolutional networks for semi-supervised learning. In: Proceedings of the AAAI conference on artificial intelligence. vol. 32 (2018)

11. Zhou, X., Lin, D., Liu, Y., Miao, C.: Layer-refined graph convolutional networks for recommendation. In: 2023 IEEE 39th International Conference on Data Engineering (ICDE). pp. 1247–1259. IEEE (2023)
12. Rong, Y., Huang, W., Xu, T., Huang, J.: Dropedge: Towards deep graph convolutional networks on node classification. arXiv preprint arXiv:1907.10903 (2019)
13. Zhu, Y., Xu, Y., Yu, F., Liu, Q., Wu, S., Wang, L.: Graph contrastive learning with adaptive augmentation. In: Proceedings of the web conference 2021. pp. 2069–2080 (2021)
14. Herlocker, J.L., Konstan, J.A., Borchers, A., Riedl, J.: An algorithmic framework for performing collaborative filtering. In: Proceedings of the 22nd annual international ACM SIGIR conference on Research and development in information retrieval. pp. 230–237 (1999)
15. Sarwar, B., Karypis, G., Konstan, J., Riedl, J.: Item-based collaborative filtering recommendation algorithms. In: Proceedings of the 10th international conference on World Wide Web. pp. 285–295 (2001)
16. Deshpande, M., Karypis, G.: Item-based top-n recommendation algorithms. ACM Transactions on Information Systems (TOIS) **22**(1), 143–177 (2004)
17. Koren, Y., Bell, R., Volinsky, C.: Matrix factorization techniques for recommender systems. Computer **42**(8), 30–37 (2009)
18. Lee, D., Seung, H.S.: Algorithms for non-negative matrix factorization. Advances in neural information processing systems **13** (2000)
19. Mao, K., Zhu, J., Xiao, X., Lu, B., Wang, Z., He, X.: Ultragcn: ultra simplification of graph convolutional networks for recommendation. In: Proceedings of the 30th ACM international conference on information & knowledge management. pp. 1253–1262 (2021)
20. Liu, F., Cheng, Z., Zhu, L., Gao, Z., Nie, L.: Interest-aware message-passing gcn for recommendation. In: Proceedings of the web conference 2021. pp. 1296–1305 (2021)
21. Bo, D., Wang, X., Shi, C., Shen, H.: Beyond low-frequency information in graph convolutional networks. In: Proceedings of the AAAI conference on artificial intelligence. vol. 35, pp. 3950–3957 (2021)
22. Wang, G., Ying, R., Huang, J., Leskovec, J.: Direct multi-hop attention based graph neural network. arXiv preprint arXiv:2009.14332 p. 137 (2020)
23. Zhao, L., Akoglu, L.: Pairnorm: Tackling oversmoothing in gnns. arXiv preprint arXiv:1909.12223 (2019)
24. He, K., Zhang, X., Ren, S., Sun, J.: Deep residual learning for image recognition. In: Proceedings of the IEEE conference on computer vision and pattern recognition. pp. 770–778 (2016)
25. Yang, Y., Wu, Z., Wu, L., Zhang, K., Hong, R., Zhang, Z., Zhou, J., Wang, M.: Generative-contrastive graph learning for recommendation. In: Proceedings of the 46th International ACM SIGIR Conference on Research and Development in Information Retrieval. pp. 1117–1126 (2023)
26. Louizos, C., Welling, M., Kingma, D.P.: Learning sparse neural networks through l_0 regularization. arXiv preprint arXiv:1712.01312 (2017)
27. Vaswani, A., Shazeer, N., Parmar, N., Uszkoreit, J., Jones, L., Gomez, A.N., Kaiser, Ł., Polosukhin, I.: Attention is all you need. Advances in neural information processing systems **30** (2017)
28. Rendle, S., Freudenthaler, C., Gantner, Z., Schmidt-Thieme, L.: Bpr: Bayesian personalized ranking from implicit feedback. arXiv preprint arXiv:1205.2618 (2012)

29. Zhang, J., Hao, B., Chen, B., Li, C., Chen, H., Sun, J.: Hierarchical reinforcement learning for course recommendation in moocs. In: Proceedings of the AAAI conference on artificial intelligence. vol. 33, pp. 435–442 (2019)
30. Liang, D., Krishnan, R.G., Hoffman, M.D., Jebara, T.: Variational autoencoders for collaborative filtering. In: Proceedings of the 2018 world wide web conference. pp. 689–698 (2018)
31. Paszke, A., Gross, S., Massa, F., Lerer, A., Bradbury, J., Chanan, G., Killeen, T., Lin, Z., Gimelshein, N., Antiga, L., et al.: Pytorch: An imperative style, high-performance deep learning library. Advances in neural information processing systems **32** (2019)
32. Glorot, X., Bengio, Y.: Understanding the difficulty of training deep feedforward neural networks. In: Proceedings of the thirteenth international conference on artificial intelligence and statistics. pp. 249–256. JMLR Workshop and Conference Proceedings (2010)
33. Kingma, D.P., Ba, J.: Adam: A method for stochastic optimization. arXiv preprint arXiv:1412.6980 (2014)
34. Kipf, T.N., Welling, M.: Variational graph auto-encoders. arXiv preprint arXiv:1611.07308 (2016)

Adaptive Disentangled Contrastive Collaborative Filtering

Sujie Yu[1], Junnan Zhuo[1], Lvying Chen[1], Hailian Yin[1], and Bohan Li[1,2,3(✉)]

[1] Nanjing University of Aeronautics and Astronautics, Nanjing, China
{ysj0618,junnanzhuo,cly.ly,yinhailian,bhli}@nuaa.edu.cn
[2] Key Laboratory of Brain-Machine Intelligence Technology, Ministry of Education, Nanjing, China
[3] Key Laboratory of Intelligent Decision and Digital Operations, Ministry of Industrial and Information Technology, Beijing, China

Abstract. Graph neural networks (GNNs) have demonstrated powerful performance in modeling high-order connectivity for collaborative filtering (CF) tasks. However, several challenges still remain unexplored in existing solutions: i) Diverse latent intents may influence users to choose certain items, which is widely ignored in representation learning; ii) The unreasonable combination of intent factors may ignore minor differences between users to some extent, which exacerbates the over-smoothing issue. To address above issues, we propose a novel Adaptive Disentangled Contrastive Collaborative Filtering (ADCCF) framework to distill fine-grained information from the entangled self-supervised signals. Specifically, we leverage cross-layer contrastive learning to guide the encoding process. Besides, ADCCF adaptively control the distance between each feature in the deep vector space, which can alleviate the over-smoothing problem. Extensive experiments on three real-world datasets indicate that our ADCCF outperforms various state-of-the-art recommendation methods.

Keywords: Recommender Systems · Collaborative Filtering · Disentangled Representation Learning · Contrastive Learning

1 Introduction

Information overload is becoming increasingly serious in recent years, complicating the task of capturing genuine user preferences. To mitigate this issue, recommender systems learn the interests of users from their historical interactions and provide personalized recommendations [1]. CF is a classical algorithm that focuses on predicting user preferences for various items based on their historical behaviors, including browsing, clicking, and purchasing. Early works depend on the user-item rating matrix to calculate preferences, such as matrix factorization (MF) [7].

As graph learning has rapidly developed recently, many research leverage GNNs to model the interaction data as graph and iteratively propagate the neighborhood information, which is known as graph collaborative filtering [5]. This technique notably enhances the proficiency of the model in capturing high-order collaborative signals. Specifically, PinSage [27] and NGCF [18] employ graph convolutional networks (GCNs) to facilitate embedding propagation. Meanwhile, LightGCN [4] considers that removing non-linear activation and feature transformation can reduce the complexity of the model.

While these approaches show good performance, there are still several issues with the current graph collaborative filtering approach: (1) neglect of intent factors: Most previous studies have overlooked that interaction behaviors of users may be influenced by many latent intents. Take Fig. 1 as an example, the user may purchase lipstick because he intends to prepare a birthday gift for his friend, but he has no interest in it. Simultaneously, he might need to buy some cakes, driven by his upcoming picnic plan. Correctly capturing disentangled factors can help understand the recommendation strategy more clearly. (2) hazards of over-smoothing representation: As the number of graph propagation layers increases, the message-passing mechanism in GNNs makes all connected node features converge to the same value, which complicates the task of distinguishing different vectors [11]. However, high-order information is crucial to address the data sparsity challenge [12], so the over-smoothing issue should not be underestimated.

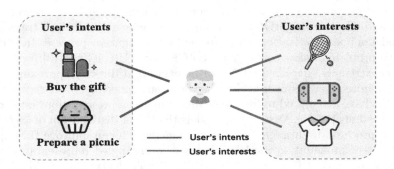

Fig. 1. Example of diverse latent intents affecting user interaction behaviors.

Given the aforementioned limitations, we propose a novel Adaptive Disentangled Contrastive Collaborative Filtering (ADCCF) framework to enrich representations while addressing the over-smoothing problem. Specifically, our model combine local collaborative relations and global intent dependencies to extract supervised signals. However, similar users may have essentially the same intent representation through the clustering algorithm. Meanwhile, irrationally combining global disentangled representations with local collaborative signals causes our model to ignore minor differences between users. To address these issues, we introduce a novel adaptive weight difference module to control the node-wise similarity, which can combat the interference of over-smoothing issue. In addition, we leverage a cross-layer contrastive learning to effectively fine-tune the uniformity of our learned representations and improve the quality of embeddings. In summary, this paper has the following major contributions:

- To capture disentangled representations, we propose a novel recommendation framework through modeling the user-item interactions from both local and global perspectives.
- ADCCF innovatively introduces an adaptive weight difference module to greatly alleviate the over-smoothing problem. The designed cross-layer contrastive learning schema helps our model get a more even distribution of node representations.
- We conduct extensive experiments on three real-world datasets, and ADCCF is consistently better than many state-of-the-art methods. Additionally, the ablation study is provided to demonstrate the effectiveness of each module in ADCCF.

2 Related Work

2.1 Graph-Based Recommender Systems

The remarkable performance of GNNs in representation learning has sparked significant interest in improving recommendation performance. For example, NGCF [18] treats user-item interactions as bipartite graph structure to capture high-order connectivity between them. To make the GCNs framework more concise and effective, LightGCN [4] proposes to remove the feature transformation and nonlinear activation parts from the propagation process. In order to solve the problem of slow convergence on CF tasks, UltraGCN [13] further proposes an extremely simplified GCNs formulation that directly approximates the limit of infinite-layer graph convolutions via a constraint loss. More recently, several works have attempted to introduce disentangled representations into recommender systems [14,16]. DGCF [19] models the intent distribution of each interaction to produce disentangled representations, which can improve the explainability of recommendation models. To learn disentangled representations with clear meanings, KDR [14] utilizes knowledge graph (KG) to guide implicit disentangled representation learning, and aligns the learned representations with semantic information derived from KG. Because different behaviors have different intent distributions, IDCL [20] models a user-item-concept graph to simultaneously learn interpretable intents and behavior distributions. However, most existing disentangled recommender systems ignore the fact that unreasonable combination of additional modeling factors may exacerbate the over-smoothing issue. To address this challenge, we propose a new intent-aware model which can adaptively control the node-wise difference.

2.2 Contrastive Learning for Recommendation

Contrastive learning aims to distinguish the similarities and differences between objects, capturing the feature representation consistency under different views [3]. Recent research works have attempted to explore its application in recommendations. For example, SGL [21] changes the graph structure by node dropout, edge dropout, and random walk to generate the contrastive views. Meanwhile, HCCF [25] reinforces the representation quality by capturing local and global collaborative relations with a hypergraph-enhanced cross-view contrastive learning architecture. To make the method simpler, SimGCL [29] attempts to produce semantically related pairs for contrastive learning by adding uniform noises to the embedding space. Given the shortcomings of manually generating contrastive views, AutoCF [23] designs a masked graph autoencoder to aggregate global information, which can automatically perform data augmentation for recommendation. Although these methods have achieved good results, they cannot capture the latent intent factors behind user-item interactions, which limits the recommendation performance of the model.

3 Preliminaries

In our recommendation scenario, let $\mathcal{U} = \{u_1, u_2, ..., u_M\}$ and $\mathcal{V} = \{v_1, v_2, ..., v_N\}$ denote the user set and item set, where M and N represent the number of users and items, respectively. A user-item interaction matrix $\mathcal{A} \in \mathbb{R}^{M \times N}$ is generated based on the historical interaction data, where $\mathcal{A}_{u,i} = 1$ if user u adopted item i before, otherwise $\mathcal{A}_{u,i} = 0$. The CF algorithm has the assumption that two users are more likely to share similar interests if they behave similarly. The objective of this work is to accurately predict the likelihood $\hat{y}_{u,i}$ that a candidate user will adopt an item based on the observed user-item interactions.

4 Methodology

In this section, we introduce the ADCCF framework in four parts: (1) Local Collaborative Relation Encoding which utilizes the graph message passing mechanism and cross-layer contrastive learning to capture high-quality local collaborative signals; (2) Disentangled Intent Representation Learning which obtains the multi-intent factors via the clustering algorithm; (3) Adaptive Weight Difference Module which can dynamically control similarity between nodes; and (4) Model Training. The overall architecture of ADCCF is illustrated in Fig. 2.

4.1 Local Collaborative Relation Encoding

High-Order Information Encoding and Aggregation. Following the common CF paradigm, we generate embedding vectors e_u and e_i for user u and item i, respectively. To propagate embeddings efficiently, we follow LightGCN [4] and discard the nonlinear activation and feature transformation in the propagation process with the following form:

$$e_u^{l+1} = \sum_{i \in \mathcal{N}_u} \frac{1}{\sqrt{|\mathcal{N}_u| \cdot |\mathcal{N}_i|}} e_i^l, \quad e_i^{l+1} = \sum_{u \in \mathcal{N}_i} \frac{1}{\sqrt{|\mathcal{N}_i| \cdot |\mathcal{N}_u|}} e_u^l, \tag{1}$$

where $\mathcal{N}_u$ and $\mathcal{N}_i$ denote the neighbor items/users set of u and i. This can counterbalance the effects of uneven node degrees. To refine the user/item representations, we employ multiple embedding propagation layers based on GNNs and sum node embeddings across all layers to obtain the final embeddings as follows:

$$e_u = \sum_{l=0}^{L} e_u^l, \quad e_i = \sum_{l=0}^{L} e_i^l. \tag{2}$$

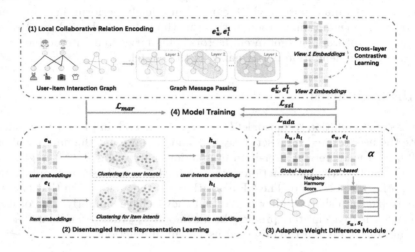

Fig. 2. The overall framework of our proposed ADCCF.

Cross-Layer Contrastive Learning. However, prevalent popularity bias frequently exists in practical recommendation scenarios. To address this issue and promote more evenly distributed user and item representations, ADCCF utilizes contrastive learning to provide self-supervised learning (SSL) signals, which makes the representations more robust. Traditional data augmentation approaches (e.g., randomly node dropout, edge dropout, or hypergraphs) may corrupt interaction graph structures and make the model more complex. Consequently, inspired by XSimGCL [28], we design cross-layer contrastive learning in ADCCF, which can enhance supervised signals more simply and efficiently.

As mentioned in [17], there is a balance when using contrastive learning where the degree of mutual information between correlated views is appropriate. So we try to compare different layers of embeddings that share some common information but differ in aggregated neighbors. Specifically, we choose to compare the embeddings in the first layer, i.e., the original embeddings, and those in the final layer. They represent the results before and after aggregating the neighborhood information, which conform to the balance theory.

According to the existing self-supervised CF paradigms, as [5,26] proposed, we treat embeddings of the same user/item in different layers as positive pairs, and the views of different users/items as negative pairs. With the generated embeddings, we get the positive and negative sample pairs: $\{e_u^0, e_u^L\}$ and $\{e_u^0, e_{u'}^L\}$. Referring to [15], we define the self-supervised contrastive loss based on mutual information maximization as follows:

$$\mathcal{L}_{ssl}^{user} = \sum_{u \in \mathcal{U}} -\log \frac{\exp(s(e_u^0, e_u^L)/\tau_1)}{\sum_{u' \in \mathcal{U}, u \neq u'} \exp(s(e_u^0, e_{u'}^L)/\tau_1)}. \quad (3)$$

where $s(\cdot)$ denotes the cosine similarity function, which measures the similarity between two embeddings, and τ_1 is the temperature in softmax, which is the hyperparameter that controls the smoothness of the softmax curve. We compute the contrastive loss for the item side as $\mathcal{L}_{ssl}^{item}$ in a similar way. Combining these two losses, we obtain the objective function for the self-supervised task, denoted by $\mathcal{L}_{ssl} = \mathcal{L}_{ssl}^{user} + \mathcal{L}_{ssl}^{item}$.

4.2 Disentangled Intent Representation Learning

Users have various intents when they interact with items, such as preferences for specific colors or the purpose of preparing gifts for friends. To capture these different intents more accurately, we assume that there are K latent intent prototypes $\{c_u^k\}_{k=1}^K$ and $\{c_v^k\}_{k=1}^K$ from the perspective of users and items. The intents on the item side can be interpreted as the context of the item, for example, a depressed user may prefer items with a "healing" context.

To learn the intent prototypes c_u, we assume that users with similar intents tend to exhibit similar behaviors. Based on this idea, we perform K-means clustering via faiss [6] over all user representations to obtain K clusters $(C_1, C_2, \ldots, C_K)$. After that, the center of these clusters can be regarded as the intent prototypes with global context, which is as follows:

$$c_u^k = \frac{1}{|C_k|} \sum_{e_u \in C_k} e_u \quad (4)$$

Users in the same cluster can be considered as sharing the same primary intent. To reflect the diversity of intents, ADCCF leverages the relevance score between the local embedding e_u and each intent prototype c_u to compute the final intent representation, which can be denoted as:

$$h_u = \sum_k^K c_u^k \frac{\exp(e_u^\top c_u^k)}{\sum_{k'}^K \exp(e_u^\top c_u^{k'})} \quad (5)$$

similar operations are employed to compute the final disentangled representations h_i from the perspective of items.

For each user u, two distinct embeddings are derived from different perspectives, namely e_u with local dependencies and h_u with global dependencies.

These embeddings contain information with different emphases, and unreasonable combination will impair performance to a certain degree. Specifically, the overemphasis on local embeddings prevents the recommendation approach from capturing fine-grained interaction patterns between users and items. Meanwhile, the overemphasis on disentangled representations ignores the minor difference between users, which exacerbates the over-smoothing problem. Therefore, we utilize a combination weight α to correlate these two embeddings, using the following design:

$$s_u = (1-\alpha)e_u + \alpha h_u, \quad s_i = (1-\alpha)e_i + \alpha h_i \qquad (6)$$

4.3 Adaptive Weight Difference Module

To alleviate over-smoothing problems caused by unreasonable combination of disentangled representations, our model proposes to dynamically regulate the embedding learning. When the central node is more harmonious with the neighboring nodes, it will be more similar to other nodes. Therefore, ADCCF adaptively locates which nodes are more likely being over-smoothed by comparing the coherence between the embedding and its neighborhood information. In particular, we accomplish the task of universally maximizing the node-wise difference by assigning distinct weights to nodes. Our goal is to minimize the following loss function:

$$\mathcal{L}_{\mathrm{ada}} = \sum_{u \in \mathcal{U}} \omega_u \cdot g(u, \mathcal{U}+\mathcal{V}) + \sum_{i \in \mathcal{V}} \omega_i \cdot g(i, \mathcal{U}+\mathcal{V}) \qquad (7)$$

It consists of two parts: user-nodes similarity and item-nodes similarity. Here, $g(\cdot)$ denotes the similarity function between each node and all other nodes. The first term $g(u, \mathcal{U}+\mathcal{V})$ represents that ADCCF manages the similarity between the embedding of user u and the embedding of every node in $\mathcal{U} \cup \mathcal{V}$, which is calculated as follows:

$$g(u, \mathcal{U}+\mathcal{V}) = \log \sum_{u' \in \mathcal{U}} \exp(s_u^\top s_{u'}/\tau_2) + \log \sum_{i \in \mathcal{V}} \exp(s_u^\top s_i/\tau_2) \qquad (8)$$

where the hyperparameter τ_2 is used to adjust the neighbor harmony score. In Eq. (7), $\omega_u \in \mathbb{R}$ controls the weight of node-wise similarity for user u, which is computed as follows:

$$\omega_a = \begin{cases} 1-\varepsilon & \text{if } s_a > 0.5 \\ 1+\varepsilon & \text{otherwise} \end{cases} \qquad (9)$$

where $0 < \varepsilon < 1$ is a hyperparameter which can control the weight, and s_a denotes the harmony score of the node a concerning its neighborhood information, i.e., the similarity between the original embedding and its purely adjacent information. Formally, it can be derived as follows:

$$s_a = \mathrm{sig}(e_a^\top \sum_{a' \in \mathcal{N}_a^k} \frac{e_{a'}}{|\mathcal{N}_a^k| \cdot ||e_a|| \cdot ||e_{a'}||}) \qquad (10)$$

where $e_a, e_{a'} \in \mathbb{R}^d$ denote the initial user/item embeddings, and sig(·) represents the sigmoid activation function. We explore the neighbor harmony score of the central node a by aggregating its k-order adjacent information using mean-pooling or summation. The larger neighbor harmony score signifies the greater structural consistency between the node and its neighboring nodes.

Through this module, ADCCF can assess the alignment of each node with its neighborhood information, dynamically work towards universally maximizing the node-wise difference, and help our model alleviate the over-smoothing issue.

4.4 Model Training

During the training stage, we employ the fused embeddings s_u, s_i obtained from Eq. (6) to predict the probability of user u interacting with item i via dot-product: $\hat{y}_{u,i} = s_u^\top s_i$. Larger interaction preference score indicates that user u is more likely to interact with item i. Given these notations, we adopt the pairwise marginal loss $\mathcal{L}_{\text{mar}}$ to optimize our ADCCF for the recommendation task:

$$\mathcal{L}_{\text{mar}} = \sum_{u \in \mathcal{U}} \sum_{i \in \mathcal{N}_u} \sum_{i' \notin \mathcal{N}_u} \max(0, 1 - \hat{y}_{u,i} + \hat{y}_{u,i'}) \tag{11}$$

where i' represents the negative samples with no observed interactions. After defining these three loss functions, we integrate them into a unified objective:

$$\mathcal{L} = \mathcal{L}_{\text{mar}} + \lambda_1 \mathcal{L}_{\text{ssl}} + \lambda_2 \mathcal{L}_{\text{ada}} + \lambda_3 \|\Theta\|_2^2 \tag{12}$$

where $\lambda_1, \lambda_2, \lambda_3$ are all tunable parameters and Θ is all the learnable model parameters.

5 Experiments

In this section, we conduct various experiments to demonstrate the effectiveness of ADCCF and address the following research questions: **RQ1:** How does our proposed ADCCF perform compared to various state-of-the-art recommendation methods? **RQ2:** What are the impacts of different key components in ADCCF model? **RQ3:** Is ADCCF effective in alleviating the problems of data sparsity and over-smoothing? **RQ4:** How do different settings of hyperparameters affect the model performance?

5.1 Experimental Settings

Datasets. We evaluate our model performance on three real-world datasets. **Gowalla:** This dataset is collected from the social platform Gowalla and contains check-in information from users worldwide from Jan to Jun, 2010. **Yelp:** It comprises ratings of venues from users on the Yelp platform from Jan to Jun, 2018. **Amazon:** It contains information about rating behaviors of users over books on the Amazon platform in 2013. Table 1 shows the statistics of each dataset.

Table 1. Statistics of datasets.

Dataset	Users	Items	Interactions	Density
Gowalla	25,557	19,747	294,983	5.85×10^{-4}
Yelp	42,712	26,822	182,357	1.59×10^{-4}
Amazon	76,469	83,761	966,680	1.51×10^{-4}

Evaluation Protocols and Metrics. To mitigate the sampling bias, we measure the prediction accuracy using the all-rank procedure [8]. We adopt two metrics *Recall@N* and *NDCG@N* to quantify the performance of all methods.

Baselines. We compare our ADCCF with the following baseline methods:

- **BiasMF** [7]: This is a classical MF method that introduces bias information into the learnable embedding vectors.
- **PinSage** [27]: It combines random walks and graph convolutions to effectively model user and item information, which incorporates both graph structure and node features.
- **NGCF** [18]: It models user-item interactions as a bipartite graph and leverages GCNs to explore high-order connectivity in the graph.
- **LightGCN** [4]: It improves GCNs message propagation by eliminating feature transformation and non-linear activation. This can simplify the training process while improving recommendation performance.
- **DGCF** [19]: This method explores finer granularity of user intents by learning the distribution over intents for each interaction and generates disentangled representations.
- **DGCL** [9]: It uses self-supervision signals to learn disentangled representations and then proposes a factor-wise discrimination objective.
- **SGL** [21]: It devises three different operators to generate contrastive views, which offers additional signals for representation learning.
- **NCL** [10]: It introduces the concepts of structural neighbors and semantic neighbors and uses contrastive learning to mine the high-order connections between nodes.
- **SimRec** [24]: This approach combines knowledge distillation and contrastive learning to preserve the global collaborative signals. It also addresses the over-smoothing issue with representation recalibration.

Parameter Settings. We implement ADCCF and all baselines with SSLRec, a unified open-source SSL framework for recommendation. For a fair comparison, we set the embedding size to 32 and the batch size to 1024. Also, we apply the Xavier initializer and the Adam optimizer with the learning rate of 0.001. Grid search is adopted to tune the hyperparameters.

Table 2. Performance comparison on different datasets in terms of Recall and NDCG.

Datasets	Gowalla				Yelp				Amazon			
Metrics	R@20	R@40	N@20	N@40	R@20	R@40	N@20	N@40	R@20	R@40	N@20	N@40
BiasMF	0.0867	0.1269	0.0579	0.0695	0.0198	0.0307	0.0094	0.0120	0.0324	0.0578	0.0211	0.0293
PinSage	0.1235	0.1882	0.0809	0.0994	0.0510	0.0743	0.0245	0.0315	0.0486	0.0773	0.0317	0.0402
NGCF	0.1757	0.2586	0.1135	0.1367	0.0681	0.1019	0.0336	0.0419	0.0551	0.0876	0.0353	0.0454
LightGCN	0.2230	0.3181	0.1433	0.1670	0.0761	0.1175	0.0373	0.0474	0.0868	0.1285	0.0571	0.0697
DGCF	0.2055	0.2929	0.1312	0.1555	0.0700	0.1072	0.0347	0.0437	0.0617	0.0912	0.0372	0.0468
DGCL	0.2068	0.2887	0.1307	0.1542	0.0704	0.1063	0.0336	0.0431	0.0609	0.0899	0.0363	0.0462
SGL	0.2332	0.3251	0.1509	0.1780	0.0803	0.1226	0.0398	0.0502	0.0874	0.1312	0.0569	0.0704
NCL	0.2283	0.3232	0.1478	0.1745	0.0806	0.1230	0.0402	0.0505	0.0955	0.1409	0.0623	0.0764
SimRec	0.2434	0.3399	0.1592	0.1865	0.0823	0.1251	0.0414	0.0519	0.1067	0.1535	0.0734	0.0879
Ours	**0.2467**	**0.3434**	**0.1601**	**0.1876**	**0.0847**	**0.1298**	**0.0427**	**0.0537**	**0.1112**	**0.1573**	**0.0759**	**0.0913**

5.2 Overall Performance Comparison (RQ1)

We compare our model with various baselines on three datasets to validate the effectiveness of ADCCF. The results are shown in Table 2.

- Compared with all baseline methods, ADCCF achieves the best performance on each dataset. We attribute the significant advantages of ADCCF to the following aspects: (1) ADCCF mitigates the data sparsity problem in the disentangling module by modeling from a fine-grained level of intents. (2) The cross-layer contrastive learning helps the model learn more evenly distributed representations, consequently improving training efficiency. (3) Our proposed adaptive weight difference module effectively mitigates the over-smoothing problem and further enhances the robustness of the model.
- From the above evaluation results, most SSL-based methods (e.g., SGL, NCL) outperform the traditional GNNs-based methods (e.g., NGCF, LightGCN), which suggests that SSL improves the generalization ability of CF models by learning the structure and distribution of the data. However, compared with these SSL-based methods, ADCCF performs even better, suggesting that it is important to disentangle intent factors behind user-item interactions.
- ADCCF has a substantial improvement compared with other disentangled baselines (e.g., DGCF, DGCL), suggesting that integrating latent multi-intent encoding and contrastive learning can improve the representation learning ability. Although DGCL attempts to use contrastive learning to learn the disentangled graph representation, factor-level contrastive learning is more susceptible to noise. At the same time, existing disentangled learning models neglect the over-smoothing problem when combining intent factors, leading to worse vector embeddings.

5.3 Ablation Study (RQ2)

In this section, to explore the effectiveness of different components on performance improvement, we consider the following three variants of ADCCF:

Table 3. Ablation study on key components of ADCCF.

Dataset	Gowalla		Yelp		Amazon	
Variant	Recall	NDCG	Recall	NDCG	Recall	NDCG
-Disen	0.2460	0.1586	0.0726	0.0360	0.1100	0.0723
-SSL	0.2368	0.1526	0.0831	0.0414	0.0918	0.0594
-Ada	0.2445	0.1576	0.0827	0.0409	0.1098	0.0722
ADCCF	**0.2465**	**0.1598**	**0.0847**	**0.0427**	**0.1112**	**0.0759**

- **Effect of Disentangled Encoding:** The variant *-Disen* stands for removing the disentangled intent encoding module. The results in Table 3 show that removing this module decreases recommendation accuracy, which suggests that learning the disentangled representation from a fine-grained perspective can better capture the global dependencies between users and items.
- **Effect of Cross-layer Contrastive Learning:** The variant *-SSL* stands for removing the SSL task across layers. As shown in Table 3, it leads to a performance decrease, confirming that the cross-layer contrastive learning can smoothly adjust the uniformity of learned representations and improve the quality of these representations.
- **Effect of Adaptive Weight Difference Module:** The variant *-Ada* disables the adaptive weight difference module. The results show that unreasonable combination of local and global embeddings may lead to over-smoothing representations. However, this module can universally maximize the node-wise difference, which contributes to better representation learning ability.

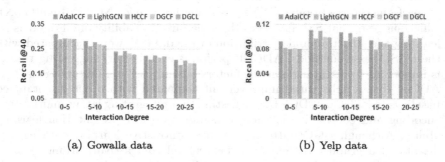

(a) Gowalla data (b) Yelp data

Fig. 3. Performance analysis of ADCCF and baseline methods under different sparsity-level users.

5.4 Study of Robustness (RQ3)

Robustness in Mitigating Data Sparsity. To investigate whether ADCCF is robust to data sparsity issue, we classify users into different groups based on their interaction frequencies, i.e., [0,5), [5,10), [10,15), [15,20), [20,25). Then we compare ADCCF with several baselines. The results are depicted in Fig. 3, and we can summarize the following observations: (1) ADCCF fundamentally shows excellent performance on datasets for both inactive and active users, especially in the group with extremely sparse supervision signals. This indicates the fact that ADCCF can better distill the supervision signals from local and global views to enrich the representations. (2) ADCCF brings the greatest benefit in the group with extremely sparse supervision signals, which demonstrates that ADCCF is still robust in special scenarios, and it can offer high-quality recommendations using sparse interaction data.

Robustness in Overcoming Over-Smoothing. Leveraging the proposed adaptive weight difference module, ADCCF alleviates the over-smoothing problem. Therefore, to validate whether ADCCF successfully mitigates the over-smoothing issue, we compare the Mean Average Distance (MAD) [2] between ADCCF and two variants -*Disen* and -*Ada*, which measures the smoothness of the graph representation. As shown in Table 4, the MAD scores of both variants are relatively low. This phenomenon indicates that unreasonable combination with disentangled information will exacerbate over-smoothing problems to a certain extent. Meanwhile, ADCCF can mitigate hazards by effectively controlling node similarities through the adaptive weight difference module.

Table 4. Comparative evaluation of graph smoothness degree of variants on different datasets (measured by MAD).

Dataset	Gowalla		Yelp		Amazon	
Variant	User	Item	User	Item	User	Item
-Disen	0.9443	0.8885	0.9195	0.8424	0.8974	0.9716
-Ada	0.9265	0.8563	0.8575	0.9329	0.8555	0.9717
ADCCF	**0.9980**	**0.999**	**0.9972**	**0.9939**	**0.9945**	**0.9906**

5.5 Influence of Hyperparameters (RQ4)

In this section, we investigate the effects of some hyperparameters on ADCCF, including the number of latent intents K, cross-layer contrastive temperature τ_1, and combination weight α. The results are shown in Fig. 4, where the y-axis represents the extent to which the model performance is altered under other settings compared to the best performance with the optimal settings.

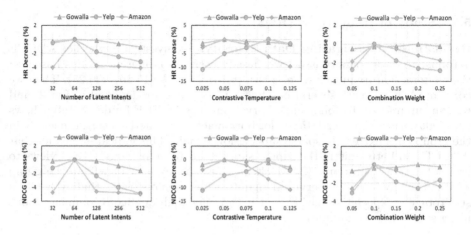

Fig. 4. Hyperparameter sensitivity analysis of ADCCF.

- **Number of Latent Intents** K: We search for the number of latent intents K in the range of $\{32, 64, 128, 256\}$. As shown in Fig. 4, the model performance improves with the number of latent intents increasing. Nevertheless, the model performance reaches the peak when $K = 64$, which suggests that considering too many intents will amplify the hazards of noises during the clustering process and further cause the phenomenon of overfitting.
- **Contrastive Temperature** τ_1: This hyperparameter in cross-layer contrastive learning controls the degree to which the model focuses on negative samples [22], and an appropriate temperature value can help the model perform better in contrastive learning. From the results in Fig. 4, it can be observed that $\tau_1 = 0.05$ works best in Gowalla and Amazon, while $\tau_1 = 0.1$ is a reasonable choice in Yelp. This phenomenon indicates that the performance improves when ADCCF gives more attention to complex samples.
- **Combination Weight** α: In order to distill the supervision signals from local and global perspectives in a cross-view way, we choose the values of combination weight α from the range of $\{0.05, 0.1, 0.15, 0.2, 0.25\}$. Through experiments, we find that our model attains optimal performance when $0.1 \leq \alpha \leq 0.2$. It proves that too much disentangled information will add additional noise, while too little will hinder the ability to capture genuine user preferences.

6 Conclusion

In this work, we propose ADCCF, a self-supervised recommendation framework based on disentangled intent factors. Specifically, we jointly introduce local dependencies and global disentangled representations in the recommender system while utilizing cross-layer contrastive learning to learn more evenly distributed representations. In addition, the adaptive weight difference module in

ADCCF accomplishes the task of universally maximizing the node-wise difference through neighbor harmony scores, which effectively mitigates the over-smoothing problem due to the unreasonable combination of intents. Extensive experimental results on various datasets show that ADCCF is more superior than other state-of-the-art methods. A promising direction for future work is to utilize our ADCCF in cross-domain recommendation scenarios, so as to improve the model performance for cold-start recommendations.

Acknowledgements. This work is supported in part by the "14th Five-Year Plan" Civil Aerospace Pre-Research Project of China under Grant No. D020101.

References

1. Bai, Y., Zhang, Y., Lu, J., Chang, J., Zang, X., Niu, Y., Song, Y., Feng, F.: Labelcraft: Empowering short video recommendations with automated label crafting. In: Proceedings of the 17th ACM International Conference on Web Search and Data Mining. pp. 28–37 (2024)
2. Chen, D., Lin, Y., Li, W., Li, P., Zhou, J., Sun, X.: Measuring and relieving the over-smoothing problem for graph neural networks from the topological view. In: Proceedings of the AAAI conference on artificial intelligence. vol. 34, pp. 3438–3445 (2020)
3. Chen, M., Huang, C., Xia, L., Wei, W., Xu, Y., Luo, R.: Heterogeneous graph contrastive learning for recommendation. In: Proceedings of the sixteenth ACM international conference on web search and data mining. pp. 544–552 (2023)
4. He, X., Deng, K., Wang, X., Li, Y., Zhang, Y., Wang, M.: Lightgcn: Simplifying and powering graph convolution network for recommendation. In: Proceedings of the 43rd International ACM SIGIR conference on research and development in Information Retrieval. pp. 639–648 (2020)
5. Jiang, Y., Huang, C., Huang, L.: Adaptive graph contrastive learning for recommendation. In: Proceedings of the 29th ACM SIGKDD conference on knowledge discovery and data mining. pp. 4252–4261 (2023)
6. Johnson, J., Douze, M., Jégou, H.: Billion-scale similarity search with gpus. IEEE Transactions on Big Data **7**(3), 535–547 (2019)
7. Koren, Yehuda, B.R., Volinsky, C.: Matrix factorization techniques for recommender systems. Computer **42**(8), 30–37 (2009)
8. Krichene, W., Rendle, S.: On sampled metrics for item recommendation. In: Proceedings of the 26th ACM SIGKDD international conference on knowledge discovery & data mining. pp. 1748–1757 (2020)
9. Li, H., Wang, X., Zhang, Z., Yuan, Z., Li, H., Zhu, W.: Disentangled contrastive learning on graphs. Adv. Neural. Inf. Process. Syst. **34**, 21872–21884 (2021)
10. Lin, Z., Tian, C., Hou, Y., Zhao, W.X.: Improving graph collaborative filtering with neighborhood-enriched contrastive learning. In: Proceedings of the ACM web conference 2022. pp. 2320–2329 (2022)
11. Liu, Y., Xuan, H., Li, B., Wang, M., Chen, T., Yin, H.: Self-supervised dynamic hypergraph recommendation based on hyper-relational knowledge graph. In: Proceedings of the 32nd ACM International Conference on Information and Knowledge Management. pp. 1617–1626 (2023)

12. Ma, W., Wang, Y., Zhu, Y., Wang, Z., Jing, M., Zhao, X., Yu, J., Tang, F.: Madm: A model-agnostic denoising module for graph-based social recommendation. In: Proceedings of the 17th ACM International Conference on Web Search and Data Mining. pp. 501–509 (2024)
13. Mao, K., Zhu, J., Xiao, X., Lu, B., Wang, Z., He, X.: Ultragcn: ultra simplification of graph convolutional networks for recommendation. In: Proceedings of the 30th ACM international conference on information & knowledge management. pp. 1253–1262 (2021)
14. Mu, S., Li, Y., Zhao, W.X., Li, S., Wen, J.R.: Knowledge-guided disentangled representation learning for recommender systems. ACM Transactions on Information Systems (TOIS) **40**(1), 1–26 (2021)
15. Oord, A.v.d., Li, Y., Vinyals, O.: Representation learning with contrastive predictive coding. arXiv preprint arXiv:1807.03748 (2018)
16. Ren, X., Xia, L., Zhao, J., Yin, D., Huang, C.: Disentangled contrastive collaborative filtering. In: Proceedings of the 46th International ACM SIGIR Conference on Research and Development in Information Retrieval. pp. 1137–1146 (2023)
17. Tian, Y., Sun, C., Poole, B., Krishnan, D., Schmid, C., Isola, P.: What makes for good views for contrastive learning? Adv. Neural. Inf. Process. Syst. **33**, 6827–6839 (2020)
18. Wang, X., He, X., Wang, M., Feng, F., Chua, T.S.: Neural graph collaborative filtering. In: Proceedings of the 42nd international ACM SIGIR conference on Research and development in Information Retrieval. pp. 165–174 (2019)
19. Wang, X., Jin, H., Zhang, A., He, X., Xu, T., Chua, T.S.: Disentangled graph collaborative filtering. In: Proceedings of the 43rd international ACM SIGIR conference on research and development in information retrieval. pp. 1001–1010 (2020)
20. Wang, Y., Wang, X., Huang, X., Yu, Y., Li, H., Zhang, M., Guo, Z., Wu, W.: Intent-aware recommendation via disentangled graph contrastive learning. arXiv preprint arXiv:2403.03714 (2024)
21. Wu, J., Wang, X., Feng, F., He, X., Chen, L., Lian, J., Xie, X.: Self-supervised graph learning for recommendation. In: Proceedings of the 44th international ACM SIGIR conference on research and development in information retrieval. pp. 726–735 (2021)
22. Wu, J., Chen, J., Wu, J., Shi, W., Wang, X., He, X.: Understanding contrastive learning via distributionally robust optimization. Advances in Neural Information Processing Systems **36** (2024)
23. Xia, L., Huang, C., Huang, C., Lin, K., Yu, T., Kao, B.: Automated self-supervised learning for recommendation. In: Proceedings of the ACM Web Conference 2023. pp. 992–1002 (2023)
24. Xia, L., Huang, C., Shi, J., Xu, Y.: Graph-less collaborative filtering. In: Proceedings of the ACM Web Conference 2023. pp. 17–27 (2023)
25. Xia, L., Huang, C., Xu, Y., Zhao, J., Yin, D., Huang, J.: Hypergraph contrastive collaborative filtering. In: Proceedings of the 45th International ACM SIGIR conference on research and development in information retrieval. pp. 70–79 (2022)
26. Yang, Y., Wu, Z., Wu, L., Zhang, K., Hong, R., Zhang, Z., Zhou, J., Wang, M.: Generative-contrastive graph learning for recommendation. In: Proceedings of the 46th International ACM SIGIR Conference on Research and Development in Information Retrieval. pp. 1117–1126 (2023)
27. Ying, R., He, R., Chen, K., Eksombatchai, P., Hamilton, W.L., Leskovec, J.: Graph convolutional neural networks for web-scale recommender systems. In: Proceedings of the 24th ACM SIGKDD international conference on knowledge discovery & data mining. pp. 974–983 (2018)

28. Yu, J., Xia, X., Chen, T., Cui, L., Hung, N.Q.V., Yin, H.: Xsimgcl: Towards extremely simple graph contrastive learning for recommendation. IEEE Transactions on Knowledge and Data Engineering (2023)
29. Yu, J., Yin, H., Xia, X., Chen, T., Cui, L., Nguyen, Q.V.H.: Are graph augmentations necessary? simple graph contrastive learning for recommendation. In: Proceedings of the 45th international ACM SIGIR conference on research and development in information retrieval. pp. 1294–1303 (2022)

Security and Privacy Issues

Enhancing IoT Security: Hybrid Machine Learning Approach for IoT Attack Detection

Alavikunhu Panthakkan[1(✉)], S. M Anzar[2], Dina J. M. Shehada[1], Wathiq Mansoor[1], and Hussain Al Ahmad[1]

[1] College of Engineering and IT, University of Dubai, Dubai, UAE
apanthakkan@ud.ac.ae
[2] Department of Electronics and Communication, TKM College of Engineering, Kollam, India 691005
anzarsm@tkmce.ac.in
https://ud.ac.ae/ , https://www.tkmce.ac.in

Abstract. The rapid proliferation of the Internet of Things (IoT) has revolutionized connectivity across various sectors, but it has also introduced significant security vulnerabilities. As IoT networks expand, they become prime targets for increasingly sophisticated cyberattacks, underscoring the critical need for effective intrusion detection systems (IDS) capable of protecting these environments. In response, this paper introduces a novel hybrid machine learning framework, the *Concatenated Ensemble Model*, which combines the strengths of Decision Trees, Gradient Boosting, and AdaBoost to improve the detection and classification of IoT-based attacks. The model is rigorously evaluated on multiple IoT datasets, with a focus on identifying anomalous activities using performance metrics such as accuracy, precision, recall, F1 score, and Cohen's Kappa. Our results demonstrate significant improvements in detection rates, showcasing the model's robustness and adaptability in diverse IoT environments. This research not only advances the state-of-the-art in IoT security but also provides valuable insights into the practical implementation of ensemble learning techniques for intrusion detection. By enhancing both the precision and resilience of attack detection, the proposed approach contributes to a more secure IoT landscape, addressing the growing challenges of cyber threats in this domain.

Keywords: Cyber-attacks · Attack Surface · Concatenated Ensemble Model · AdaBoost · Fraudulent Activities · IoT Security

1 Introduction

The Internet of Things (IoT) refers to a vast network of interconnected physical devices that communicate and exchange data over the internet [8,17]. With the rapid advancement of technology, IoT applications have expanded into numerous domains, including smart homes, healthcare, transportation, industrial automation, and beyond [2]. This integration of IoT devices has transformed the way we interact with our environment, providing unprecedented levels of convenience, efficiency, and innovation.

However, the rapid proliferation of IoT devices has brought significant concerns regarding the security and privacy of connected systems [7]. As more devices become part of the IoT ecosystem, the risk of cyber-attacks, data breaches, and intrusions grows exponentially. These threats exploit the inherent vulnerabilities of IoT devices, many of which are designed with limited computing power and minimal security features [5]. Common attacks include denial-of-service (DoS), man-in-the-middle, and unauthorized data access, which can compromise the integrity of critical systems.

To address these evolving threats, machine learning-based Intrusion Detection Systems (IDS) have emerged as a promising solution for enhancing the security of IoT networks [18]. Traditional security measures, such as firewalls and encryption, are often inadequate in the IoT context due to the heterogeneous nature of the devices and their large-scale deployment. Machine learning-based IDS can process and analyze large volumes of IoT data in real-time, identifying patterns and anomalies that signal potential security breaches [15]. These systems can detect both known attack signatures and novel, previously unseen threats, making them a powerful tool in safeguarding IoT environments.

Despite the advances in machine learning-based IDS, existing systems face several challenges, including improving detection accuracy, reducing false positives, and efficiently handling the resource constraints of IoT devices. Addressing these issues is critical to ensuring that IoT systems remain secure as cyber-attacks become more sophisticated.

In this context, this paper proposes a novel hybrid machine learning framework called the *Concatenated Ensemble Model*, which combines Decision Trees, Gradient Boosting, and AdaBoost algorithms to enhance the accuracy, efficiency, and resilience of attack detection in IoT networks. Through a comprehensive evaluation using various IoT datasets and performance metrics, we demonstrate the model's capability in identifying and classifying cyber threats, contributing to the ongoing efforts to strengthen IoT security.

The structure of this paper is as follows: Sect. 2 presents a detailed review of the literature, focusing on hybrid machine learning approaches for intrusion detection. Section 3 describes the background work that underpins the development of the proposed model. In Sect. 4, the results and performance analysis are discussed. Finally, Sect. 5 concludes the paper by summarizing the key contributions and findings of this research.

2 Literature Review

The rapid growth of the Internet of Things (IoT) has led to an increased demand for robust intrusion detection systems (IDS) capable of protecting IoT networks from a variety of cyberattacks. Machine learning and hybrid approaches have emerged as prominent solutions for enhancing IoT security. This section reviews key contributions from the literature that have explored hybrid machine learning approaches for IoT intrusion detection.

Sattari et al. [17] proposed a hybrid deep learning model to detect bottlenecks in IoT networks, integrating various neural networks to improve prediction accuracy. This study demonstrated the potential of hybrid models to optimize IoT performance, particularly in detecting bottlenecks, though the focus was not directly on intrusion detection. Emeç and Özcanhan [8] presented a hybrid deep learning model specifically for IoT intrusion detection, leveraging convolutional neural networks (CNNs) and recurrent neural networks (RNNs) to enhance the detection of malicious activities. Their work emphasizes the effectiveness of combining these models for increased accuracy in complex IoT environments.

Bangui and Buhnova [5] proposed a lightweight intrusion detection system (IDS) designed for edge computing networks using deep forest and bio-inspired algorithms. Their approach optimizes the resource efficiency of IDS systems in IoT, making it particularly suitable for edge computing scenarios, though the focus remains on lightweight solutions rather than comprehensive hybrid models. Altunay and Albayrak [2] introduced a hybrid CNN-LSTM (Convolutional Neural Network and Long Short-Term Memory) model for industrial IoT (IIoT) networks. Their approach demonstrated high performance in detecting intrusions, particularly in IIoT environments, by leveraging the strengths of both CNN and LSTM models for time-series data and spatial feature extraction.

Subbarayalu et al. [19] developed a hybrid network intrusion detection system for smart IoT environments. This system combined machine learning algorithms such as Support Vector Machines (SVM) and decision trees to improve intrusion detection rates. The study highlights the utility of hybrid models in environments where both resource efficiency and high detection accuracy are critical. In the context of lightweight IDS solutions, Mendonça et al. [12] explored the application of deep learning algorithms in a lightweight IDS for the Industrial Internet of Things (IIoT). Their findings emphasize the need for efficient IDS models that can operate within the constraints of IIoT devices while maintaining high detection accuracy.

Sarhan et al. [16] proposed a hierarchical blockchain-based federated learning (HBFL) framework for collaborative IoT intrusion detection. By distributing model training across devices without sharing sensitive data, their framework aims to enhance privacy and security in distributed IoT networks, offering a novel approach to federated learning in the context of IoT security. Campos et al. [6] conducted a comprehensive review on the use of federated learning for intrusion detection in IoT networks, identifying key challenges in terms of privacy, communication overhead, and the potential of model poisoning in federated learning environments. Their review highlights the importance of developing secure federated learning frameworks for IoT.

Gassais et al. [9] introduced a multi-level host-based IDS for IoT networks, employing a hierarchical approach to detect intrusions at different layers of the IoT network. This method improves detection by isolating threats across the network stack, but the complexity of implementation could limit its scalability in large IoT networks. Recent advances in hybrid models also include Awotunde et al. [4], who proposed a multi-level random forest model combined with a fuzzy inference system for IoT intrusion detection. Their approach integrates both rule-based and statistical methods to enhance decision-making in detecting intrusions, further demonstrating the versatility of hybrid systems.

The use of recurrent neural networks (RNNs) for anomaly detection in IoT networks was explored by Ullah and Mahmoud [20], who focused on designing an efficient RNN-based model for detecting abnormal patterns in IoT traffic. Their results demonstrated improved accuracy and reduced false positives in anomaly detection. Kumar et al. [11] proposed a distributed ensemble-based IDS using fog computing to protect IoT networks. Their model distributes detection tasks across fog nodes, leveraging ensemble learning to improve detection rates in real-time applications. Similarly, Illy et al. [10] explored the use of ensemble learning techniques in fog-to-things environments, reinforcing the significance of distributed systems in ensuring IoT security.

Other notable contributions include Singh et al. [18], who designed an ensemble model for securing IoT networks using a combination of machine learning techniques, and Om Kumar et al. [14], who developed a recurrent kernel convolutional neural network (RK-CNN) for intrusion detection in IoT networks, showcasing the benefits of hybrid deep learning models.

3 Research Gap and Motivation

Despite advancements in hybrid machine learning models for IoT intrusion detection, key research gaps remain. Many models struggle with scalability and real-time performance in large, distributed IoT environments. Additionally, lightweight models often compromise detection accuracy to meet resource constraints, highlighting the need for solutions that balance efficiency and accuracy.

Federated learning shows promise for privacy-preserving IoT security but presents challenges related to communication overhead and robustness against attacks. Moreover, a comprehensive hybrid solution integrating ensemble learning, deep learning, and rule-based methods into a unified framework is still lacking. Addressing these gaps is critical for developing robust, scalable, and adaptive IoT security systems.

The literature underscores advancements in detection accuracy and efficiency through deep learning, hybrid approaches, and lightweight models. However, challenges in scaling, reducing false positives, and adapting to evolving threats persist. These must be resolved to create more effective IoT intrusion detection systems.

In response, our research proposes a hybrid machine learning framework that integrates Decision Trees, Gradient Boosting, and AdaBoost to improve detection accuracy, reduce false positives, and ensure real-time processing. This framework addresses issues of scalability, resource constraints, and adaptability, offering a promising solution for IoT security against current and emerging threats.

4 Background Work

In this section, we review the key machine learning techniques employed in the proposed model, specifically focusing on Decision Trees, Gradient Boosting, and AdaBoost. These algorithms form the foundation of many ensemble learning models due to their individual strengths in handling complex classification tasks. By understanding the underlying principles of these methods, we can appreciate how they complement each other to improve classification accuracy, resilience, and interpretability in Intrusion Detection Systems (IDS) for IoT networks. Below, we explore each of these techniques in detail, discussing their respective advantages and applications.

4.1 Decision Tree

Decision trees are robust, interpretable machine learning models commonly used for classification. They recursively partition the feature space based on input variables, making them effective for identifying decision boundaries in complex datasets. A major advantage of decision trees is their inherent interpretability, which allows practitioners to understand and visualize the decision-making process–particularly valuable in fields like healthcare and finance.

Decision trees are versatile, handling both numerical and categorical data, and are capable of modeling non-linear relationships and feature interactions. They also provide insights into feature importance, aiding in feature selection and dimensionality reduction in high-dimensional datasets. Additionally, they are resilient to outliers and can handle missing data with minimal preprocessing, making them practical for real-world applications.

4.2 Gradient Boosting

Gradient Boosting is a powerful ensemble learning technique, particularly effective for binary classification. It builds an ensemble of weak learners, typically decision trees, with each model correcting the errors of the previous one. XGBoost, a popular implementation, efficiently handles large datasets and consistently delivers high predictive performance, making it widely used in competitive machine learning.

One of Gradient Boosting's strengths is its ability to capture complex relationships in data, refining predictions iteratively to improve accuracy. It is also highly effective for imbalanced datasets, assigning higher weights to misclassified instances and improving the identification of minority classes. Additionally, Gradient Boosting models offer interpretability through feature importance analysis, which helps domain experts validate models and understand key variables driving predictions.

4.3 AdaBoost

AdaBoost, short for Adaptive Boosting, is an effective ensemble learning technique for binary classification. It combines the outputs of multiple weak learners–often simple decision trees–into a single, stronger predictive model by assigning higher weights to misclassified instances. This ensures that each subsequent model focuses on the most challenging cases, progressively improving performance.

A key strength of AdaBoost is its ability to enhance weak classifiers, making them more effective at handling complex relationships in the data. The method is also resistant to overfitting, as it places less emphasis on well-classified examples, improving its ability to generalize to new, unseen data–an essential trait in real-world applications. AdaBoost's adaptability and strong generalization have proven effective across various domains, including face detection, object recognition, and medical diagnosis, making it a popular choice for practitioners addressing binary classification tasks.

5 Methodology

This paper presents a novel approach for IoT attack classification through a concatenated machine learning model. This methodology enhances overall performance by leveraging the predictions from multiple base models and combining them using a meta-model. The stacking technique used in this method capitalizes on the strengths of diverse models such as Decision Trees, Gradient Boosting, and AdaBoost, improving the system's robustness and accuracy for IoT attack detection. This ensemble learning method combines the predictions of multiple base models to improve the system's robustness and accuracy, ultimately leading to better overall performance.

5.1 IoT Attack Classification Framework

The proposed concatenated model for IoT attack classification follows a two-level stacking architecture, as illustrated in Fig. 1. The methodology consists of two primary stages:

- **First-Level Models:** The IoT attacks dataset is preprocessed and provided as input to the base models, which include:
  - *Decision Tree (DT)*: A simple, interpretable model that captures decision boundaries effectively.
  - *Gradient Boosting (GB)*: A robust ensemble technique that builds trees sequentially, correcting errors from previous iterations.
  - *AdaBoost (AB)*: Another ensemble learning method that iteratively boosts weak learners by focusing on previously misclassified instances.

  Each model is trained separately on the IoT dataset. After training, each base model generates predictions on the test set.

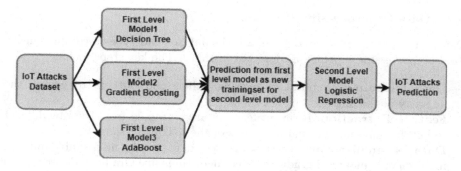

Fig. 1. Proposed Concatenated ML Model

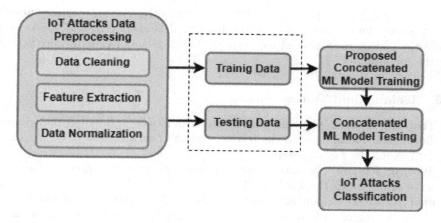

Fig. 2. Schematic diagram of proposed ML Model

- **Concatenation and Meta-Model (Second-Level Model):** The predictions from the first-level models (Decision Tree, Gradient Boosting, and AdaBoost) are concatenated into a new training set, which is then passed to a second-level meta-model. In this case, the *Logistic Regression* model is employed as the meta-classifier. The meta-model is responsible for learning how to combine the predictions from the first-level models to make a final classification decision. This stacked approach helps to capture complex patterns and dependencies that individual models might miss, leading to improved prediction performance.

Figure 1 shows the overall stacking process, with predictions from each base model forming a new dataset that serves as the input for the second-level Logistic Regression model. This method effectively consolidates the strengths of each base model and mitigates their individual weaknesses, thus leading to a more robust classification system.

5.2 Data Preprocessing Pipeline

Before training the models, the IoT attacks dataset undergoes a comprehensive data preprocessing pipeline, as depicted in Fig. 2. The key steps include:

- **Data Cleaning:** Removal of missing values and outliers to ensure the integrity of the dataset.
- **Feature Extraction:** Relevant features are extracted from the raw dataset to better represent the characteristics of the IoT attacks.
- **Data Normalization:** Features are scaled to ensure uniformity, improving the performance of gradient-based algorithms like Gradient Boosting and Logistic Regression.

Once the dataset is preprocessed, it is split into training and testing subsets. The training data is used to train the base models (DT, GB, and AdaBoost), while the testing data is used to evaluate the models and generate predictions for the meta-model.

5.3 Training and Testing Phases

- **Training Phase:** The base models are individually trained on the training data. Their predictions on the testing data are concatenated to create a new dataset for the meta-model.
- **Testing Phase:** The concatenated predictions from the first-level models serve as input to the Logistic Regression meta-model, which produces the final IoT attack classification. The meta-model is responsible for making the ultimate prediction based on the combined strengths of the first-level classifiers.

5.4 IoT Attack Classification Using the Stacking Model

The concatenated machine learning model aims to provide high accuracy and robust detection of IoT attacks. By stacking diverse models, the system compensates for the weaknesses of individual models and provides more reliable predictions. The final classification is based on the output of the Logistic Regression meta-model, which learns from the collective predictions of the Decision Tree, Gradient Boosting, and AdaBoost models. The proposed methodology leverages the ensemble learning strategy of stacking to improve IoT attack detection. By integrating various base models into a comprehensive framework, the proposed concatenated model outperforms individual classifiers, ensuring greater robustness and precision in IoT security applications.

5.5 Model Parameters Employed

In the proposed concatenated machine learning model, we configured the following key parameters: Decision Trees with a maximum depth of 5, a minimum samples split of 2, a minimum samples leaf of 1, and 'gini' as the criterion; Gradient Boosting with 100 estimators, a learning rate of 0.1, a maximum depth of 3, and a subsample ratio of 1.0; and AdaBoost with 50 estimators, a learning rate of 1.0, and a base estimator set as a DecisionTreeClassifier with a maximum depth of 1. These parameter choices offer a balanced foundation for the model, providing an optimal trade-off between complexity and performance, and can be further fine-tuned to accommodate specific dataset characteristics and requirements.

6 RESULTS AND ANALYSIS

6.1 Experimental Set-up

Our novel IoT Attack Detection System leverages the cloud infrastructure of Google Colab to facilitate the efficient development and execution of Python-based concatenated machine learning models. By utilizing Colab's access to high-performance GPUs and TPUs, seamless integration with Google Drive for data management, and pre-installed libraries, we streamline the implementation process. The availability of these advanced computing resources significantly reduces model training time, making this platform highly suitable for large-scale IoT attack detection applications involving complex models and extensive datasets. Specifically, we utilized NVIDIA T4 GPUs, known for their energy-efficient design and AI-optimized performance based on the Turing architecture, to accelerate the training process on the ToN-IoT dataset.

6.2 ToN-IoT Dataset

The ToN-IoT dataset is a valuable asset for cybersecurity researchers aiming to strengthen the security of Internet of Things (IoT) networks. It contains real-world data meticulously collected from various IoT devices and networks, providing a comprehensive view of both normal and malicious activities. In addition to network traffic, the dataset includes sensor data, operating system logs, and network details, giving a holistic perspective of IoT ecosystem behavior.

The dataset covers a wide range of cyber threats, from basic Denial of Service (DoS) attacks to more sophisticated exploits and botnet operations, reflecting a realistic environment within a simulated Industrial Internet of Things (IIoT) network. Its balanced distribution of normal and attack samples ensures an ideal foundation for developing intrusion detection systems capable of accurately distinguishing between legitimate and malicious activities.

Designed to support research in intrusion detection and prevention systems, the ToN-IoT dataset offers rich insights into evolving IoT threats, fostering the advancement of AI-powered cybersecurity solutions. It consists of network traffic captures with 43 distinct features categorized into flow, basic, time, and additional attributes. Each flow is clearly labeled as either normal or one of ten different attack types, providing a structured format for model training and evaluation.

With a near-equal distribution of normal (54.54%) and attack (45.45%) samples, the dataset mirrors realistic traffic patterns from actual IoT networks, further enhancing its applicability to real-world cybersecurity challenges. The ToN-IoT dataset plays a crucial role in facilitating research on IoT intrusion detection and attack mitigation, contributing significantly to the ongoing efforts to improve IoT security.

6.3 Performance Evaluation Metrics

Machine learning models are typically evaluated using various metrics to quantify their performance across different aspects [3]. One common method of assessment is the confusion matrix, which compares the predicted classes with the actual classes, offering a detailed view of the model's classification outcomes [13]. The confusion matrix includes four key terms: True Positives (TP), True Negatives (TN), False Positives (FP), and False Negatives (FN).

In this study, we employed the confusion matrix alongside several evaluation metrics–accuracy, precision, recall, F1-score, and Cohen's kappa–to gain comprehensive insights into the model's performance.

Accuracy. measures the overall correctness of a model by calculating the ratio of correctly predicted instances (both positive and negative) to the total number of instances in the dataset [1]. While accuracy offers a quick and intuitive overview of the model's performance, it can be misleading for imbalanced datasets. In such cases, the model may appear to perform well by correctly predicting the majority class, even though its performance on the minority class may be poor.

$$Accuracy = \frac{(TP + TN)}{(TP + TN + FP + FN)} \quad (1)$$

Precision. is a key metric in binary classification, measuring the accuracy of positive predictions. It is calculated as the ratio of true positives to the sum of true positives and false positives. Precision highlights the proportion of correctly identified positive instances out of all instances predicted as positive. A high precision value indicates that the model makes accurate positive predictions with few false positives, which is particularly important in scenarios where incorrect positive predictions can lead to significant consequences, such as in medical diagnoses.

$$Precision = \frac{TP}{(TP + FP)} \quad (2)$$

Recall. is a binary classification metric that measures a model's ability to correctly identify all relevant positive instances. It is calculated as the ratio of true positives to the sum of true positives and false negatives. A high recall indicates that the model successfully captures a large proportion of actual positive instances, minimizing false negatives. This metric is especially important in situations where failing to detect positive cases can have serious consequences, such as in fraud detection or medical screening.

$$Recall = \frac{TP}{(TP + FN)} \quad (3)$$

F1 Score. is a commonly used metric in binary classification that provides a balanced assessment by combining both precision and recall into a single measure. It is particularly useful for evaluating models on imbalanced datasets. The F1 Score calculates the harmonic mean of precision and recall, offering a comprehensive view of the model's ability to make accurate positive predictions while also capturing all relevant positive instances. The F1 Score ranges from 0 to 1, with a value of 1 representing perfect performance.

$$F1Score = \frac{2 * (Precision * Recall)}{(Precision + Recall)} \quad (4)$$

Cohen's Kappa (κ). is a statistical metric used to assess the level of agreement between predicted and actual classifications in binary classification tasks, beyond what would be expected by chance. It accounts for the possibility of random agreement, providing a more robust measure than simple accuracy, especially in imbalanced datasets. The value of Cohen's Kappa ranges from -1 to 1, with 1 representing perfect agreement, 0 indicating agreement no better than chance, and negative values suggesting disagreement.

The formula for Cohen's Kappa is given by:

$$\kappa = \frac{P_o - P_e}{1 - P_e} \quad (5)$$

where P_o is the observed agreement between the predicted and actual classifications. P_e is the expected agreement due to chance.

6.4 Results Interpretation and Discussion

The confusion matrices presented in Figs. 3 provide a detailed evaluation of the classification performance for different machine learning models: Decision Tree, Gradient Boosting, AdaBoost, and the proposed Concatenated ML model. These confusion matrices offer insight into how well each model distinguishes between normal and attack traffic in IoT networks by showcasing the distribution of true positives (correctly predicted attacks), true negatives (correctly predicted normal traffic), false positives (normal traffic misclassified as attacks), and false negatives (attacks misclassified as normal traffic).

In Fig. 3(a), the confusion matrix for the Decision Tree classifier shows that 17,109 normal instances are correctly classified, but 1,566 normal instances are incorrectly identified as attacks. For attack detection, 31,330 instances are correctly identified as attacks, while 1,530 attack instances are misclassified as normal traffic. This shows the Decision Tree performs reasonably well but has some limitations in correctly classifying attacks and normal traffic.

Figure 3(b) illustrates the confusion matrix for the Gradient Boosting classifier. Here, 16,946 normal instances are accurately classified, but 1,729 instances are misclassified as attacks. For attacks, 31,266 instances are correctly identified, while 1,594 attack instances are misclassified as normal traffic. The Gradient

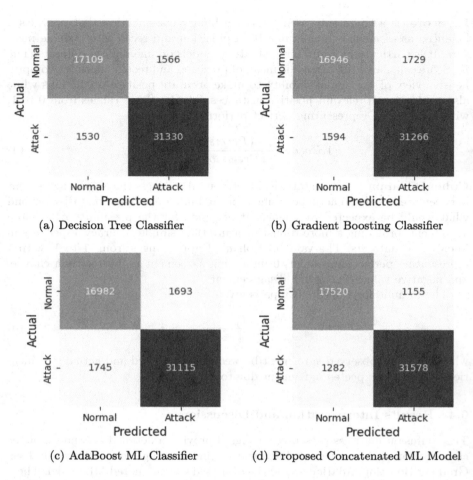

Fig. 3. Confusion Matrices Comparing the Performance of Different Models

Boosting model performs similarly to the Decision Tree but shows slightly more misclassifications of normal traffic.

Figure 3(c), the AdaBoost classifier confusion matrix reveals that 16,982 normal instances are correctly classified, with 1,693 misclassified as attacks. For attack detection, 31,115 instances are correctly identified, while 1,745 are misclassified as normal traffic. The AdaBoost model has higher false-negative rates compared to the previous models, indicating more misclassification of attacks as normal.

Figure 3(d) presents the confusion matrix for the proposed Concatenated ML model. This model demonstrates a marked improvement, with 17,520 normal instances correctly classified and only 1,155 misclassified as attacks. For attack detection, 31,578 instances are correctly identified, and only 1,282 are misclassified as normal traffic. The Concatenated ML model shows the fewest misclassifications overall, indicating its superior performance in distinguishing between normal and attack traffic. The performance metrics summarized in Table 1 further confirm the effectiveness of the Concatenated ML model. It achieves the highest accuracy of 95.28%, surpassing the Decision Tree (93.99%), Gradient Boosting (93.54%), and AdaBoost (93.33%). Precision, recall, and F1-Score metrics consistently reflect the Concatenated ML model's enhanced ability to correctly classify both normal and attack instances with minimal misclassification. With a precision and recall of 95.27%, the model minimizes false positives and false negatives effectively, leading to a balanced and highly accurate classification system.

Table 1. Performance Evaluation of ML Classifiers

Classifier	Accuracy	Precision	Recall	F1-Score	Cohen's Kappa
Decision Tree	93.99	93.99	93.99	93.99	86.99
Gradient Boosting	93.54	93.55	93.55	93.55	86.02
AdaBoost	93.33	93.33	93.33	93.33	85.57
Concatenated ML	95.28	95.27	95.27	95.27	89.78

The *Cohen's Kappa* statistic provides an additional measure of the model's performance, taking into account the agreement between predicted and actual classifications. The Concatenated ML model achieves the highest Cohen's Kappa score of 89.78, indicating substantial agreement and strong predictive reliability. In contrast, the Decision Tree, Gradient Boosting, and AdaBoost models exhibit lower Kappa values, reinforcing the superior performance of the Concatenated ML model.

In conclusion, the confusion matrices and performance metrics indicate that the proposed Concatenated ML model outperforms the individual classifiers in terms of accuracy, precision, recall, and overall classification reliability. By leveraging the strengths of multiple base models through stacking, the Concatenated ML model enhances the detection of IoT attacks while minimizing misclassification errors, making it a robust solution for real-time IoT security applications.

7 Conclusion

This paper presents a substantial contribution to the field of IoT intrusion detection by thoroughly evaluating the proposed concatenated machine learning model. By leveraging the strengths of multiple base classifiers–Decision Tree, Gradient Boosting, and AdaBoost–combined in a stacked ensemble with a Logistic Regression meta-model, we achieved enhanced classification performance. The empirical results demonstrate that the concatenated model significantly outperforms individual models in terms of accuracy, precision, recall, F1-Score, and Cohen's Kappa, with the latter showing a strong agreement between predicted and actual classifications, surpassing simple accuracy metrics.

Our proposed model effectively mitigates the risk of overfitting, as the diverse strengths of each base model counterbalance the limitations of the others. The confusion matrices and performance metrics confirm the model's superior ability to accurately classify both normal and attack instances, minimizing false positives and false negatives. With an accuracy of 95.28% and Cohen's Kappa of 89.78, the concatenated model provides a robust and reliable solution for IoT intrusion detection, offering balanced precision and recall that is critical for high-stakes applications.

These findings underscore the potential of the concatenated approach as a versatile and efficient tool for binary classification tasks in IoT security. This work paves the way for the development of more advanced and effective intrusion detection systems, tailored specifically to the evolving threats in IoT environments.

References

1. Abhiram, A., Anzar, S., Panthakkan, A.: Deepskinnet: a deep learning model for skin cancer detection. In: 2022 5th International Conference on Signal Processing and Information Security (ICSPIS). pp. 97–102. IEEE (2022)
2. Altunay, H.C., Albayrak, Z.: A hybrid cnn+ lstm-based intrusion detection system for industrial iot networks. Engineering Science and Technology, an International Journal **38**, 101322 (2023)
3. Anzar, S., Sathidevi, P.: An efficient pso optimized integration weight estimation using d-prime statistics for a multibiometric system. International Journal on Bioinformatics and Biosciences **2**(3), 1504–1511 (2012)
4. Awotunde, J.B., Ayo, F.E., Panigrahi, R., Garg, A., Bhoi, A.K., Barsocchi, P.: A multi-level random forest model-based intrusion detection using fuzzy inference system for internet of things networks. International Journal of Computational Intelligence Systems **16**(1), 31 (2023)
5. Bangui, H., Buhnova, B.: Lightweight intrusion detection for edge computing networks using deep forest and bio-inspired algorithms. Comput. Electr. Eng. **100**, 107901 (2022)
6. Campos, E.M., Saura, P.F., González-Vidal, A., Hernández-Ramos, J.L., Bernabe, J.B., Baldini, G., Skarmeta, A.: Evaluating federated learning for intrusion detection in internet of things: Review and challenges. Comput. Netw. **203**, 108661 (2022)
7. Choukri, W., Lamaazi, H., Benamar, N.: Abnormal network traffic detection using deep learning models in iot environment. In: 2021 3rd IEEE Middle East and North Africa COMMunications Conference (MENACOMM). pp. 98–103. IEEE (2021)
8. Emeç, M., Özcanhan, M.H.: A hybrid deep learning approach for intrusion detection in iot networks. Advances in Electrical and Computer Engineering **22**(1), 3–12 (2022)
9. Gassais, R., Ezzati-Jivan, N., Fernandez, J.M., Aloise, D., Dagenais, M.R.: Multi-level host-based intrusion detection system for internet of things. Journal of Cloud Computing **9**(1), 62 (2020)

10. Illy, P., Kaddoum, G., Moreira, C.M., Kaur, K., Garg, S.: Securing fog-to-things environment using intrusion detection system based on ensemble learning. In: 2019 IEEE wireless communications and networking conference (WCNC). pp. 1–7. IEEE (2019)
11. Kumar, P., Gupta, G.P., Tripathi, R.: A distributed ensemble design based intrusion detection system using fog computing to protect the internet of things networks. J. Ambient. Intell. Humaniz. Comput. **12**(10), 9555–9572 (2021)
12. Mendonça, R.V., Silva, J.C., Rosa, R.L., Saadi, M., Rodriguez, D.Z., Farouk, A.: A lightweight intelligent intrusion detection system for industrial internet of things using deep learning algorithms. Expert. Syst. **39**(5), e12917 (2022)
13. Mohammed Anzar, S.T., Sathidevi, P.S.: On combining multi-normalization and ancillary measures for the optimal score level fusion of fingerprint and voice biometrics. EURASIP Journal on Advances in Signal Processing **2014**, 1–17 (2014)
14. Om Kumar, C., Marappan, S., Murugeshan, B., Beaulah, P.M.R.: Intrusion detection model for iot using recurrent kernel convolutional neural network. Wireless Personal Communications **129**(2), 783–812 (2023)
15. Panthakkan, A., Anzar, S., Mansoor, W.: Enhancing iot security: A machine learning approach to intrusion detection system evaluation. In: 2023 IEEE International Conference and Expo on Real Time Communications at IIT (RTC). pp. 19–23. IEEE (2023)
16. Sarhan, M., Lo, W.W., Layeghy, S., Portmann, M.: Hbfl: A hierarchical blockchain-based federated learning framework for collaborative iot intrusion detection. Comput. Electr. Eng. **103**, 108379 (2022)
17. Sattari, F., Farooqi, A.H., Qadir, Z., Raza, B., Nazari, H., Almutiry, M.: A hybrid deep learning approach for bottleneck detection in iot. IEEE Access **10**, 77039–77053 (2022)
18. Singh, R., Sharma, K.P., Awasthi, L.K.: A machine learning-based ensemble model for securing the iot network. Cluster Computing pp. 1–15 (2024)
19. Subbarayalu, V., Surendiran, B., Arun Raj Kumar, P.: Hybrid network intrusion detection system for smart environments based on internet of things. The Computer Journal **62**(12), 1822–1839 (2019)
20. Ullah, I., Mahmoud, Q.H.: Design and development of rnn anomaly detection model for iot networks. IEEE Access **10**, 62722–62750 (2022)

Boosting Adversarial Transferability by Uniform Scale and Mix Mask Method

Tao Wang, Qianmu Li[✉], Zhichao Lian, Zijian Ying, Fan Liu, and Shunmei Meng

Nanjing University of Science and Technology, Nanjing, China
qianmu@njust.edu.cn

Abstract. Enhancing transferability of adversarial examples generated from surrogate models has garnered significant attention in recent research. This property allows these examples to deceive other black-box models. Input augmentation has emerged as a promising approach to improve adversarial transferability, focusing on two aspects: incorporating additional data and transforming the input data itself. However, the first category of approaches, which linearly mix external information, suffers from reduced effectiveness in blending features. On the other hand, the second category of approaches often encounters the issue of gradient redundancy, as they generate very similar inputs across multiple scales. To overcome these challenges, we introduce a novel framework called Uniform Scale and Mix Mask Method (US-MM) for generating adversarial examples. The Mix Mask method refines external information into masks, enabling a nonlinear mixing process. Meanwhile, the Uniform Scale approach explores the boundaries of perturbation using a linear factor, effectively minimizing the negative impact of scale copies. Empirical evaluations on ImageNet ILSVRC 2012 validation set demonstrate that US-MM achieves obviously better transfer attack success rate compared to state-of-the-art methods. Ablation experiments are also conducted to validate the effectiveness of each component in US-MM.

Keywords: Adversarial Attack · Transfer Attack

1 Introduction

Deep neural networks (DNNs) have demonstrated remarkable success in computer vision tasks. However, recent research has revealed that adding imperceptible perturbations to images can deceive these models [5], posing serious security concerns. For example, such attacks can compromise personal safety in autonomous driving or cause issues with facial recognition systems. One concerning phenomenon is that adversarial examples crafted to deceive one model can also fool other models, even if they have different architectures and parameters. This transferability of adversarial examples has garnered widespread attention. Attackers can exploit this property without having detailed knowledge of the target model, making it a black-box attack. While these adversarial example generation methods were initially proposed for white-box attacks where the attacker has complete knowledge of the target model, they can still suffer from overfitting issues with the source model. This can result in weak transferability when these adversarial examples are used to attack other black-box models.

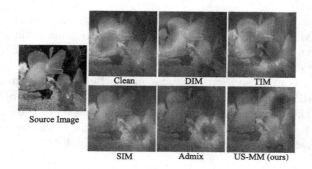

Fig. 1. A collection of heatmaps about the source image and adversarial examples crated by DIM [30], TIM [2], SIM [12], Admix [26] and our proposed US-MM. These methods are single-transformation-based attack methods. The red regions are of significance in model prediction. (Color figure online)

For boosting the transferability of adversarial examples in black-box setting, various approaches have been proposed. For example, the methods based on input transformation try to introduce transformed images [2,12,30], e.g. scale transformation, to craft adversarial examples. Current works indicate that adding additional information from those transformed images can enhance the adversarial transferability [16,33]. These methods achieved leadership in computational performance and attack effectiveness.

However, in the current transformation-based attacks, there is a lack of focus on the importance of transformation factors. For instance, the mixup strategy, which is used in some attack methods, introduces information from images of other classes is to further improve transferability. However, this mixup strategy is limited by its linear approach, which weakly adds information without adaptation. Additionally, this linear mix can also damage some pixel regions in the source image, leading to limited transferability of the adversarial examples.

In the other hand, transforming the input data itself by scaling is proved to play a role in enhancing the transferability of adversarial examples. However, due to the exponential scale strategy which is often used in current methods, this approach may lead to a problem of gradient redundancy when the scale copies become more. As a result, it is need to choose a careful scale way to achieve optimal performance.

To address these issues, we propose a novel and flexible attack method called the Uniform Scale and Mix Mask Method (US-MM). The MM component improves upon the mixup strategy by using mix masks for multiplication instead of addition. The US component overcomes the limitations of exponential scaling by uniformly sampling scale values within an interval. Each component can directly improve the transferability of adversarial examples, and the combination of the two can further enhance performance. The visualization results about several single-transformation-based attack methods are shown in Fig. 1. Our contributions are summarized as follows:

- We propose a novel non-linear mixup strategy, namely Mix Mask Method(MM), to overcome the issue of damaged regions in the source image.
- We propose a novel method, namely Uniform Scale Method (US), to eliminate the adverse effects from quantities of scale copies.
- We conduct comparison experiments on the benchmark dataset ImageNet ILSVRC 2012 validation set, showing that the proposed US-MM method achieves higher transfer attack success rates compared to the baselines.
- We conduct ablation experiments to validate that our methods can improve the transferability of adversarial examples individually.

2 Related Work

Attack methods can be categorized into two types based on the attacker's knowledge of the victim model: white-box attacks [17] and black-box attacks. In white-box attacks, the attacker has access to all information about the target model, such as its architecture and parameters, to generate adversarial examples. In contrast, black-box attacks only allow attackers to have query permissions.

2.1 White-Box Adversarial Attack

Fast Gradient Sign Method (FGSM) [5] sets the optimization objective to maximize the loss of classification function and makes the sign of input gradient as noise for benign image. Basic Iterative Method (BIM) [10] extends the idea of FGSM by applying multiple iterations of small perturbations to achieve better white-box attack performance.

Although these methods can achieve nearly 100% success rates in white-box attack setting, when tested on other black-box models, the adversarial examples show weak transferability due to overfitting with the source model.

2.2 Black-Box Adversarial Attack

It is challenging in black-box attack scenarios because attackers are absolutely ignorant of victim model but only obtain model output. There are two sorts of black-box attack algorithms. One is query-based attacks [19,21] while the other is transfer-based attacks. Query-based attacks design query samples purposefully and optimize adversarial noise based on query results. However, it is impractical in physical world because of the huge amount of query operations.

Instead, based on the phenomenon that adversarial examples generated for one model might mislead another model, transfer-based attacks works by attacking a local surrogate model. To enhance transferability, existing transfer-based attacks usually utilize several avenues to craft adversarial examples.

Optimizer-based Attacks. Dong et al. [1] use momentum to help escape from poor local minima in multiple iterations, denoted as MI-FGSM. Wang et al. [27] estimate momentum on several samples which crafted on previous gradient's direction repeatedly. Wang et al. [25] utilize the average value of gradient difference between source image and surrounding samples to swap adversarial examples, achieving higher transferability.

Mid-layer-based Attacks. Huang et al. [9] shift the adversarial noise to enlarge the distance of specific layer in DNNs between the benign image and adversarial examples. Ganeshan et al. [4] design a novel loss function which reduces the activation of supporting current class prediction and enhances the activation of assisting other class prediction.

Ensemble-model Attacks. Li et al. [11] introduce dropout layers into source model and acquire ghost networks by setting different parameters. In each iteration, the surrogate model is selected randomly from network collection, known as longitudinal ensemble. Xiong et al. [31] tune the ensemble gradient in order to reduce the variance of ensemble gradient for each single gradient.

Input Transformation Based Attacks. Xie et al. [30] propose the first attack method based on input transformation. They resize the input image randomly and expand it to a fixed size by filling pixels. Dong et al. [1] use a set of translated images to optimize adversarial examples. To reduce computation complexity, they apply convolution kernel to convolve the gradient. Lin et al. [12] assume the scale-invariant property of DNNs and propose an attack method working by calculating the gradient by several scaled copies, denoted as SIM. Wang et al. [26] observe that introducing the information of images in other categories during generating examples can improve the transferability significantly. They propose an Admix method which mixes source image and the images with different labels. Wang et al. [28] divide image into several regions and apply various transformations onto the blocks while retaining the structure of image.

3 Preliminary

In this section, we will first define the notations used for generating adversarial examples. Then, we will discuss the limitations of the current transform strategies and state the motivation behind our work.

3.1 Problem Settings

Adversarial attack tries to create an adversarial example x^{adv} from a benign image x. The victim model, typically a deep learning model, is denoted as f, with parameters θ. The output of the victim model is $f(x,\theta) \in R^K$, where K is the number of classes. The true label of x is represented as y, and the loss function of f is denoted as $J(f(x;\theta),y)$. The objective of the attack method is to generate an adversarial example within a specified constraint, such that the victim model misclassifies it. This can be achieved by generating x^{adv}, which satisfies $\|x^{adv} - x\|_p < \epsilon$, and results in $f(x^{adv};\theta) \neq f(x;\theta)$. The most common constraint used in adversarial attacks is the L_∞ norm.

3.2 Motivation

The mixup strategy is used to enhance adversarial transferability by introducing features from other classes. For example, Admix [26] uses a set of admixed images to calculate the average gradient. It first randomly choose several images from other categories. Then, for every sampled image x', Admix gets the mixed image $\tilde{x}$ by adopting a linear mix method on original image x as follows:

$$\tilde{x} = \gamma \cdot x + \eta' \cdot x' = \gamma \cdot (x + \eta \cdot x'), \tag{1}$$

where γ and η' are the mix portions of the original image and sampled image in the admixed image respectively satisfying $0 \leq \eta' < \gamma \leq 1$ and η is computed by $\eta = \eta'/\gamma$.

However, there are two problems with this mixup way to introduce external information. The first problem is that for a random pixel $P_{x'}$ in x', even if η decreases its value, it is still unpredictable whether the pixel value will be greater than the corresponding pixel P_x in x. This means that the condition $P_x < \eta \cdot P_{x'}$ will always be true, unless η is very close to 0. However, mixed image will lose efficacy when η is too small. As a result, the mixed image x_{mixed} will have a larger portion from the image of another category in some pixel positions, disrupting the feature information of the original image x.

The second issue is that the pixel values in x' are always positive, which only increases input diversity in the positive direction. This results in a limited range of mixing options. It may be more effective to mix the image in a negative direction as well.

At the same time, a kind of exponential scale strategy, which is used in the Scale-Invariant Method (SIM) [12], utilizes the gradient from multiple scaled images to generate adversarial examples with more adversarial transferability. The core of exponential scale strategy is using scaling transformation $S(x)$ to modify the input image, which is represented as follows:

$$S_i(x) = x/2^i, \tag{2}$$

where i is the indicator of scale copies.

SIM set the number of scale copies m to limit the number of scale copies, with a higher m representing more general feature information. However, when m is greater than a certain degree, performance of SIM starts to decline, as shown in Fig. 2. This is because the scaled images tend to be similar when i is a large integer. These analogous inputs swap nearly same gradients, which

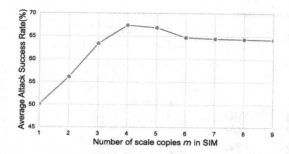

Fig. 2. Average attack success rates (%) of SIM when attacking five pretrained models. The examples are crafted on Inc-v3 model.

occupy a large proportion of the final gradient, leading to a problem of gradient redundancy. Therefore, we believe that using the gradient of such scaled images to generate adversarial noise can reduce transferability, and sampling images uniformly is necessary.

4 Methodology

In this section, we present our proposed methods, the Mix Mask Method and Uniform Scale Method, and provide an algorithm for a better understanding of the proposed approach.

4.1 Mix Mask Method

To address the limitations of linear mixup, we propose the Mix Mask Method (MM), which works by generating a mix mask from an image of a different category and applying it to the input image. This method has two main improvements: first, the transformation range is related to the per-pixel value of the source image, and second, the transformation contains both positive and negative directions. In the first step of MM, a mix mask is generated according to the mix image using the following equation:

$$M_{mix} = (1-r) \cdot \mathbf{1} + 2r \cdot x', \qquad (3)$$

where M_{mix} is the mask, r is the mix range size, x' is the mix image and $\mathbf{1}$ is an all one matrix same shape with x'. Because the images are normalized to [0, 1], the value of each element in M_{mix} is mapped into [1−r, 1+r].

Then, mask M_{mix}, which contains the information of the mix image, can be utilized to influence the source image x, which is

$$x^m = M_{mix} \odot x, \qquad (4)$$

where x^m is the mixed image generated by the source image and the mask, and $\odot$ is element-wise product.

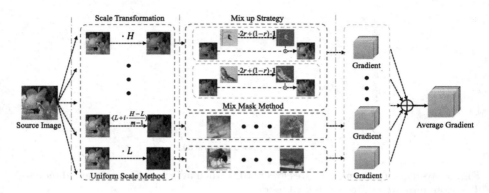

Fig. 3. Illustration of Uniform Scale and Mix Mask Method (US-MM).

In MM, the transformation range of per pixel in source image is limited within a symmetric interval by applying mix mask, which means a kind of reliable and bidirectional transformation measure. Then MM can introduce features from other categories of images more effectively than linear ways.

4.2 Uniform Scale Method

The most straightforward solution to address the gradient redundancy with exponential scale strategy is to utilize the uniform scale with a convert function $U_i(x, m_{us})$ to generate scale copies, which is

$$U_i(x, m_{us}) = \frac{i}{m_{us} - 1} \cdot x, \tag{5}$$

where x is the input image and m_{us} is a positive integer to present the number of uniform scale copies.

However, some problems may arise when the value of i is in two extreme cases. When $i = 0$, the scaled input is changed to a black image, generating the meaningless gradient information. When $i = m_{us} - 1$, the transformed image is the same as the input, having no contribution to increasing input diversity. To overcome these problems, we define a lower bound L to avoid generating black image and a upper bound H to increase input diversity further.

Thus, the final form of the Uniform Scale Method (USM) is as follows:

$$U_i(x, m_{us}, L, H) = (L + i * \frac{H - L}{m_{us} - 1}) \cdot x, \tag{6}$$

where x is the input image and m_{us} is a positive integer to present the number of uniform scale copies. Particularly, we stipulate that $U_i(x, m_{us}, L, H) = H \cdot x$ when $m_{us} = 1$.

Algorithm 1. Uniform Scale and Mix Mask Method

Input: A classifier f with parameter θ and loss function J
Input: A benign image x with ground-truth label y
Input: The maximum perturbation ϵ and number of iterations T.
Input: Number of uniform scale copies m_{us}, scale lower bound L, scale upper bound H,
Input: Number of mix images m_{mix}, mix range size r
Output: An adversarial example x^{adv}

1: $\alpha = \frac{\epsilon}{T}; x_0^{adv} = x$
2: **for** $t = 0$ to $T - 1$ **do**
3:    $G = 0$
4:    **for** $i = 0$ to $m_{us} - 1$ **do**
5:       $x_i^{scaled} = U_i(x_t^{adv}, m_{us}, L, H)$
6:       **for** $j = 0$ to $m_{mix} - 1$ **do**
7:          Get a mix mask M_{mix} by Eq.(3)
8:          $x_{i,j}^m = M_{mix} \odot x_i^{scaled}$
9:          $x_{i,j}^m = Clip(x_{i,j}^m, 0, 1)$
10:         $G = G + \nabla_{x_{i,j}^m} J(f(x_{i,j}^m; \theta), y)$
11:       **end for**
12:    **end for**
13:    $x_{t+1}^{adv} = x_t^{adv} + \alpha * sign(G)$
14: **end for**
15: **return** $x^{adv} = x_T^{adv}$

4.3 Algorithm of US-MM Method

US-MM method contains two parts, scale transformation and mix up strategy. The structure of our US-MM method is exhibited in Fig. 3. The pseudo-code of the process of Uniform Scale and Mix Mask Method is summarized in Algorithm 1. Note that our US component can replace traditional exponential scale strategy in any appropriate situation and it is easy to integrate our MM component into other transfer-based attacks.

5 Experiments

In this section, we conduct experiments to verify the effectiveness of our proposed approach. We first specify the setup of the experiments. Then, we report the results about attacking several pretrained models and defense models with baseline methods and US-MM. Finally, we do ablation study to explore the role of hyper-parameters. We also display the effectiveness of our two methods.

5.1 Experiment Setup

Dataset. We evaluate all methods on 1000 images from ILSVRC 2012 validation set [18] provided by Lin et al. [12].

Models. We choose five pretrained models based on ImageNet to generate adversarial examples, i.e. Inception-v3 (Inc-v3) [23], VGG16 [20], ResNet50 (Res50) [7], DenseNet121 (Dense121) [8] and ViT-Base (ViT-B) [3]. We evaluate the examples on five models, i.e. Inception-v4 (Inc-v4), Inception-ResNet-v2 (IncRes-v2) [22], ConvNeXt-Tiny (ConvNeXt-T) [14], DeiT-Tiny (DeiT-T) [24] and Swin-Tiny (Swin-T) [13]. To further show the effectiveness of our US-MM, we consider two adversarially trained models, namely Inc-v3$_{adv}$ and IncRes-v2$_{adv}$. Four adversarial defense methods are also tested at the same time, i.e. R&P [29], FD [15], Bit-Red [32] and JPEG [6]. All these models can be found in[1].

Baselines. Because our US-MM is a kind of single-transformation-based attack method, we choose four classic single-transformation-based attack methods as the baselines, i.e. DIM [30], TIM [2], SIM [12] and Admix [26]. We also consider integrating our method into SIA [28] in our ablation studies, while SIA is a collection of several single-transformation-based attack methods. All attacks are integrated into MI-FGSM [1], which is the most classic method to improve transferability.

Attack Setting. We follow the most settings in [26]. We set the maximum perturbation ϵ to 16 and number of iteration T to 10. For MI-FGSM, we make momentum delay factor $\mu = 1.0$. We set the probability of input transformation $p = 0.7$ in DIM and use Gaussian kernel with size of 7×7 in TIM. Based on our study about SIM, we set scale copies $m = 5$ to achieve the best performance. For Admix, except for the same setting as SIM, we set the number of mix images to 3 and mix ratio $\eta = 0.2$.

5.2 Attacking Transferability

In this section, we apply our baseline and proposed attack methods on four pretrained convolutional neural networks and ViT-B, then test them on five pretrained models. For all baseline methods, We use the optimal parameter settings for them separately. To ensure the same computational complexity as Admix, we set the number of uniform scale copies to $m_{us} = 5$ and the number of mix images to $m_{mix} = 3$ in US-MM. For other parameters of US-MM, we set the scale lower bound $L = 0.1$, scale upper bound $H = 0.75$, and mix range size $r = 0.5$. We then collect the results of the model outputs for the generated adversarial examples and count the number of images that are incorrectly classified. The attack success rate is defined as the proportion of these images to the entire dataset. The experimental results are shown in Table 1.

It can be observed that our proposed method, US-MM, achieves the best performance in almost all situations. In the one case where it does not have the highest attack success rate, it is very close to the best. Additionally, when thoroughly examining the results, it can be seen that US-MM has a significant improvement compared to the second-best attack success rate, with an average increase of 8.4%.

---

[1] https://github.com/huggingface/pytorch-image-models.

5.3 Attacking Defense Models

Recently, adversarial training and input pre-processing have been used as defense methods to avoid the adversarial attacks. In this section, we evaluate our baseline and proposed attack methods on two adversarially trained models and four input pre-processing based defense methods. The part of experimental results are shown in Table 2 where the test adversarial examples are crafted on Inc-v3 and the defense methods are used in IncRes-V2.

It can be observed that our US-MM still has the highest attack success rate compared to other baseline methods, especially in attacking adversarially trained models. Besides, US-MM suffers minimal attack performance degradation when testing on the input pre-processing based defense methods.

Table 1. The attack success rates (%) against five models by baseline attacks and our method. The best results are marked in bold.

Model	Attack	Inc-v4	IncRes-v2	ConvNeXt-T	Deit-T	Swin-T
Inc-v3	DIM	39.1	34.4	14.4	27.3	17.6
	TIM	36.4	30.1	14.9	35.5	16.0
	SIM	67.3	64.5	25.5	42.6	30.8
	Admix	81.6	81.9	37.2	50.9	41.0
	US-MM	**92.8**	**93.0**	**52.6**	**62.7**	**56.2**
VGG16	DIM	42.8	30.3	25.3	37.6	30.2
	TIM	43.8	31.0	21.8	47.1	31.4
	SIM	78.6	65.5	45.6	54.2	47.2
	Admix	88.1	76.9	54.8	62.0	56.3
	US-MM	**92.4**	**83.1**	**64.7**	**67.9**	**66.2**
Res50	DIM	41.2	37.7	25.7	42.5	33.0
	TIM	42.8	37.3	24.5	49.0	30.9
	SIM	76.7	73.9	44.5	57.3	50.3
	Admix	87.4	85.3	56.1	67.2	62.4
	US-MM	**92.9**	**91.4**	**69.9**	**76.2**	**74.5**
Dense121	DIM	46.7	41.0	28.6	44.4	34.7
	TIM	48.3	40.8	29.2	52.0	36.7
	SIM	79.0	75.5	49.7	63.8	59.4
	Admix	89.4	86.6	64.2	71.9	70.6
	US-MM	**94.6**	**92.9**	**75.5**	**81.4**	**80.1**
ViT-B	DIM	27.0	21.4	31.4	54.7	42.7
	TIM	26.5	21.0	28.6	51.5	35.3
	SIM	13.9	13.2	22.8	51.7	34.4
	Admix	52.5	50.8	65.2	**89.8**	78.9
	US-MM	**59.0**	**56.8**	**70.7**	89.6	**83.2**

Table 2. The attack success rates (%) against two adversarially trained models and four defense methods by baseline attacks and our method. The best results are marked in bold.

Model	Attack	Inc-v3$_{adv}$	IncRes-v2$_{adv}$	R&P	FD	Bit-Red	JPEG
Inc-v3	DIM	20.7	8.3	33.3	32.9	33.2	31.3
	TIM	29.0	21.6	37.0	43.6	29.0	30.4
	SIM	41.2	18.1	62.6	58.4	64.4	63.0
	Admix	55.4	25.0	79.0	72.5	80.7	77.7
	US-MM	**67.9**	**29.8**	**89.5**	**81.3**	**91.5**	**90.3**

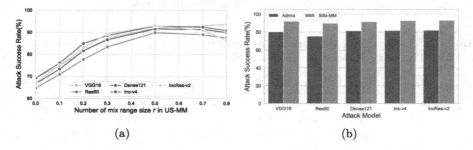

Fig. 4. (a) Attack success rates (%) of US-MM when attacking other five pretrained models for different mix range size r. (b) Attack success rates (%) of SIM-MM and Admix when attacking five pretrained models.

5.4 Ablation Studies

In this section, we study the attack performance with different value of hyperparameters and verify the effectiveness of our two methods. We conduct a total of six ablation experiments where the first three are related to MM and the last three are about US. The default experiment setting about ablation studies is that the Inc-v3 is the victim model while tested models are five pretrained models, i.e. VGG16, Res50, Dense121, Inc-v4 and IncRes-v2. Attack methods are realized based on MI-FGSM.

Mix Range Size. To investigate the relationship between attack success rate and mix range size r, we conduct experiments with getting r from 0 to 0.8. We set $L = 0.1$ and $H = 0.75$ as above studies. As shown in Fig. 4a, the attack success rate increases rapidly when r is set from 0 to 0.5. Then the transferability seems have a little decrease when r becomes bigger than 0.5. It seems that a smaller r value results in a smaller transformation magnitude, which leads to a crafted image with a similar gradient as the original image. On the other hand, a larger r value can destroy the features of the original image and introduce harmful gradient information. This highlights the importance of finding a balance between the two for optimal adversarial transferability.

SIM-MM vs. Admix. For demonstrating the advantage of MM, we integrate MM to SIM as SIM-MM and conduct the comparison between SIM-MM and Admix. For Admix, m_1 and m_2 are set to 5 and 3 respectively. To maintain the same computational complexity with Admix, SIM-MM is done in the setting of $m = 5$ and $m_{mix} = 3$. Experimental results are shown in Fig. 4b. It can be observed that with the same scale strategy, SIM-MM shows a much better attack performance than Admix on five test models.

SIA vs. SIA+Admix vs. SIA+MM. Unlike the single-transformation-based attack methods mentioned above, SIA [28], which integrates multiple data augmentation methods such as rotation or adding noise, shows more excellent attack performance. However, we notice that SIA did not consider the mix method. To further demonstrate the effectiveness of our MM method, we conduct experiments about SIA, SIA+Admix and SIA+MM. For reducing the impact of randomness, we choose three victim models, i.e. Inc-v3, Res50 and ViT-B. We choose ConvNeXt-T as the test model because of its high robustness. The results are shown in Table 3. It can be observed that MM is more effective than Admix when integrated into SIA.

Table 3. The attack success rates (%) against ConvNeXt-T by SIA, SIA+Admix and SIA+MM. The best results are marked in bold.

Model	SIA	SIA+Admix	SIA+MM
Inc-v3	69.6	69.8	**70.7**
Res50	85.5	85.3	**86.0**
ViT-B	87.4	87.0	**87.7**

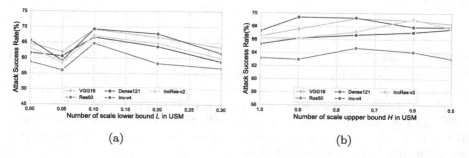

(a) (b)

Fig. 5. (a) Attack success rates (%) of USM when attacking other five pretrained models for different scale lower bound L. (b) Attack success rates (%) of USM when attacking other five pretrained models for different scale upper bound H.

Lower Bound of Scale. Considering scale copies are usually not too closed to the source image, we set upper bound to 0.75, which is $H = 0.75$. We test different lower bound values from 0 to 0.3. Experimental results are shown in Fig. 5a. It can be observed from Fig. 5a. that USM achieves the obviously max attack success rate when L is set to 0.1 under the condition of $H = 0.75$.

Upper Bound of Scale. Following the study for the lower bound, experiment exploring upper bound set the lower bound to 0.1, which refers to $L = 0.1$. We test different upper bound values from 0.5 to 1. Experimental results are shown in Fig. 5b. It can be observed from Fig. 5b that most curves are relate even, which means H might not have great influence on adversarial transferability in a certain extent. It seems that there is not an obviously outperformed upper bound value according to Fig. 5b. However, $H = 1$ seems not a good choice because the attack success rate is lower compared with other values. $H = 1$ means USM calculate the gradient of raw adversarial examples and it seems to occur the overfitting problem. This might be the reason why adversarial transferability is not good enough when $H = 1$. Based on these results, it seems that $H = 0.75$ is a good choice when $L = 0.1$.

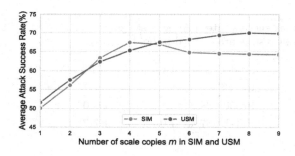

Fig. 6. Average attack success rates (%) of USM and SIM when attacking five pre-trained models.

USM vs. SIM. To validate the effect of USM, we compare USM with SIM in the setting of different m, which denotes the number of scale copies. We set $L = 0.1$ and $H = 0.75$ for USM in the experiment based on the performance found in the above two ablation experiments. In Fig. 6, it can be observed that USM has the similar attack performance with SIM when m increases from 1 to 5. However, when m keeps increasing, SIM achieves a reduced attack success rate while our USM still has an increasing attack success rate.

6 Conclusion

In this paper, we propose a novel adversarial example generation method, namely US-MM method. MM Method improves the mixup strategy from linear addition to a mix mask method considering value range for both source image and mix image. MM Method also considers the impact from mix image in both positive and negative direction. US Method refines the scale changing with uniforming the scale changes within the scope between the upper bound and lower bound. The results of the comparison experiment clearly demonstrate the superior performance of our proposed method in attack transferability. Ablation experiment explores the influence of the hyper-parameters and verifies the effectiveness of both MM Method and US Method. In the future, we attempt to theoretically analyze the characteristics of adversarial transferability from three directions: model space, sample space, and feature space, and provide more detailed explanations.

Acknowledgments. This work was supported by Frontier Technologies R&D Program of Jiangsu, under Grant BF2024071.

References

1. Dong, Y., Liao, F., Pang, T., Su, H., Zhu, J., Hu, X., Li, J.: Boosting adversarial attacks with momentum. In: Proceedings of the IEEE conference on computer vision and pattern recognition. pp. 9185–9193 (2018)
2. Dong, Y., Pang, T., Su, H., Zhu, J.: Evading defenses to transferable adversarial examples by translation-invariant attacks. In: Proceedings of the IEEE/CVF Conference on Computer Vision and Pattern Recognition. pp. 4312–4321 (2019)
3. Dosovitskiy, A., et al.: An image is worth 16x16 words: Transformers for image recognition at scale. arXiv preprint arXiv:2010.11929 (2020)
4. Ganeshan, A., BS, V., Babu, R.V.: Fda: Feature disruptive attack. In: Proceedings of the IEEE/CVF International Conference on Computer Vision. pp. 8069–8079 (2019)
5. Goodfellow, I.J., Shlens, J., Szegedy, C.: Explaining and harnessing adversarial examples. arXiv preprint arXiv:1412.6572 (2014)
6. Guo, C., Rana, M., Cisse, M., Van Der Maaten, L.: Countering adversarial images using input transformations. arXiv preprint arXiv:1711.00117 (2017)
7. He, K., Zhang, X., Ren, S., Sun, J.: Deep residual learning for image recognition. In: Proceedings of the IEEE conference on computer vision and pattern recognition. pp. 770–778 (2016)
8. Huang, G., Liu, Z., Van Der Maaten, L., Weinberger, K.Q.: Densely connected convolutional networks. In: Proceedings of the IEEE conference on computer vision and pattern recognition. pp. 4700–4708 (2017)
9. Huang, Q., Katsman, I., He, H., Gu, Z., Belongie, S., Lim, S.N.: Enhancing adversarial example transferability with an intermediate level attack. In: Proceedings of the IEEE/CVF international conference on computer vision. pp. 4733–4742 (2019)
10. Kurakin, A., Goodfellow, I., Bengio, S.: Adversarial machine learning at scale. arXiv preprint arXiv:1611.01236 (2016)
11. Li, Y., Bai, S., Zhou, Y., Xie, C., Zhang, Z., Yuille, A.: Learning transferable adversarial examples via ghost networks. In: Proceedings of the AAAI Conference on Artificial Intelligence. vol. 34, pp. 11458–11465 (2020)
12. Lin, J., Song, C., He, K., Wang, L., Hopcroft, J.E.: Nesterov accelerated gradient and scale invariance for adversarial attacks. arXiv preprint arXiv:1908.06281 (2019)

13. Liu, Z., Lin, Y., Cao, Y., Hu, H., Wei, Y., Zhang, Z., Lin, S., Guo, B.: Swin transformer: Hierarchical vision transformer using shifted windows. In: Proceedings of the IEEE/CVF international conference on computer vision. pp. 10012–10022 (2021)
14. Liu, Z., Mao, H., Wu, C.Y., Feichtenhofer, C., Darrell, T., Xie, S.: A convnet for the 2020s. In: Proceedings of the IEEE/CVF conference on computer vision and pattern recognition. pp. 11976–11986 (2022)
15. Liu, Z., Liu, Q., Liu, T., Xu, N., Lin, X., Wang, Y., Wen, W.: Feature distillation: Dnn-oriented jpeg compression against adversarial examples. In: 2019 IEEE/CVF Conference on Computer Vision and Pattern Recognition (CVPR). pp. 860–868. IEEE (2019)
16. Long, Y., Zhang, Q., Zeng, B., Gao, L., Liu, X., Zhang, J., Song, J.: Frequency domain model augmentation for adversarial attack. In: European Conference on Computer Vision. pp. 549–566. Springer (2022)
17. Madry, A., Makelov, A., Schmidt, L., Tsipras, D., Vladu, A.: Towards deep learning models resistant to adversarial attacks. arXiv preprint arXiv:1706.06083 (2017)
18. Russakovsky, O., Deng, J., Su, H., Krause, J., Satheesh, S., Ma, S., Huang, Z., Karpathy, A., Khosla, A., Bernstein, M., et al.: Imagenet large scale visual recognition challenge. Int. J. Comput. Vision **115**, 211–252 (2015)
19. Shi, Y., Han, Y., Tian, Q.: Polishing decision-based adversarial noise with a customized sampling. In: Proceedings of the IEEE/CVF Conference on Computer Vision and Pattern Recognition. pp. 1030–1038 (2020)
20. Simonyan, K., Zisserman, A.: Very deep convolutional networks for large-scale image recognition. arXiv preprint arXiv:1409.1556 (2014)
21. Su, J., Vargas, D.V., Sakurai, K.: One pixel attack for fooling deep neural networks. IEEE Trans. Evol. Comput. **23**(5), 828–841 (2019)
22. Szegedy, C., Ioffe, S., Vanhoucke, V., Alemi, A.: Inception-v4, inception-resnet and the impact of residual connections on learning. In: Proceedings of the AAAI conference on artificial intelligence. vol. 31 (2017)
23. Szegedy, C., Vanhoucke, V., Ioffe, S., Shlens, J., Wojna, Z.: Rethinking the inception architecture for computer vision. In: Proceedings of the IEEE conference on computer vision and pattern recognition. pp. 2818–2826 (2016)
24. Touvron, H., Cord, M., Douze, M., Massa, F., Sablayrolles, A., Jégou, H.: Training data-efficient image transformers & distillation through attention. In: International conference on machine learning. pp. 10347–10357. PMLR (2021)
25. Wang, X., He, K.: Enhancing the transferability of adversarial attacks through variance tuning. In: Proceedings of the IEEE/CVF Conference on Computer Vision and Pattern Recognition. pp. 1924–1933 (2021)
26. Wang, X., He, X., Wang, J., He, K.: Admix: Enhancing the transferability of adversarial attacks. In: Proceedings of the IEEE/CVF International Conference on Computer Vision. pp. 16158–16167 (2021)
27. Wang, X., Lin, J., Hu, H., Wang, J., He, K.: Boosting adversarial transferability through enhanced momentum. arXiv preprint arXiv:2103.10609 (2021)
28. Wang, X., Zhang, Z., Zhang, J.: Structure invariant transformation for better adversarial transferability. In: Proceedings of the IEEE/CVF International Conference on Computer Vision. pp. 4607–4619 (2023)
29. Xie, C., Wang, J., Zhang, Z., Ren, Z., Yuille, A.: Mitigating adversarial effects through randomization. arXiv preprint arXiv:1711.01991 (2017)
30. Xie, C., Zhang, Z., Zhou, Y., Bai, S., Wang, J., Ren, Z., Yuille, A.L.: Improving transferability of adversarial examples with input diversity. In: Proceedings of the

IEEE/CVF conference on computer vision and pattern recognition. pp. 2730–2739 (2019)
31. Xiong, Y., Lin, J., Zhang, M., Hopcroft, J.E., He, K.: Stochastic variance reduced ensemble adversarial attack for boosting the adversarial transferability. In: Proceedings of the IEEE/CVF Conference on Computer Vision and Pattern Recognition. pp. 14983–14992 (2022)
32. Xu, W., Evans, D., Qi, Y.: Feature squeezing: Detecting adversarial examples in deep neural networks. arXiv preprint arXiv:1704.01155 (2017)
33. Zhang, J., Huang, J.t., Wang, W., Li, Y., Wu, W., Wang, X., Su, Y., Lyu, M.R.: Improving the transferability of adversarial samples by path-augmented method. In: Proceedings of the IEEE/CVF Conference on Computer Vision and Pattern Recognition. pp. 8173–8182 (2023)

A Differential Privacy Decision Forest Algorithm for Reducing the Effect of Noise

Runfei Liu[1,2], Mingze Chu[1,2], Yuming Jiang[1,2], Xuefeng Ding[1,2], Yuncheng Shen[3], and Dasha Hu[1,2(✉)]

[1] College of Computer Science, Sichuan University, Chengdu 610065, China
2022223040050@stu.scu.edu.cn
[2] Big Data Analysis and Fusion Application Technology Engineering Laboratory of Sichuan Province, Chengdu 610065, China
hudasha@scu.edu.cn
[3] College of Physics and Information Engineering, Zhaotong University, Zhaotong 657000, China
Shenyuncheng@ztu.edu.cn

Abstract. Decision tree is widely used as a classification model, but it is susceptible to a significant privacy issue. Researchers have integrated differential privacy into decision trees, yielding positive outcomes. However, to satisfy differential privacy, the noise will inevitably have a negative effect on its performance. We proposes a differential privacy decision forest algorithm (DPDF-REN). First, the termination division criteria of nodes are established. A node will terminate the division, when its division does not yield better classification results, but instead increases the effect of noise. Second, a node depth-based privacy budget allocation strategy (NDPS) allocates more privacy budget to deeper nodes. This strategy gradually reduces the noise in order to alleviating the signal-to-noise ratio imbalance. Finally, a leaf node reliability-based ensemble (LR-En) is proposed. Leaf node reliability measures the likelihood of being in the maximum instance count class. The ensemble reduces the effect of noise by assigning weights to each decision tree based on the leaf node reliability. We conducted a series of experiments on several public datasets. The experimental results show that the DPDF-REN algorithm has better classification performance compared to other algorithms.

Keywords: Differential privacy · Decision tree · Noise reduction · Privacy budget allocation

1 Introduction

Decision tree is a data mining algorithm for classification [6], which is constructed by selecting the best attributes for data division at internal nodes and generating class at leaf nodes. However, the transparency of decision trees [19] exposes serious privacy issues. Assuming that there are two adjacent data sets used to train two trees, which differ in one record at most. It is possible for an adversary to obtain sensitive individual information from the database by comparing the result of the counts [9].

In recent years, differential privacy [2] has shown great advantages in solving the problem of privacy protection. Differential privacy has rigorous mathematical definitions and proofs that provide formal privacy guarantees. Even if the attacker has all the background information except the target information, differential privacy can still provide reliable privacy protection. Researchers have applied differential privacy to decision trees by adding random noise to the dividing process of the internal nodes and to the output of the leaf nodes in order to preserve privacy. The combination of differential privacy and decision trees has made progress in several directions, such as edge computing [21], data publishing [18], and federated learning [12] [15].

But there are still some problems in the current research on differential privacy decision trees. 1) The division of nodes does not always lead to better classification results, but instead increases the effect of noise. 2) There is a serious imbalance of signal-to-noise ratio in differential privacy decision trees. 3) The performance of differential privacy decision trees is difficult to evaluate. We proposes a differential privacy decision forest algorithm (DPDF-REN) for reducing the effect of noise to address the above three problems and minimize this negative effect.

Firstly, the termination division criteria of nodes are established. The termination criteria are to make the nodes that are mostly in the same class or the nodes with signal-to-noise ratio less than 1 terminate the division and become leaf nodes.

Secondly, a node depth-based privacy budget allocation strategy (NDPS) alleviates the signal-to-noise ratio imbalance. NDPS adds less noise when the instance counts is reduced, allocating more privacy budget to deeper nodes.

Thirdly, a leaf node reliability-based ensemble (LR-En) is proposed. Leaf node reliability utilizes instance counts to measure the likelihood that leaf node's class is the maximum instance count class to evaluate the performance of differential privacy decision trees. Assign weights to each decision tree based on leaf node reliability.

The rest of the paper is organized as follows. In Sect. 2, related work on differential privacy decision trees and differential privacy decision forests is reviewed. In Sect. 3, the background knowledge required for our work is presented. In Sect. 4, our proposed DPDF-REN algorithm is presented. In Sect. 5, experiments are conducted and the results are analyzed. Finally, the paper is summarized in Sect. 6.

2 Related Work

Allocating privacy budget to allow decision tree to achieve better performance is a challenge. Liu et al. [14] designed a privacy budget allocation strategy that divides the total privacy budget into equal budget shares. The closer the internal nodes are to the root node, the smaller their budget share, while the leaf nodes are given the largest budget share. In [1], Dong et al. proposed a linear privacy budget allocation strategy that increases the privacy budget linearly with the increase of layers. However, neither of them considered the difference between the Laplace mechanism and the exponential mechanism. An alternative perspective of privacy budget allocation strategy was designed by Niu et al. [19]. They argued that larger noise can be added when there is a large difference between the instance counts in the largest and next largest classes in a leaf node, and vice versa.

Decision forest tend to have better performance [19]. Jagannathan et al. [10] first introduced differential privacy into decision forests, and they used bootstrap sampling to obtain a training set of decision trees. Unfortunately, each additional tree further divides the privacy budget. In order to query each tree in parallel, Fletcher and Islam [5] [7] attempted to construct each tree using disjoint subsets of data. In addition to saving on privacy budget, constructing decision trees with disjoint data subsets also reduces the correlation between decision trees. Subsequently, Xin et al. [22] proposed an algorithm for integrating greedy decision trees. They also took the approach of training decision trees using disjoint training sets.

Assigning appropriate weights to different differential privacy decision trees benefits the performance of decision forests. Niu et al. [19] used quantum genetic algorithms to assign weights. Liu et al. [13] adjusted instance weights to construct complementary forests. Guan et al. [8] decided which trees to include in the final forest by determining whether removing each decision tree will reduce the error rate of the forest. While these methods achieve nice results, they need to rely on publicly available test datasets to evaluate the performance of decision trees or forests, or else they violate differential privacy. Li et al. [11] and Wang et al. [20] assigned weights to decision trees via out-of-bag balance error rates after noise addition, but this approach consumes additional privacy budget and relies too much on out-of-bag data.

3 Preliminaries

Definition 1. (ε−Differential Privacy [2]) *Datasets x and y are adjacent if they differ in only one record. Suppose f is a random function, R_f is any possible outcome of f, all data $x, y \in D^n$, ε is a privacy budget parameter, and $Pr\,[E]$ denotes the risk of disclosure of event E. If f satisfies ε-differential privacy, then f satisfies the following inequality*

$$Pr\left[f\left(x\right)\epsilon R^f\right] \leq exp\left(\varepsilon\right) \cdot Pr\left[f\left(y\right)\epsilon R_f\right] \tag{1}$$

Theorem 1. (Sequential Composition [16]) *Having $M = \{M_1, M_2, \ldots, M_m\}$ is a sequence of privacy mechanisms. If each M_i satisfies ε_i−differential privacy and M is performed sequentially on a dataset, M will satisfy $\sum_i^m \varepsilon_i$− differential privacy.*

Theorem 2. (Parallel Composition [16]) *Having $M = \{M_1, M_2, \ldots, M_m\}$ is a sequence of privacy mechanisms. If each M_i satisfies ε_i-differential privacy on disjoint subsets of the dataset, M will satisfy $\max(\{\varepsilon_1, \varepsilon_2, \ldots, \varepsilon_m\})$-differential privacy.*

Definition 2. (Laplace Mechanism [3]) *For a query M, M satisfies ε- differential privacy if and only if the output $M(x) = f(x) + L$, where $f: f \to R$, $L \sim Laplace(\Delta f/\varepsilon)$ is a random variable extracted from the Laplace distribution with mean 0 and scale $\Delta f/\varepsilon$. Δf will be described below.*

Definition 3. (Exponential Mechanism [17]) *There is a utility function $u: u(z,x) \to R$, where the larger the value of u, the higher the probability of $z \in Z$ output. A query f satisfies ε-differential privacy when the output z is proportional to $exp(\frac{\varepsilon u(z,x)}{2\Delta u})$, that is*

$$Pr(f(x) = z) \propto exp(\frac{\varepsilon u(z,x)}{2\Delta u}) \tag{2}$$

In the above definition, Δf denotes the sensitivity of the query function f and Δu denotes the sensitivity of the utility function u. The sensitivity of a function is the maximum difference in the output of the function for two adjacent datasets x and y.

Definition 4. (Sensitivity [2]) *The sensitivity of a query f is*

$$GS(f) = \max_{x,y: \|x-y\|_1 \leq 1} \|f(x) - f(y)\|_1 \tag{3}$$

4 DPDF-REN Algorithm

In this section, Differential Privacy Decision Forest Algorithm for Reducing the Effect of Noise (DPDF-REN) is proposed. First, a differential privacy decision tree is constructed and a privacy budget allocation strategy in the decision tree is designed. Then, an ensemble of the decision tree is proposed.

4.1 Constructing Differential Privacy Decision Tree

Termination Division Criteria. In traditional decision tree, a node should terminate division when the instances in the node are same class. To satisfy differential privacy, the condition that the instances in a node are same class can hardly be satisfied. Nodes terminate division when most instances are same class. A threshold μ is introduced to gauge what proportion of the maximum class's instance count is reached when the node terminates dividing. The termination criterion (1) is thus proposed.

(1) $\frac{X_{max}}{X} > \mu$, where X_{max} represents the instance count of the maximum class. X represents the instance count in the node, which is the total sum of the

instance counts of all classes. μ is the threshold.

A node with a signal-to-noise ratio of less than 1 represents that noise has outweighed the signal, and continued division of the node does not yield better results, but rather increases the effect of the noise. The termination criterion (2) is thus proposed.

(2) $\frac{X}{|C|\sigma} < 1$, where $\frac{X}{|C|\sigma}$ is the signal-to-noise ratio of nodes [5]. $|C|$ is the number of classes in C. σ is the standard deviation of the Laplace noise.

Nodes that satisfy criterion (2) are already overwhelmed by noise [5], causing it to disable its classification ability. The class of the parent node is used in place of the class of the node to restore the classification ability for it.

Overall, we constructs a differential privacy decision tree, as shown in Algorithm 1.

Algorithm 1 DP_TREE

Input: The training dataset D; Privacy budget for Laplace mechanism ε_1; Privacy budget for exponential mechanism ε_2; Privacy budget allocation parameter λ; Threshold for node termination division μ; Maximal tree depth h; The current tree depth d_T; The class of parent node p; The class set C; The complete set of all attributes U;

Output: Decision tree T satisfies ε-differential privacy;

1: **function** DP_TREE($D, \varepsilon_1, \varepsilon_2, \lambda, \mu, h, d_T, p, C, U$)
2:    **for** each c in C **do**
3:       calculate the number of samples x_c of c in D;
4:       $X_c = \text{count}(x_c) + \text{Lap}(\frac{1}{\varepsilon_1})$;
5:       **if** $X_c > X_{max}$ or $max = \emptyset$ **then**
6:          $max = c$;
7:       **end if**
8:    **end for**
9:    $X = \sum_{c \in C} X_c$;
10:   $\sigma = \frac{\sqrt{2}}{\varepsilon_1}$;
11:   **if** $\frac{X}{|C|\sigma} < 1$ **then**
12:      create a node with class p;
13:   **else**
14:      create a node with class max;
15:   **end if**
16:   **if** $\frac{X_{max}}{X} > \mu$ or $\frac{X}{|C|\sigma} < 1$ or $d_T \geq h$ **then**
17:      set the node as a leaf node;
18:      **return** ;
19:   **end if**
20:   select the split attribute A by using exponential mechanism $\exp(\varepsilon_2)$;
21:   **for** each a in $U(A)$ **do**
22:      divide the value of attribute A in D with a value of a into D_a;
23:      DP_TREE($D_a, \lambda \varepsilon_1, \varepsilon_2, \lambda, \mu, h, d_T + 1, X_{\max}, C, U$);
24:   **end for**
25:   **return** $Tree$;
26: **end function**

Privacy Budget Allocation Strategy. During the construction of the decision tree, the dataset is subjected to division attribute queries and count queries. Division attribute queries and count queries are differentially privatized by utilizing the exponential and Laplace mechanisms, respectively. As the number of layers in the decision tree increases, the count of instances allocated in the nodes decreases. Since the exponential mechanism is not affected by the count of node instances [8], it is appropriate to allocate an equal privacy budget for the exponential mechanism at the internal nodes. Specifically, the privacy budget allocated for the exponential mechanism is $\varepsilon_{2i} = \frac{\varepsilon_2}{h-1}$, where h represents the maximal decision tree depth.

If the same privacy budget is allocated to different layers for the Laplace mechanism, it will cause a significant imbalance in the signal-to-noise ratio of the decision tree. A node depth-based privacy budget allocation strategy (NDPS) is designed for the Laplace mechanism makes the privacy budget increase with the increase of layers of the decision tree.

We introduces the parameter λ to represent the ratio of the instance count of the upper node to the instance count of the lower node between neighboring layers. Since the signal in each layer of nodes is reduced according to λ, the noise added to each node should also be reduced according to the parameter λ. $|C|\sigma$ denotes the effect of Laplace noise under ε, $|C|\sigma = \frac{\sqrt{2}|C|}{\varepsilon}$. To ensure that the signal-to-noise ratio is balanced, the privacy budget relationship between the different layers is

$$\frac{\varepsilon_{11}}{\sqrt{2}|C|} = \frac{\varepsilon_{12}}{\lambda\sqrt{2}|C|} = \frac{\varepsilon_{13}}{\lambda^2\sqrt{2}|C|} = \ldots = \frac{\varepsilon_{1h}}{\lambda^{h-1}\sqrt{2}|C|} \qquad (4)$$

In addition, NDPS for each layer needs to meet

$$\varepsilon_1 = \varepsilon_{11} + \varepsilon_{12} + \varepsilon_{13} + \ldots + \varepsilon_{1h} \qquad (5)$$

Combining equations (4)(5) yields

$$\sum_{j=0}^{h-1} \lambda^j \varepsilon_{11} = \varepsilon \qquad (6)$$

From equation (6), NDPS is

$$\varepsilon_{1i} = \frac{\varepsilon_1}{\sum_{j=0}^{h-1} \lambda^j} \lambda^{i-1} \qquad (7)$$

Privacy Analysis.

Lemma 1. *The differential privacy decision tree in Algorithm 1 satisfies ε- differential privacy.*

Proof. Division attribute queries and count queries are differentially privatized by utilizing the exponential and Laplace mechanisms, respectively. The division process after the division attribute query divides the dataset only through the division attribute, which guarantees differential privacy through the post-processing property of DP [4]. According to Theorem 2, nodes in the same layer can share the same privacy budget since the datasets of these nodes are not intersected. The privacy budget in the decision tree is allocated on a layer-by-layer basis. From the privacy budget allocation strategy outlined in the previous section, it is evident that each layer allocates a privacy budget of $\varepsilon_{2i} = \frac{\varepsilon_2}{h-1}$ for the exponential mechanism and a privacy budget of $\varepsilon_{1i} = \frac{\varepsilon_1}{\sum_{j=0}^{h-1} \lambda^j} \lambda^{i-1}$ for the Laplace mechanism. The Laplace mechanism is to be used for layers 1 to h, while the exponential mechanism is to be used for layers 1 to $h-1$. The total privacy budget for the differential privacy decision tree is shown in equation (8).

$$\sum_{i=1}^{h} \varepsilon_{1i} + \sum_{i=1}^{h-1} \varepsilon_{2i} = \frac{\varepsilon_1}{\sum_{j=0}^{h-1} \lambda^j} \times \sum_{i=1}^{h} \lambda^{i-1} + \frac{\varepsilon_2}{h-1} \times (h-1) = \varepsilon_1 + \varepsilon_2 = \varepsilon \quad (8)$$

Thus, the differential privacy decision tree in Algorithm 1 satisfies ε-differential privacy.

4.2 Leaf Node Reliability-Based Ensemble

We propose average leaf node reliability to evaluate the performance of each differential privacy decision tree by utilizing instance counts from the training phase.

$$ave(T) = \frac{\sum_{l \in leaf} R_l}{|leaf|} \quad (9)$$

where $leaf$ is the set of leaf nodes of the decision tree T, $|leaf|$ is the number of leaf nodes of the decision tree T, and R_l is the reliability of leaf node l. R is defined as follows.

$$R = \frac{\sum_{c \in C} (X_c - X_{max})^2}{|C|} \quad (10)$$

where C represents the set of classes, X_c is the instance count of class c, X_{max} is the instance count of the maximum class, $|C|$ is the number of classes in C.

The average leaf node reliability of all decision trees is normalized to serve as the weights of the decision trees. Decision trees together with their weights form a decision forest.

The algorithm for the complete DPDF-REN is shown in Algorithm 2.

Lemma 2. *DPDF-REN satisfies $\varepsilon-$differential privacy.*

Proof. 4.1.3 It has been shown that the differential privacy decision tree in Algorithm 1 satisfies $\varepsilon-$differential privacy. In the DPDF-REN algorithm, ensemble is performed based on instance counts obtained in the training phase that satisfies differential privacy, which is guaranteed by post-processing. At the same time, different decision trees are trained using disjoint training datasets, so the decision forest algorithm still satisfies $\varepsilon-$differential privacy according to Theorem 2.

Algorithm 2 DPDF-REN

Input: The dataset D; Privacy budget ε; The class set C; The complete set of all attributes U; The number of decision trees τ;
Output: Decision forest F satisfies ε-differential privacy;
1: **function** DPDF-REN($D, \varepsilon, C, U, \tau$)
2:    generate training subsets $D = \{D_1, D_2, \ldots, D_\tau\}$ based on sampling without replacement method;
3:    divide the ε into ε_1 and ε_2;
4:    sets the values of the parameters h, λ, μ;
5:    $T = \{\}$;
6:    **for** D_i in D **do**
7:       $T_i =$ DP_TREE($D_i, \varepsilon_1, \varepsilon_2, \lambda, \mu, h, 1, None, C, U$);
8:       $ave(T_i) = \frac{\sum_{l \in leaf_i} R_l}{|leaf_i|}$;
9:       $T = T \cup T_i$;
10:   **end for**
11:   $F = \{\}$;
12:   **for** T_i in T **do**
13:       $F = F \cup [T_i, \frac{ave(T_i)}{\sum_{j=1}^{\tau} ave(T_j)}]$;
14:   **end for**
15:   **return** F;
16: **end function**

5 Experiments

A series of experiments are conducted on real datasets from the UCI Machine Learning Library[1] to evaluate DPDF-REN algorithm. Specific information can be found in Table 1. To standardize the experimental setup, the data were preprocessed as follows. First, data records containing null values are removed. Then, the datasets is divided into training: validation: test set equal to 6:2:2 or training: test set equal to 8:2 for different algorithm. The accuracy of the model will vary each time without changing any parameters. Therefore, in all experiments, the same experiment was averaged over ten runs.

In the DPDF-REN algorithm there are two hyperparameters μ and λ. After testing, the DPDF-REN algorithm can achieve relatively good results when $\mu = 0.9$ and $\lambda = 1.5$. Therefore, in the following experiments, $\mu = 0.9$, $\lambda = 1.5$.

5.1 Comparison with Other Decision Forests

The noiseless decision forest is the baseline for this experiment. Compared to DPDF-REN, the noiseless decision forest is integrated using equal weights for different decision trees, except for the difference of whether noise is added or not. The difference from the baseline can be used to measure the negative effect

---
[1] http://archive.ics.uci.edu/ml.

Table 1. Specific information about dataset.

Dataset	Number of samples	Number of attributes	Number of classes
Adult	48842	14	2
Mushroom	8124	22	2
Pima	768	9	2
Car	1728	6	4

of DPDF-REN on accuracy due to satisfying differential privacy. Apart from baseline, DPDF-REN was also compared with TpDPRF [13], SNR [5]. To ensure that the settings of these two parameters do not bias the DPDF-REN algorithm, we set the maximum tree depth and the number of trees to 5 and 30, respectively, with reference to the experiments done in [13]. The SNR algorithm sets the maximum tree depth and the number of trees adaptively.

In this section the comparison of the classification accuracy of each algorithm is discussed under different privacy budgets, which are set to values from 0.2 to 1. The experimental results are shown in Fig. 1.

The results in Fig. 1 show that DPDF-REN outperforms the TpDPRF and SNR algorithms on all datasets, but still falls short of the noiseless baseline. This gap generally decreases as the privacy budget increases. In addition, a small portion of the curve decreases as the privacy budget increases, unlike what would be expected. This phenomenon is due to the randomness of the noise. Under the effect of randomness, the amount of noise added to the decision tree does not increase strictly with the increase of the privacy budget. However, it can be observed that most of the curves show an increasing trend.

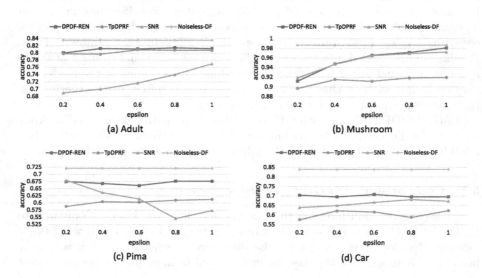

Fig. 1. Comparison of DPDF-REN with other decision forests.

5.2 Effect of Privacy Budget Allocation Strategy on DPDF-REN

To show that NDPS is effective, NDPS is compared with the uniform allocation baseline in Algorithm 1 to analyze the change of signal-to-noise ratio with the increase of layers of the decision tree. The signal-to-noise ratio for each layer is shown in equation (11).

$$STNR_i = \frac{\sum_n^{layer_i} \sum_{c \in C} x_{nc}}{|C|\sigma layer_i} \quad (11)$$

where $layer_i$ is the number of nodes in layer i, C represents the set of classes, $|C|$ represents the number of class in C, x_{nc} is the count of instances with class c in the nth node on layer i, σ is the standard deviation of the Laplace noise.

Experiments are run on the four datasets while keeping the parameters constant, except for changing the privacy budget allocation strategy.

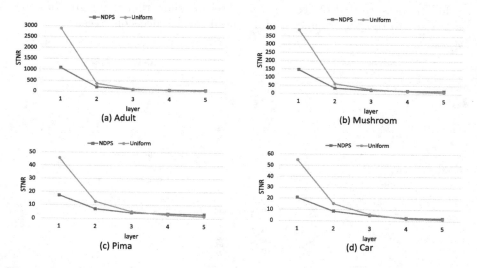

Fig. 2. Comparison of signal-to-noise ratio for different number of layers in decision tree. ε is fixed to 1.

As can be seen from Fig. 2, the uniform allocation strategy has a higher signal-to-noise ratio in the lower layers of the decision tree, and NDPS has a higher signal-to-noise ratio in the deeper layers of the decision tree. Meanwhile, NDPS has more balanced signal-to-noise ratio in the whole decision tree. These happen because the total privacy budget is certain, the uniform allocation strategy allocates equal privacy budget at each layer, while NDPS allocates more privacy budget with the increase of layers. The signal-to-noise ratio of the uniform allocation strategy decreases significantly due to the decrease in instance counts but constant noise. In contrast, NDPS reduces noise as the instance count decreases. Although the ratio of instance count to noise reduction is not exactly the same in NDPS, it ensures a relatively balanced signal-to-noise ratio.

5.3 Effect of Ensemble on DPDF-REN

In differential privacy decision trees, leaf nodes affected by noise may incorrectly select class that does not have the maximum true instance count, which are called noisy nodes. The number of noisy nodes is directly related to the negative effect of noise in the decision tree. We propose a metric to evaluate the effect of noise by evaluating the weights that the noisy nodes have in the ensemble model.

$$NNE = \sum_{i=1}^{\tau} \left(\frac{L_i}{N_i} w_i \right) \qquad (12)$$

where τ is the number of decision trees, L_i and N_i are the number of noisy nodes and total leaf nodes in decision tree i, w_i is the weight of decision tree i. A smaller NNE indicates that the noise has less effect on the model.

Figure 3 compares the trend of accuracy and NNE to verify the relationship between noise effect and accuracy in the decision tree. The trend of NNE is roughly opposite to that of accuracy. Such results show that the degree of noise effect is key to the performance of differential privacy decision trees.

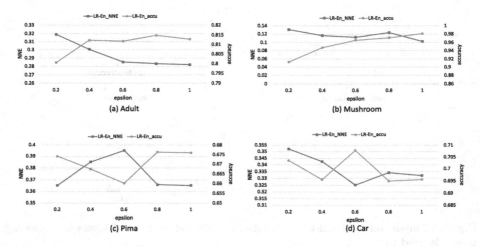

Fig. 3. NNE and accuracy of LR-En under different privacy budgets.

In Fig. 4, LR-En is compared with DPT_1 and DPT_2 in terms of accuracy. DPT_1 and DPT_2 denote the one-tree-only method and the simple majority voting method, respectively. DPT_1 and DPT_2 are completely identical to LR-En except for the ensemble. The optimal experimental results were achieved by LR-En, which further validates the effectiveness of LR-En.

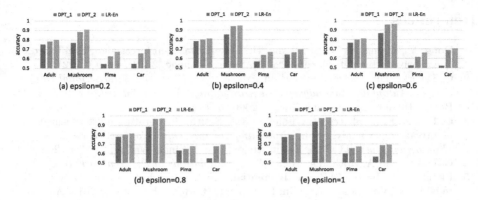

Fig. 4. Effect of ensemble on accuracy.

6 Conclusion

We proposes a differential privacy decision forest algorithm (DPDF-REN) for reducing the effect of noise. In this algorithm, the termination division criteria for nodes are first established to ensure the validity of data classification while reducing the privacy violation by noise. Secondly, a node depth-based privacy budget allocation strategy (NDPS) is designed to gradually increase the privacy budget allocated to a node as layers increase to ensure that nodes with fewer instance count are added with less noise. NDPS controls the privacy budget allocated for each query to make the signal-to-noise ratio of each decision tree more balanced, thus allowing more nodes to make the correct decision. Then secondly, a leaf node reliability-based ensemble (LR-En) is introduced. LR-EN reduces the effect of noise in the decision forest by optimizing the decision tree weights based on the leaf node reliability. Finally, the performance of DPDF-REN algorithm is compared with other decision forest algorithms, and the effectiveness of NDPS and LR-En is verified respectively. Experimental results show that NDPS has a more balanced signal-to-noise ratio in decision trees, LR-En reduces the weight of the noisy nodes in the ensemble model and DPDF-REN effectively reduces the effect of noise and shows the highest classification accuracy in the classification problem.

Currently, only data centralized scenarios are considered in this paper. In future work, more data scenarios, such as distributed data scenarios, will be further considered, which will make the DPDF-REN algorithm closer to practical applications.

Acknowledgments. This research was funded by the National Natural Science Foundation of China (No. 62262074 and No. U2268204).

References

1. Dong, Y., Zhang, S., Xu, J., Wang, H., Liu, J.: Random forest algorithm based on linear privacy budget allocation. J. Database Manage. **33**(2), 1–19 (2022)
2. C. Dwork, "Differential privacy," in *Proceedings of the 33rd International Conference on Automata, Languages and Programming - Volume Part II*, ser. ICALP'06. Berlin, Heidelberg: Springer-Verlag, 2006, p. 1–12
3. C. Dwork, F. McSherry, K. Nissim, and A. Smith, "Calibrating noise to sensitivity in private data analysis," in *Proceedings of the Third Conference on Theory of Cryptography*, ser. TCC'06. Berlin, Heidelberg: Springer-Verlag, 2006, p. 265–284
4. C. Dwork and A. Roth, *The Algorithmic Foundations of Differential Privacy*, 2014
5. Fletcher, S., Islam, M.Z.: A differentially private random decision forest using reliable signal-to-noise ratios. In: Pfahringer, B., Renz, J. (eds.) AI 2015: Advances in Artificial Intelligence, pp. 192–203. Springer International Publishing, Cham (2015)
6. Han, J., Kamber, M., Pei, J.: Data Mining: Concepts and Techniques, 3rd edn. Morgan Kaufmann Publishers Inc., San Francisco, CA, USA (2011)
7. Fletcher, S., Islam, M.Z.: Differentially private random decision forests using smooth sensitivity. Expert Syst. Appl. **78**, 16–31 (2017)
8. Guan, Z., Sun, X., Shi, L., Wu, L., Du, X.: A differentially private greedy decision forest classification algorithm with high utility. Computers & Security **96**, 101930 (2020)
9. S. Fletcher and M. Z. Islam, "Decision tree classification with differential privacy: A survey," *ACM Comput. Surv.*, vol. 52, no. 4, aug 2019
10. Jagannathan, G., Pillaipakkamnatt, K., Wright, R.N.: A practical differentially private random decision tree classifier. Trans. Data Privacy **5**(1), 273–295 (2012)
11. X. Li, B. Qin, Y. Luo, and D. Zheng, "A differential privacy budget allocation algorithm based on out-of-bag estimation in random forest," *Mathematics*, vol. 10, no. 22, 2022
12. Li, Y., Feng, Y., Qian, Q.: Fdpboost: Federated differential privacy gradient boosting decision trees. Journal of Information Security and Applications **74**, 103468 (2023)
13. Liu, J., Li, X., Wei, Q., Liu, S., Liu, Z., Wang, J.: A two-phase random forest with differential privacy. Appl. Intell. **53**(10), 13037–13051 (2022)
14. Liu, X., Li, Q., Li, T., Chen, D.: Differentially private classification with decision tree ensemble. Appl. Soft Comput. **62**, 807–816 (2018)
15. Y. Liu, Z. Fan, X. Song, and R. Shibasaki, "Fedvoting: A cross-silo boosting tree construction method for privacy-preserving long-term human mobility prediction," *Sensors*, vol. 21, no. 24, 2021
16. F. McSherry, "Privacy integrated queries: an extensible platform for privacy-preserving data analysis," *Commun. ACM*, vol. 53, no. 9, p. 89–97, sep 2010
17. F. McSherry and K. Talwar, "Mechanism design via differential privacy," in *48th Annual IEEE Symposium on Foundations of Computer Science (FOCS'07)*, 2007, pp. 94–103
18. N. Mohammed, R. Chen, B. C. Fung, and P. S. Yu, "Differentially private data release for data mining," in *Proceedings of the 17th ACM SIGKDD International Conference on Knowledge Discovery and Data Mining*, ser. KDD '11. New York, NY, USA: Association for Computing Machinery, 2011, p. 493–501
19. X. Niu and W. Ma, "An ensemble learning model based on differentially private decision tree," *Complex & Intelligent Systems*, pp. 1–14, 2023

20. C. Wang, S. Chen, and X. cheng Li, "Adaptive differential privacy budget allocation algorithm based on random forest," in *International Conference on Bio-Inspired Computing: Theories and Applications*, 2021
21. Wu, X., Qi, L., Gao, J., Ji, G., Xu, X.: An ensemble of random decision trees with local differential privacy in edge computing. Neurocomputing **485**, 181–195 (2022)
22. B. Xin, W. Yang, S. Wang, and L. Huang, "Differentially private greedy decision forest," in *ICASSP 2019 - 2019 IEEE International Conference on Acoustics, Speech and Signal Processing (ICASSP)*, 2019, pp. 2672–2676

CDGM: Controllable Dataset Generation Method for Cybersecurity

Yushun Xie[1], Haiyan Wang[4], Runnan Tan[3], Xiangyu Song[4], and Zhaoquan Gu[2,4(✉)]

[1] Shenzhen Institute for Advanced Study, University of Electronic Science and Technology of China, Shenzhen, China
yshxie@std.uestc.edu.cn
[2] School of Computer Science and Technology, Harbin Institute of Technology, Shenzhen, China
guzhaoquan@hit.edu.cn
[3] Cyberspace Institution of Advanced Technology, Guangzhou University, Guangzhou, China
1112106007@e.gzhu.edu.cn
[4] Department of New Networks, Peng Cheng Laboratory, Shenzhen, China
{wanghy01,songxy02}@pcl.ac.cn

Abstract. Cyberattacks can lead to data breaches, service disruptions, and economic losses, and may even threaten national security and social stability. Therefore researchers have proposed various methods based on public datasets to improve the intelligence and automation of cybersecurity defense techniques. However, these public datasets usually have limited coverage of the types of cyberattacks, resulting in the proposed methods being ineffective against attacks not included in the dataset. Meanwhile, cybersecurity defenders often need to study cyberattack scenarios involving specific assets that are usually not represented in public datasets. To address these challenges, we propose a new approach to cybersecurity controlled dataset generation. Our method can reproduce any cyberattack using our four-role architecture, generating customized private attack data that includes specific assets, this capability satisfies the needs of researchers. By integrating the private attack data with a cybersecurity knowledge base derived from open-source datasets, we construct a comprehensive cybersecurity dataset. Extensive experiments demonstrate that the cybersecurity dataset generated by our method is suitable for various common cybersecurity tasks, such as threat hunting, alert analysis, and knowledge reasoning.

Keywords: Cybersecurity Dataset · Data generate · Cyberattack

1 Introduction

With the development of information technology, the scale of cyberspace is gradually increasing. Cyberspace refers to the virtual environment composed of computer networks, including the Internet, local area networks (LANs), wide area networks (WANs), and the data and information transmitted within them. It plays a vital role in the modern society. As cyberspace expands into various areas of life, the impact of cyberattacks is increasingly serious. For example, the 2017 WannaCry ransomware attack [12] affected hundreds of thousands of computers worldwide, causing victims to suffer significant financial losses.

The extremely rapid evolution of technology has led to the diversification of cyberattacks, necessitating that cybersecurity defenders possess substantive technical skills, strong analytical ability, and the concept of continuous learning. Nonetheless, the increasing frequency and sophistication of cyberattacks have outpaced the availability of qualified cybersecurity professionals, creating a gap in the field of cybersecurity defense. This shortage drives researchers to develop automated and intelligent techniques to deal with cyberattacks.

Hundreds of millions of attacks exist in cyberspace and cybersecurity techniques are ever changing, which leads to a rise in security events. At the same time, Advanced Persistent Threat (APT) and other complex attacks are highly correlated in terms of attack patterns, vulnerabilities and assets [29], making it challenging for cybersecurity experts to get the desired necessary information from cyberspace. Currently, the researchers focus on extracting valuable information from massive data and summarizing the correlation to present the whole process of cyberattacks. However, the long persistence and highly concealed characteristics of modern cyberattacks, such as APT attacks, are highly the limitations of traditional defense techniques based on expert rules and machine learning.

Researchers have proposed various methods to enhance the performance of cybersecurity defense techniques. Xu et al. [30] used principal component analysis to analyze attack feature vectors. Hwang et al. [17] assessed three host-based datasets to evaluate and compare their combined detection capabilities across diverse attack stages and types. CSKG4APT [24] collected and analyzed fragmented information to portray the portrait of attack organization, assisting security analysts in decisions-making. Macas et al. [14] provided a comprehensively survey of adversarial attack and defense across eight principal cybersecurity application categories.

Based on public datasets, these existing methods perform strongly, but the public data often have limited coverage of cyberattack types, limiting the effectiveness of these techniques against cyberattacks not included in the datasets. Researchers frequently require specific attack data to study the particular attack scenarios involving unique assets, which public datasets can't provide. As a result, the generation of controlled cyberattack datasets has become an important research topic. For example, Lin et al. [19] built a dataset with three sources and utilized machine learning algorithms to train models.

To address the problem of low attack coverage in public datasets and satisfy researchers' needs for particular private attack data, we propose a Controllable Dataset Generation Method for Cybersecurity (CDGM), as illustrated in Fig. 1. First, we establish a four-role architecture that allows researchers to produce customized private attack data. Second, we summarize a cybersecurity knowledge base from open-source datasets to serve as the foundation. Finally, we integrate the private attack data with the cybersecurity knowledge base to generate a specific cybersecurity dataset, satisfying the unique demands of researchers and facilitating advanced research.

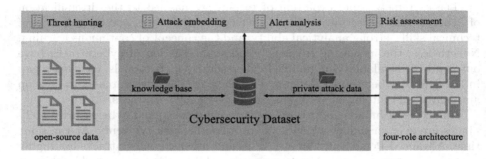

Fig. 1. CDGM: controllable dataset generation method for cybersecurity.

Our contributions are summarized as follows:

1) We propose a controllable dataset generation method for cybersecurity, named CDGM, to generate specific attack data involving unique assets. This dataset comprises two components: a cybersecurity knowledge base derived from open-source datasets and the private cybersecurity data generated from particular attack scenarios.
2) We summarize a cybersecurity knowledge base by analyzing open-source cybersecurity datasets from a top-down, abstract-to-concrete perspective. This cybersecurity knowledge base contains 9 types of entities and 11 types of relationships, providing unified knowledge standards and serving as a foundation to support private data generated from attack scenarios.
3) We establish a four-role architecture for producing private attack data, consisting of an attack server, a target server, a benign server and a data server. The private customized attack data includes specific assets, satisfying the unique needs of researchers. Additionally, the private data fully exploits the value of the cybersecurity knowledge base, enhancing its adaptability to complex cybersecurity tasks.
4) We create a specialized attack dataset by combining a cybersecurity knowledge base with private attack data. Important use cases demonstrate that this dataset can be effectively utilized for classical cybersecurity tasks such as threat hunting, alert analysis and attack knowledge reasoning.

The rest of this paper is organized as follows. Section 2 introduces related work. Section 3 describes the preliminary, which includes open-source datasets and attack scenarios. In Sect. 4, we introduce our cybersecurity dataset generation method. In Sect. 5, we conduct an empirical study on the dataset produced using our method and we conclude our work in Sect. 6.

2 Related Work

2.1 Cybersecurity Data and Ontology

Various cybersecurity data and ontologies have been developed in academia, industry, and government agencies. Standard cybersecurity data includes the Adversarial Tactics, Techniques, and Common Knowledge (ATT&CK) [2] for understanding adversary behavior used in cyberattacks; the Common Attack Pattern Enumeration and Classification (CAPEC) [3] for known attack patterns used by adversaries; the Common Weakness Enumeration (CWE) [6] for publicly known software and hardware weaknesses; the Common Vulnerabilities and Exposures (CVE) [5] for publicly disclosed vulnerabilities; and the Common Platform Enumeration (CPE) [4] for describing IT products and platforms.

The cybersecurity ontology describes the cybersecurity concepts and their relationships within the cybersecurity domain. Structured Threat Information eXpression (STIX) [10] is a standardized format for exchanging cyber threat intelligence, covering almost all security fields. Based on the STIX standard, some works created various cybersecurity ontologies. For instance, STUCCO [18] ontology integrates different structured and unstructured data resources, establishing a professional ontology for cybersecurity. Unified Cybersecurity Ontology (UCO) [28] aligns heterogeneous ontology schemas, providing a common standard for understanding cybersecurity knowledge. UCO 2.0 [23] extends some classes and relationships defined by STIX 2.0 [11] to update UCO. BRON [16] gains comprehensive insights by linking ATT&CK [2] with CAPEC [3], CWE [6], CVE [5], while enabling bi-directional paths to trace contextual information.

2.2 Attack Scenarios Data

Cyberattacks occur frequently in cyberspace, making Intrusion Detection Systems (IDSs) essential for identifying malicious activities in networks and systems. To improve the performance of IDS, several well-developed datasets for evaluation have been released, such as HDFS [27], KDD99 [22], CICIDS2017 [26], etc.

HDFS [27] is a distributed file system designed to store and process large datasets across clusters of computers using Hadoop, containing approximately 3% anomalous behavior data. KDD99 [22], created as part of the KDD Cup 1999 competition, is widely used in intrusion detection and network security. It records four types of attacks and contains 41 network traffic features. CICIDS2017 [26], provided by the Canadian Institute for Cybersecurity (CIC), includes network traffic data generated in simulated network environments for analyzing and detecting various cybersecurity attacks. Recent studies [15,20,21,25] have also proposed advanced feature characterization, representation, and extraction techniques.

Despite the availability of numerous datasets and methods, existing datasets fail to cover the full range of cybersecurity attacks and lack differentiation. In this work, we propose CDGM to meet the specialized needs of researchers.

3 Preliminary

3.1 Open-Source Datasets

Cybersecurity knowledge is typically collected, structured and published by industry experts, research institutions and government agencies such as the National Institute of Standards and Technology (NIST) [9] and MITRE [8]. The types of knowledge we selected include weaknesses, vulnerabilities, attack patterns and TTPs. Online documentation is available for all open-source datasets.

ATT&CK [2]. The ATT&CK is a framework for categorizing and understanding threat behavior in cyberattacks. It provides a standardized and structured way to describe the TTPs employed by attackers throughout the various stages of the cyber kill chain. The dataset is designed to help security professionals gain a comprehensive insight into the tactics and techniques commonly used by attackers allowing them to fully understand threat behavior and evaluate security defenses.

CAPEC [3]. The CAPEC is a structured list of common attack patterns. CAPEC entries include a detailed description of the attack patterns, including their objectives, prerequisites, execution flow, and potential impact. Each attack pattern is assigned a unique identifier and categorized based on various attributes such as attack type, attack phase and impact type. It plays a crucial role in enhancing cybersecurity by providing the details of attack methods and techniques.

CWE [6]. The CWE is a community-developed list of common software and hardware weaknesses that can lead to security vulnerabilities. Based on the characteristics and impact, each weakness is classified into categories and subcategories. Each weakness is assigned a unique identifier and is described with detailed information, including examples, potential consequences, and mitigation strategies. The primary goal of CWE is to provide a common language for discussing, identifying, and addressing security weaknesses in software and hardware systems.

CVE [5]. The CVE is a widely adopted standard for uniquely identifying and tracking publicly disclosed vulnerabilities in software and hardware products. CVE entries contain detailed information about vulnerabilities, including descriptions, affected products, severity ratings, and references to related resources such as patches or advisories. Each vulnerability is assigned a unique identifier in the format "CVE-YYYY-NNNN". It can improve the overall cybersecurity environment by encouraging security researchers to reference and share information about vulnerabilities.

CPE [4]. The CPE is a standardized method for describing and identifying software and hardware platforms. CPE entries consist of a standardized format comprising three main components: part, product and version. It provides a structured naming scheme that enables organizations to uniquely identify and categorize applications, hardware and operating systems. Therefore, it facilitates interoperability and integration across different security tools and platforms.

3.2 Attack Scenarios

Attack scenarios refer to various approaches used to compromise computer systems, networks, or their data. These malicious behaviors are typically designed and executed by attackers aiming to obtain confidential information, disrupt system functionality, steal private funds, or sell personal information. The cybersecurity attack scenarios exhibit diversity and dynamism, evolving continuously with technological advancements, so it is necessary to continuously improve the corresponding defense strategy.

General attack scenarios can be divided into three stages: initial access, establish control and execute commands. Initial access is exploiting a discovered vulnerability or weakness to gain access to a target system or network, which may involve using malware, phishing attacks, password cracking, etc. Establish control is building a persistent control point within the target system, usually by installing a backdoor, implanting malware, elevating privileges, etc. Executing commands is performing malicious activities, such as stealing data, modifying system configurations, conducting extortion, etc.

Among the many types of cybersecurity attacks, the three most typical attacks are phishing attacks, distributed denial of service (DDoS) attacks and malware attacks. Phishing attacks involve sending deceptive emails or messages to trick victims into disclosing sensitive information. DDoS attacks flood target systems with a massive volume of traffic, rendering services unavailable. Malware attacks distribute malicious software like viruses or ransomware to infect victims' devices, allowing attackers to steal data, control systems, or encrypt files surreptitiously. These threats highlight the importance of implementing robust security measures to safeguard against cyberattacks.

4 Proposed Method

Our proposed method generates a unique cybersecurity dataset tailored to researchers' specific demands and applicable to various downstream tasks. The dataset consists of two main parts: a cybersecurity knowledge base derived from open-source datasets and the private attack traffic data generated from specific cyberattacks. The cybersecurity knowledge base server is the foundation, storing a large amount of fundamental cybersecurity data that supports applications involving specific attack traffic data. The private traffic data, representing the apex, is obtained through our established four-role architecture. The dataset, integrating the cybersecurity knowledge base and private attack data, provides specialized services for various cybersecurity tasks.

4.1 Cybersecurity Knowledge Base

We analyze typical open-source datasets to extract valuable information as entities, adopting a top-down and abstract-to-concrete perspective. Specifically, we analyze ATT&CK, CAPEC, CWE, CVE and CPE.

There are three types of entities contained in ATT&CK: tactics, techniques and sub-techniques. Tactics serve as the parent class of techniques, and techniques serve as the parent class of sub-techniques. Notably, tactics may be directly connected to specific sub-techniques, differently through the technique. CAPEC, as an enumeration and classification of attack patterns, can be mapped to ATT&CK and related to CWE, thus serving as a bridge between ATT&CK and CWE. CWE describes weaknesses in software and hardware, with each CWE representing a separate entity. Additionally, CAPEC and CWE exhibit a hierarchical structure, encompassing specific relationships within them. A combination of one or more CWEs may manifest as a vulnerability that exposes a CVE. Each CVE operates as a distinct entity and affects multiple CPEs, thus facilitating the connection between CWE and CPE. CPE lists all the platforms affected by CVE, categorizing them into applications, hardware, and operating systems.

In summary, we identify 9 types of entities: Tactic, Technique, SubTechnique, AttackPattern, Weakness, Vulnerability, Application, Hardware and OS. Leveraging these entity types, we process the open-source datasets, yielding a total of 1,026,520 entities. Further statistics details are given in Table 1.

Table 1. Statistics for entities.

Dataset	Entity type	Number of entities	Explanation
ATT&CK	Tactic	14	The high-level objectives pursued by attackers. The columns of the ATT&CK matrix
	Technique	196	The specific methods used by attackers. The row elements of the ATT$CK matrix
	SubTechnique	411	The subdivisions or variants of Technique
CAPEC	AttackPattern	559	The common attack patterns i.e. CAPEC-id
CWE	Weakness	933	The common software and hardware weaknesses i.e. CWE-id
CVE	Vulnerability	215574	The common vulnerabilities and exposures i.e. CVE-id
CPE	Application	691059	The affected applications
	Hardware	54033	The affected hardware
	OS	65178	The affected operating systems

Note that there are too many CPE entities, so we extract "Product" and "Version" of the configuration.

Isolated entities are insufficient for conveying the semantic information of knowledge. Hence, we introduce external relationship and internal relationship to establish connections between entities. External relationship describes connections across different open-source datasets, while internal relationships denote connections within the same open-source dataset. Detailed definitions are given in Table 2.

Table 2. Statistics for relationships.

	Relationship Type	Source	Target
External relationship	Interact	Vulnerability	Weakness
		Weakness	Vulnerability
	Affect	Vulnerability	Application
		Vulnerability	Hardware
		Vulnerability	OS
	RelatedPattern	Weakness	AttackPattern
	RelatedWeakness	AttackPattern	Weakness
	RelatedAttack	AttackPattern	Tactic
			Technique
			SubTechnique
Internal relationship	ChildOf, CanPrecede Required, PeerOf CanAlsoBe, CanFollow	Weakness	Weakness
		AttackPattern	AttackPattern
		Tactic	Technique
		Tactic	SubTechnique
		Technique	SubTechnique

4.2 Private Attack Data Generated by the Four-Role Architecture

Although several researches have proposed datasets for attack scenarios, such as HDFS, KDD99 and CICIDS2017, the diverse nature of cyberattacks renders existing datasets insufficient to cover all attack scenarios comprehensively. To address this limitation, we establish a four-role architecture designed to reproduce private attack data, enabling researchers to replicate specific attack data based on their configurations.

Figure 2 presents an overview of the four-role architecture, which contain 4 essential parts: an attack server, a target server, a benign server and a data server. The attack server sends commands and launches attacks on the victims, while the target server becomes the subject of the attack. Positioned between the attack server and the target server, the benign server provides common network services such as SMS mail service and web applications. Finally, the data server aggregates traffic data from all incoming and outgoing connections. All the server systems operate on a Unix-based platform and are interconnected via a central switch.

Upon successful implementation of the attack scenario, we employ existing tools like tcpdump to collect both benign and malicious traffic data, which is stored in the data server. To convert the traffic data into cybersecurity knowledge, we replay data stored in the data server and analyze it using IDS devices. These devices generate alert logs containing multiple kinds of cybersecurity

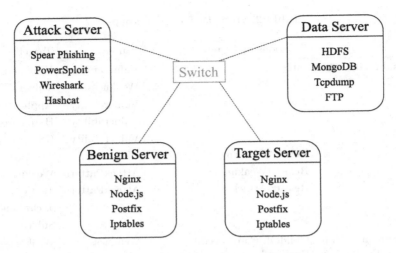

Fig. 2. Four-role architecture of generated private attack data.

knowledge. Users can filter the cybersecurity knowledge contained in the alert logs, and we extract the knowledge related to 9 types of entities based on the designed cybersecurity ontology.

We integrate the private attack data with a cybersecurity knowledge base to conduct a specific cybersecurity dataset. Utilizing this dataset, we can query details for various entities. For example, we can search the information about "Weakness: CWE-204 (Observable Response Discrepancy)" [7]. This weakness occurs when a product provides different responses to incoming requests, inadvertently revealing internal state information to unauthorized actors outside of the intended control sphere. Such discrepancies are particularly prevalent during authentication processes, where variations in failed-login messages may allow attackers to ascertain the validity of usernames. These exposures can arise from inadvertent bugs or intentional design flaws.

Furthermore, querying the dataset produced using our method reveals that "Weakness: CWE-204" is related to 12 vulnerabilities such as "Vulnerability: CVE-2002-2094", "Vulnerability: CVE-2004-0778", "Vulnerability: CVE-2004-1428" with the relationship "Interact". Additionally, we discover that "AttackPattern: CAPEC-331", "AttackPattern: CAPEC-332", "AttackPattern: CAPEC-541" and "AttackPattern: CAPEC-580" are linked to "Weakness: CWE-204" via the "RelatedPattern" relationship. Moreover, these four attack patterns are connected to "Technique: T1082" and "SubTechnique: T1592.002" in ATT&CK. Furthermore, "Weakness: CWE-204" and "Weakness: CWE-203", as well as "AttackPattern: CAPEC-331" and "AttackPattern: CAPEC-332", can be connected through "ChildOf". Figure 3 shows some of the paths that link "Weakness: CWE-204" to other entities.

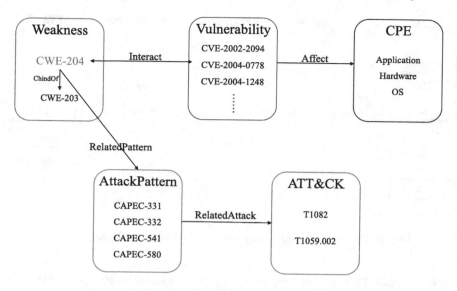

Fig. 3. The paths related to "Weakness: CWE-204".

5 Experiments

We employ the four-role architecture to reproduce the Eternal Blue attack and gather private attack data. Leveraging the cybersecurity knowledge base derived from open-source datasets, we generate a scenario-specific cybersecurity dataset tailored to Eternal Blue.

Eternal Blue exploits the SMB vulnerability of the Windows system, enabling attackers to obtain the highest privilege of the system privileges directly. Numerous hackers have capitalized on this vulnerability to create ransomware, underscoring the significant and persistent threat posed by Eternal Blue. In our emulation, we use Kail and Windows as the attack servers to simulate hacking behaviors, while employing Internet Information Services (IIS) as the benign server. Additionally, a Windows 7 machine serves as the target server. Initially, the attacker leverages the WebDAV write permission of the IIS within the internal network to upload a backdoor. Subsequently, the attacker establishes forwarding with port 445 of the target server, using the IIS as a springboard. Finally, the attacker exploits CVE-2017-0146 in the Windows 7 to execute remote code, thereby gaining the highest system privileges.

5.1 Threat Hunting

Here we perform a threat hunting analysis of Eternal Blue (Vulnerability: CVE-2017-0146), and Fig. 4 shows the entities associated with this vulnerability.

CPE. The "Vulnerability: CVE-2017-0146" exists in a variety of Microsoft products, including one type of Application: "Application: server_message_

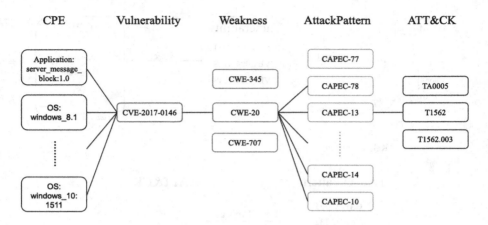

Fig. 4. The threat hunting analysis of "Weakness: CVE-2017-0146".

block:1.0" and 12 types of OS, including "OS: windows_10:1511", "OS: windows_8.1", "OS: windows_server_2008:r2" and so on.

CWE. The "Vulnerability: CVE-2017-0146" is one of the observed examples of the "Weakness: CWE-20" (Improper Input Validation): the product receives input or data, but it does not validate or incorrectly validates that the input has the properties that are required to process the data safely and correctly. Meanwhile, this weakness has internal relationships, "Weakness: CWE-20" and "Weakness: CWE-707" are linked by "ChildOf", and "Weakness: CWE-20" and "Weakness: CWE-345" are linked by "PeerOf".

CAPEC. The weakness related to "Vulnerability: CVE-2017-0146" is a feature of "AttackPattern: CAPEC-13" (Subverting Environment Variable Values): the adversary directly or indirectly modifies environment variables used by or controlling the target software. The adversary's goal is to cause the target software to deviate from its expected operation in a manner that benefits the adversary. This attack pattern has internal relationships with "AttackPattern: CAPEC-77", "AttackPattern: CAPEC-14" and "AttackPattern: CAPEC-10".

ATT&CK. The attack patterns associated with "Vulnerability: CVE-2017-0146" is closely related to "SubTechnique: T1562.003" (Impair Command History Logging): adversaries prevent the logging of command history on a compromised system, hindering forensic analysis and detection efforts. This Subtechnique belongs to "Technique: T1562" (Impair Defenses), which falls under "Tactic: TA0005" (Defense Evasion).

Based on threat hunting, we can gain insights into Eternal Blue's attack strategies and attack patterns, allowing defenders to implement protective measures at the corresponding stage. For example, "CWE-20" suggests that defenders can carry out "Input Validation" at the "Implementation" stage.

5.2 Alert Analysis

2022 Top Routinely Exploited Vulnerabilities [1] is a regularly issue technical guidance by The Cybersecurity and Infrastructure Security Agency (CISA) and the Federal Bureau of Investigation (FBI). It encourages vendors, designers, developers, and end-user organizations to implement the recommendations found within the Mitigations section of this advisory–including the following–to reduce the risk of compromise by malicious cyber actors. We count the top serious vulnerabilities in 2022 and track the weaknesses, attack patterns and platforms corresponding to each vulnerability, Table 3 shows the results for the 12 most commonly used vulnerabilities in 2022. Note that because there are fewer mapping relationships between CAPEC and ATT&CK, TTPs (Tactic, Technique and Subtechnique) are not listed.

Table 3. 2022 top routinely exploited vulnerabilities.

Vulnerability	#Weakness	#AttackPattern	#Application, Hardware and OS
CVE-2018-13379	1	5	2
CVE-2021-34473	1	1	5
CVE-2021-31207	1	5	5
CVE-2021-34523	1	10	5
CVE-2021-40539	1	10	170
CVE-2021-26084	1	37	8
CVE-2021-44228	4	55	397
CVE-2022-22954	1	3	13
CVE-2022-22960	1	3	13
CVE-2022-1388	1	5	66
CVE-2022-30190	1	1	18
CVE-2022-26134	1	37	14

Table 3 indicates that the most commonly exploited CVE is "Vulnerability: CVE-2018-13379", which affect 2 configurations of products, both exhibiting the same weakness: "Weakness: CWE-22" (Path Traversal). This weakness corresponds to 5 attack patterns. On the other hand, "Vulnerability: CVE-2021-44228", ranked seventh, affects the most products, with a total of 397, corresponding to 4 weaknesses. These weaknesses are the features of 55 different attack patterns.

Multiple vulnerabilities exploit "CWE-22" (Path Traversal), "CWE-287" (Improper Authentication), and "CWE-74" (Injection). It indicates that these weaknesses pose significant security threat. If an attack environment simulates platforms with these vulnerabilities and weaknesses, users should replace the affected platforms, or refer to the specific recommendations given in 'Potential Mitigations'.

5.3 Attack Knowledge Reasoning

We train the knowledge graph embedding model based on the customized cybersecurity dataset. Specifically, we select the CVE entities and involved CPE entities from 2018 to 2019 within the cybersecurity knowledge base. Due to the limited number of explicit links between CAPEC and ATTC_K, and the insufficient number of triples containing "CanAlsoBe", "Required", and "CanFollow" (fewer than 50), which is not enough to support the KGE model. We exclude these data from the experiment, Table 4 and Table 5 detail the statistics of entities and relationships.

We choose TransE [13] as the classical KGE model, it is a translational distance model, we believe that the performance of TransE is representative of most KGE models. Meanwhile, we use the link prediction task to evaluate TransE. Toward the link prediction task, the following two metrics are generally used to evaluate the KGE model: Mean reciprocal ranking (MRR) and Hits@n.

As can be seen from Table 6, the cybersecurity knowledge embedding using TransE performs well. Among them, the external relationship "Affect" and the internal relationship "ChildOf" achieve highest performance respectively, likely due to the sufficient amount of knowledge containing these two relationships. Note that there is a limitation that we selectively pick knowledge as the training data, but the available experimental results are sufficient to demonstrate that the

Table 4. The statistics of entities in customized cybersecurity dataset.

Entity type	Number of entities
AttackPattern	559
Weakness	933
Vulnerability	31187
Application	28353
Hardware	2228
OS	2633

Table 5. The statistics of relationships in customized cybersecurity dataset.

		Relationship type	Number of relationships
Cybersecurity knowledge base	External relationship	Interact	27317
		Affect	164577
		RelatedPattern	610
		RelatedWeakness	724
	Internal relationship	ChildOf	1601
		CanPrecede	321
		PeerOf	135
Private attack knowledge	External relationship	Interact	3148
		Affect	25176

Table 6. The performance of relationships in TransE.

	MRR	Hits@10
Interact	0.243	0.427
Affect	0.305	0.523
RelatedPattern	0.207	0.334
RelatedWeakness	0.198	0.328
ChildOf	0.421	0.761
CanPrecede	0.083	0.200
PeerOf	0.229	0.383
Total	0.289	0.457

cybersecurity dataset generated through CDGM can satisfy the task of attack knowledge reasoning.

6 Conclusion

In this paper, we proposed a controllable dataset generation method, CDGM, designed to produce a customized cybersecurity dataset tailored to various attack scenarios. The dataset comprises a cybersecurity knowledge base and private attack data. While the cybersecurity knowledge base acts as the basis for supporting the private attack data, the private attack data enhances the utility of the cybersecurity knowledge base. We established a four-role architecture to generate diverse cyberattack data with unique assets, filling the gaps left by existing public datasets. Experimental results indicate that the cybersecurity dataset constructed by our method is effectively suited for cybersecurity applications. In further work, we will explore additional techniques to refine the relationship between CAPEC and ATT&CK, thereby further enhancing the usability of the customized cybersecurity dataset.

Acknowledgments. This work was supported in part by the Major Key Project of PCL (Grant No. PCL2023A07-4), the National Natural Science Foundation of China (Grant No. 62372137), and the Guangxi Natural Science Foundation (No. 2022GXNSFBA035650).

References

1. 2022 top routinely exploited vulnerabilities, https://www.cisa.gov/news-events/cybersecurity-advisories/aa23-215a
2. Att&ck matrix for enterprise, https://attack.mitre.org/
3. Common attack pattern enumeration and classification, https://capec.mitre.org/
4. Common platform enumeration, https://cpe.mitre.org/
5. Common vulnerabilities and exposure, https://cve.mitre.org/

6. Common weakness enumeration, https://cwe.mitre.org/
7. Cwe-204 detail, https://cwe.mitre.org/data/definitions/204.html
8. Mitre, https://www.mitre.org/
9. National institute of standards and technology, https://www.nist.gov/
10. Stix 1.0 documentation, https://stixproject.github.io/documentation/
11. Stix 2.0 documentation, https://oasis-open.github.io/cti-documentation/stix/examples.html
12. Akbanov, M., Vassilakis, V.: Wannacry ransomware: Analysis of infection, persistence, recovery prevention and propagation mechanisms. Journal of Telecommunications and Information Technology **1**, 113–124 (04 2019). https://doi.org/10.26636/jtit.2019.130218
13. Bordes, A., Usunier, N., García-Durán, A., Weston, J., Yakhnenko, O.: Translating embeddings for modeling multi-relational data. In: Burges, C.J.C., Bottou, L., Ghahramani, Z., Weinberger, K.Q. (eds.) Advances in Neural Information Processing Systems 26: 27th Annual Conference on Neural Information Processing Systems 2013. Proceedings of a meeting held December 5-8, 2013, Lake Tahoe, Nevada, United States. pp. 2787–2795 (2013), https://proceedings.neurips.cc/paper/2013/hash/1cecc7a77928ca8133fa24680a88d2f9-Abstract.html
14. Carrasco, M.A.M., Wu, C., Fuertes, W.: Adversarial examples: A survey of attacks and defenses in deep learning-enabled cybersecurity systems. Expert Syst. Appl. **238**(Part E), 122223 (2024), https://doi.org/10.1016/j.eswa.2023.122223
15. Chou, D., Jiang, M.: A survey on data-driven network intrusion detection. ACM Comput. Surv. **54**(9), 182:1–182:36 (2022), https://doi.org/10.1145/3472753
16. Hemberg, E., Kelly, J., Shlapentokh-Rothman, M., Reinstadler, B., Xu, K., Rutar, N., O'Reilly, U.M.: Linking threat tactics, techniques, and patterns with defensive weaknesses, vulnerabilities and affected platform configurations for cyber hunting. CoRR **abs/1905.02497** (2021), http://arxiv.org/abs/1905.02497
17. Hwang, R., Lee, C., Lin, Y., Lin, P., Wu, H., Lai, Y., Chen, C.K.: Host-based intrusion detection with multi-datasource and deep learning. J. Inf. Secur. Appl. **78**, 103625 (2023). https://doi.org/10.1016/j.jisa.2023.103625
18. Iannacone, M.D., Bohn, S., Nakamura, G., Gerth, J., Huffer, K.M.T., Bridges, R.A., Ferragut, E.M., Goodall, J.R.: Developing an ontology for cyber security knowledge graphs. In: Proceedings of the 10th Annual Cyber and Information Security Research Conference, CISR '15, Oak Ridge, TN, USA, April 7-9, 2015. pp. 12:1–12:4. ACM (2015), https://doi.org/10.1145/2746266.2746278
19. Lin, Y., Wang, Z., Lin, P., Nguyen, V., Hwang, R., Lai, Y.: Multi-datasource machine learning in intrusion detection: Packet flows, system logs and host statistics. J. Inf. Secur. Appl. **68**, 103248 (2022). https://doi.org/10.1016/j.jisa.2022.103248
20. Martín, M.L., Carro, B., Arribas, J.I., Sánchez-Esguevillas, A.: Network intrusion detection with a novel hierarchy of distances between embeddings of hash IP addresses. Knowl. Based Syst. **219**, 106887 (2021). https://doi.org/10.1016/j.knosys.2021.106887
21. Martín, M.L., Sánchez-Esguevillas, A., Arribas, J.I., Carro, B.: Supervised contrastive learning over prototype-label embeddings for network intrusion detection. Inf. Fusion **79**, 200–228 (2022). https://doi.org/10.1016/j.inffus.2021.09.014
22. Özgür, A., Erdem, H.: A review of KDD99 dataset usage in intrusion detection and machine learning between 2010 and 2015. PeerJ Prepr. **4**, e1954 (2016). https://doi.org/10.7287/peerj.preprints.1954v1

23. Pingle, A., Piplai, A., Mittal, S., Joshi, A., Holt, J., Zak, R.: Relext: relation extraction using deep learning approaches for cybersecurity knowledge graph improvement. In: ASONAM '19: International Conference on Advances in Social Networks Analysis and Mining, Vancouver, British Columbia, Canada, 27-30 August, 2019. pp. 879–886. ACM (2019), https://doi.org/10.1145/3341161.3343519
24. Ren, Y., Xiao, Y., Zhou, Y., Zhang, Z., Tian, Z.: CSKG4APT: A cybersecurity knowledge graph for advanced persistent threat organization attribution. IEEE Trans. Knowl. Data Eng. **35**(6), 5695–5709 (2023). https://doi.org/10.1109/TKDE.2022.3175719
25. Sarhan, M., Layeghy, S., Portmann, M.: Towards a standard feature set for network intrusion detection system datasets. Mob. Networks Appl. **27**(1), 357–370 (2022). https://doi.org/10.1007/s11036-021-01843-0
26. Sharafaldin, I., Lashkari, A.H., Ghorbani, A.A.: Toward generating a new intrusion detection dataset and intrusion traffic characterization. In: Mori, P., Furnell, S., Camp, O. (eds.) Proceedings of the 4th International Conference on Information Systems Security and Privacy, ICISSP 2018, Funchal, Madeira - Portugal, January 22-24, 2018. pp. 108–116. SciTePress (2018), https://doi.org/10.5220/0006639801080116
27. Shvachko, K., Kuang, H., Radia, S., Chansler, R.: The hadoop distributed file system. In: Khatib, M.G., He, X., Factor, M. (eds.) IEEE 26th Symposium on Mass Storage Systems and Technologies, MSST 2012, Lake Tahoe, Nevada, USA, May 3-7, 2010. pp. 1–10. IEEE Computer Society (2010), https://doi.org/10.1109/MSST.2010.5496972
28. Syed, Z., Padia, A., Finin, T., Mathews, M.L., Joshi, A.: UCO: A unified cybersecurity ontology. In: Artificial Intelligence for Cyber Security, Papers from the 2016 AAAI Workshop, Phoenix, Arizona, USA, February 12, 2016. AAAI Technical Report, vol. WS-16-03. AAAI Press (2016), http://www.aaai.org/ocs/index.php/WS/AAAIW16/paper/view/12574
29. Xiao, L., Xu, D., Mandayam, N.B., Poor, H.V.: Attacker-centric view of a detection game against advanced persistent threats. IEEE Trans. Mob. Comput. **17**(11), 2512–2523 (2018). https://doi.org/10.1109/TMC.2018.2814052
30. Xu, W., Huang, L., Fox, A., Patterson, D.A., Jordan, M.I.: Detecting large-scale system problems by mining console logs. In: Fürnkranz, J., Joachims, T. (eds.) Proceedings of the 27th International Conference on Machine Learning (ICML-10), June 21-24, 2010, Haifa, Israel. pp. 37–46. Omnipress (2010), https://icml.cc/Conferences/2010/papers/902.pdf

Enhancing Privacy in Big Data Publishing: η-Inference Model

Zhenyu Chen[1], Lin yao[1], Guowei Wu[1], and Shisong Geng[2](✉)

[1] School of Software, Dalian University of Technology, Dalian, China
{yaolin,wgwdut}@dlut.edu.cn
[2] Institute of Software Chinese Academy of Sciences, Beijing, China
shisong@iscas.ac.cn

Abstract. The advent of big data has significantly advanced the development of various applications and services by leveraging the extensive informational benefits of data publishing. However, the sensitive information contained within these data requires robust privacy protection prior to sharing and publishing. Effective privacy protection demands a clear definition of privacy standards and guarantees. While differential privacy (DP) and its variants are widely considered the gold standard for privacy preservation, their practical implementation often faces challenges that can lead to undesirable outcomes. This paper highlights the limitations of DP, such as the misuse of sequential composition, the presence of dishonest entities, and a privacy threat known as stigmatization, which falls outside the scope of DP's guarantees. To address these challenges, we propose the η-inference model, which ensures information limited privacy and effectively mitigates the problems encountered with DP. Additionally, the η-inference model exhibits data incrementally invariant properties, making it particularly suitable for dynamic and distributed data publishing scenarios.

Keywords: Data publishing · privacy protection · η-inference

1 Introduction

In the era of big data, the vast benefits of data sharing are evident, particularly in enhancing knowledge-based decision-making across various domains such as data publishing, smart homes, and health tracking [13]. However, this data sharing presents a dual challenge: while data owners aim to maximize the utility of shared information, they also face significant risks of privacy breaches, especially when users accessing the data may not be entirely trustworthy. This necessitates robust privacy protection mechanisms to prevent unauthorized access to sensitive information.

In typical data publishing scenarios, the data owner begins by selecting a specific privacy definition, such as differential privacy (DP) or k-anonymity, to provide a qualitative privacy guarantee. Subsequently, a quantitative privacy guarantee is established by assigning values to privacy parameters. The critical step involves finding or designing an anonymization mechanism that ensures the privacy of the published database is adequately protected while maximizing utility, such as optimizing error rates and misclassification rates [26]. Ultimately, the database is published based on this anonymization mechanism [11,16,28,32].

However, privacy protection often comes at the cost of utility. The challenge lies in choosing a privacy definition that maintains high utility while providing adequate qualitative privacy guarantees. A strict privacy definition can unnecessarily damage utility, while a loose definition may lead to unpredictable privacy threats. Over the past decades, differential privacy has been considered the golden standard for data privacy, offering robust protection against adversaries with arbitrary background knowledge [19]. DP operates by ensuring that an adversary cannot distinguish between any two neighboring databases based on the outputs of a DP mechanism [6].

Despite its strengths, DP is not a one-size-fits-all solution. Different applications have diverse characteristics: some datasets are more sensitive than others, some applications need to protect aggregate secrets, while others must protect individual secrets, and in some cases, only certain attributes need protection [13]. To address these varied needs, researchers have developed numerous DP variants, such as Local Differential Privacy (LDP) [5], d-privacy [4], Pufferfish privacy [13], Bayesian DP [27], and Rényi DP [18], among others, to provide tailored privacy guarantees for specific applications.

Nonetheless, practical challenges remain. As noted in [30], the noise added by DP mechanisms cannot be verified by the data recipient. This can lead to additional privacy or utility issues if the data owner or service provider is not trustworthy. For instance, a data owner might add excessive noise to ensure better privacy, resulting in significant utility loss. Conversely, a service provider might reduce the noise to enhance service competitiveness, thereby compromising privacy. Furthermore, Kenny et al. [12] found that DP can introduce additional privacy risks compared to traditional definitions like l-diversity.

Given these limitations, it is clear that DP alone is insufficient to address the privacy challenges in various applications. Therefore, it is crucial to reassess the privacy guarantees needed and explore new, feasible approaches to privacy protection. This paper proposes the η-inference model as a solution, offering information limited privacy that addresses the shortcomings of DP and adapts to dynamic and distributed data publishing scenarios.

1.1 Contribution

In this paper, we first summarize the current limitations of DP and its variants in the practice of privacy protection. Then, we propose an η-inference model to ensure the information limited privacy, which can be used to address the limitations of DP.

The main contributions of this work are summarized as follows:

- We highlight the practical challenges and limitations of DP, including the misuse of sequential composition, the presence of dishonest entities, and the privacy threat of stigmatization, which DP does not address.
- We introduce a new privacy threat called stigmatization, where an adversary can cause significant harm without needing to accurately identify the true privacy information of the victim. We conclude that ensuring information limited privacy is essential to prevent stigmatization.
- We propose a new η-inference model to guarantee information limited privacy. It can effectively mitigate the problems associated with DP, providing a more robust privacy guarantee across various scenarios.

The remainder of this paper is organized as follows. In Sect. 2, we introduce the related work. In Sect. 3, we describe the model for the database and notations. The limitations of DP is introduced in Sect. 4 and η-inference is presented in Sect. 5. In Sect. 6, we show the simulation results on four real datasets. Finally, we conclude our work.

2 Related Work

Existing privacy definitions can generally be classified into two categories: syntactic privacy and mechanism privacy [7].

Syntactic Privacy. The syntactic privacy is guaranteed by controlling the form of the output to hide the privacy information in the published database with certain assumptions. For instance, k-anonymity [23] requires that each record in the published database be indistinguishable from at least k-1 other records based on quasi-identifiers. This ensures that a potential adversary cannot pinpoint an individual's identity if k-anonymity is maintained. Expanding on this concept, various other syntactic privacy definitions have been proposed to address different types of information that need to be protected. These include l-diversity [15], which ensures that sensitive attributes have at least l well-represented values, and t-closeness [14], which ensures that the distribution of sensitive attributes in any equivalence class is close to the distribution of the attribute in the overall database. Other notable definitions include (α,k)-anonymity [25], (α,β)-privacy [22], t-closeness [14], δ-presence [20].

Mechanism Privacy. Mechanism privacy, on the other hand, limits the relationship between the original and the published databases by applying privacy-preserving mechanisms. Differential privacy (DP), introduced by Dwork [6], is a prominent example. When a DP mechanism perturbs the published database, it ensures that a potential adversary cannot confidently determine whether a given piece of information comes from the original database or from the added noise. This makes it difficult for adversaries to establish a precise relationship between sensitive values in the published database and the original data, thereby protecting the privacy of the original database.

Various extensions of DP have been developed to provide different types of privacy guarantees for specific applications. These include Pufferfish privacy [13], which allows for customizable privacy definitions based on the adversary's knowledge and the data's sensitivity, and d-privacy [4], which broadens the scope of DP to different data types and applications. Other notable variants are Blowfish privacy [8], which introduces policies to specify which data correlations should be protected, and Local Differential Privacy (LDP) [5], which ensures privacy at the individual data contributor level. Additional approaches include information privacy [21], plausible deniability [1], Bayesian DP [27], Rényi DP [18], and local information privacy [10].

In summary, while syntactic privacy focuses on anonymizing data based on predefined assumptions, mechanism privacy employs mathematical mechanisms to obscure the relationship between the original and published data. Both approaches have their strengths and limitations, and ongoing research continues to explore and refine these privacy definitions to address the evolving challenges in data privacy.

2.1 The Privacy Guarantee

Based on the classification, we define two types of privacy guarantees for data publishing, described as follows:

Privacy type 1 (Source information limited privacy) *By observing the published database $\tilde{X}$, the ability of the potential adversary to obtain privacy information accurately from the original database X based on $\tilde{X}$ should be limited.*

Privacy type 2 (Information limited privacy) *By observing the published database $\tilde{X}$, the ability of the potential adversary to obtain privacy information accurately based on $\tilde{X}$ should be limited.*

Source information limited Privacy can be effectively ensured through mechanism privacy approaches, such as differential privacy. In contrast, information limited privacy cannot be adequately guaranteed by syntactic privacy methods, as these typically address only specific types of sensitive information.

3 Model for Database and Notations

As shown in Table. 1, we consider an original database X with m rows and n columns. Each row represents a record of an individual, and each column represents an attribute. For all $i \in \mathcal{I} = \{1, 2, \cdots, m\}$ and $j \in \mathcal{J} = \{1, 2, \cdots, n\}$, let $X^{(i,:)}$ and $X^{(:,j)}$ be the i-th row and j-th column, respectively. Let $X^{(\mathcal{I},:)} \equiv X^{(\mathcal{I},\mathcal{J})} \equiv X^{(:,\mathcal{J})}$. In this paper, the original database X is considered to be a sequence of m independent observations from $\mathcal{X}$, where $\mathcal{X}$ is the union of all databases. That is, each attribute in $\mathcal{X}$ contains all possible values and $\mathcal{X}$ contains all possible records. The probability distribution on the i-th observation $X^{(i,:)}$ which obtained from $\mathcal{X}$ is formulated as,

$$\theta_{X^{(i,\mathcal{J})}} = Pr(X^{(i,1)} \leftarrow \rho^{(i,1)}, \cdots, X^{(i,n)} \leftarrow \rho^{(i,n)}), \qquad (1)$$

where $\rho^{(i,j)}$ is a probability mass function if $X^{(i,j)}$ is a discrete attribute with $\sum_{\rho^{(i,j)} \in \rho^{(i,j)}} \rho^{(i,j)} = 1$, and a probability density function if $X^{(i,j)}$ is a continuous attribute with $\int_{-\infty}^{+\infty} \rho^{(i,j)}(x) dx = 1$.

Table 1. An Example of Tabular Data

Name	Gender	Job	Age	Zipcode	Salary	Disease
Alice	F	Teacher	30	11100	4500	HIV
Ben	M	Engineer	33	12200	6000	Flu
Cary	F	Engineer	45	12200	4700	SARS
David	M	Teacher	42	11100	4500	Fever
Eric	M	Doctor	32	25300	6700	Flu

Notations of the Privacy Protection Scenario. As mentioned before, the data owner first select a specific privacy definition $PD \in \mathcal{PD}$. Then, values for specific privacy parameters denoted with Greek alphabet such as ϵ or η and α et al., are assigned. After that, a mechanism M which can satisfy the chosen privacy definition is found or designed, which is denoted as $M \leftarrow \mathcal{M}(PD)$. The data owner applied the mechanism M on the original database $X \subset \mathcal{X}$ and output the database for publishing $\tilde{X}$, which is denoted as $M(X) = \tilde{X}$.

4 The Limitations of DP

Specifically, differential privacy was proposed with the idea of **Privacy type** 1, where the goal is to limit the ability of the potential adversary to infer the information from the original database X based on the output of mechanisms that satisfy differential privacy.

A standard definition of differential privacy is formalized as follows:

Definition 1 (Differential privacy) *A differential privacy definition on the random mechanism M requires the output of M on any two neighboring database $X \subset \mathcal{X}$ and $X' \subset \mathcal{X}$ differing in one element is similar. Specifically, for any $\epsilon \in \mathbb{R}^+$, the ϵ-DP on M is satisfied if the following inequality holds:*

$$\forall S \subset Range(M), Pr[M(X) \in S] \leq e^{\epsilon} Pr[M(X') \in S], \qquad (2)$$

where $Range(M)$ denotes the range of the output of M.

The definition of differential privacy means that, by observing the output of a random mechanism M satisfying differential privacy, a malicious adversary cannot distinguish the real database X from any possible neighboring database X' differs from X by one element. To ensure this characteristic, the mechanism M has to ensure the information from X is limited, typically by adding random noise that is independent of how X is sampled from $\mathcal{X}$.

4.1 Misusing of Sequential Composition

DP is appealing not only because it can resist various privacy threats but also because it possesses advantageous properties such as post-processing and sequential composition. Specifically, sequential composition is the property most susceptible to abuse, potentially leading to violations of DP. It is described as,

Theorem 1 (Sequential composition). *Given $M_1(\cdot) : \mathcal{X} \to \mathbb{S}$ satisfies ϵ_1-DP, and $M_2(\cdot,\cdot) : \mathbb{S} \times \mathcal{X} \to \mathcal{X}$ satisfies ϵ_2-DP for any $s \in \mathbb{S}$, then $M(X) = M_2(M_1(X), X)$ satisfies $(\epsilon_1 + \epsilon_2)$-DP.*

Given the following example, the sequential composition is easy likely to be abused while violate the privacy guarantee provided by DP:

Example 1. Considering Alice is going to adopt $M_1(\cdot) : \mathcal{X} \to \mathcal{X}$ that satisfies ϵ_1-DP to anonymize X into a perturbed version $\tilde{X}$ for publishing. Unfortunately, due $M_1(\cdot)$ is a random mechanism, the first-round output $M_1(X) = \tilde{X}_1$ does not yield satisfactory results. Subsequently, Alice initiates a second round of anonymization using $M_1(\cdot)$. Fortunately, $M_1(X) = \tilde{X}_2$ provides better utility compared to $\tilde{X}_1$. Consequently, Alice decides to publish $\tilde{X}_2$ and discard $\tilde{X}_1$.

It is a common phenomenon when using random mechanisms to anonymize the original database X. The issue is that, Alice in *Example* 1 introduces an additional selection mechanism between $\tilde{X}_1$ and $\tilde{X}_2$, denoted as $M_2(\cdot,\cdot) : \mathcal{X} \times \mathcal{X} \to \mathcal{X}$. Specifically, because Alice ensures that $\tilde{X}_2$ has better utility than $\tilde{X}_1$ and publishes $\tilde{X}_2$, M_2 only satisfies $\epsilon_2 = +\infty$-DP. As a result, the overall process $M(X) = M_2(M_1(X), X)$ provides $\epsilon_1 + \epsilon 2 = +\infty$-DP. Consequently, no privacy guarantee can be ensured, even if M_1 is an outstanding mechanism.

4.2 Presence of Dishonest Entities

As previously mentioned, the noise added by a DP mechanism cannot be verified by the recipient, leading to a significant trust issue because untrustworthy entities cannot be detected or defended against.

Example 2. Recall the mistake in *Example* 1: Alice published the database $\tilde{X}_2$ with $M(X) = M_2(M_1(X), X)$, which provides $+\infty$-DP. However, Alice ignored the mistake she made and she claimed that $\tilde{X}_2$ is well protected with $M_1(X)$ with ϵ_1-DP.

In this scenario, even though privacy is not adequately protected, the recipient cannot verify whether the privacy measures are correctly implemented. Consequently, a malicious service provider might purposely reduce the noise added by the DP mechanism to enhance service quality and gain a competitive market advantage. Meanwhile, a similar trust crisis can arise on the local user side.

Example 3. As mentioned in [30], local users are reluctant to share their data containing privacy information. Even though the DP in a local model is adopted, the local users may deliberately add more noise than the LDP mechanism generates to better protect the privacy of their data.

In this case, the utility of the data could be severely compromised. Although [30] claims that they can verify the noise added by local users, the verification algorithm can be bypassed by simply adding noise twice. Consequently, because the noise added by the DP mechanism cannot be verified by the recipient, it introduces extra privacy or utility issues when the publishers or local users are dishonest.

4.3 The Privacy Threat Outside DP

DP is often regarded as the golden standard for privacy protection, as it can safeguard users' privacy against adversaries with arbitrary background knowledge [19], which is consistent with the description in **Privacy type** 1. However, DP does not provide guarantees for **Privacy type** 2.

Table 2. The original database X

Name	Disease
Alice	HIV
Ben	Flu
Cary	SARS
...	...

Table 3. Perturbed database $\tilde{X}$

Name	Disease
Alice	SARS
Ben	Flu
Cary	Fever
...	...

Example 4 (Differential privacy not enough). Considering an original database X shown in Table. 2, which contains two attributes: name and disease. The data publisher is going to perturb X into $\tilde{X}$ using an exponential mechanism M which has been proven to be a random mechanism satisfying differential privacy [17]. As a result, the perturbed database $\tilde{X}$ can be viewed as a permuted version of X. In this case, the malicious adversary can determine the disease information of each individual appearing in $\tilde{X}$ with a probability of 1.0. Consequently, while the malicious adversary cannot obtain privacy information from X with high probability, they can obtain privacy information from $\tilde{X}$ with high probability.

This introduces a stigmatization threat, which falls outside the scope of DP.

Example 5 (Stigmatization). According to Example 4, the malicious adversary can determine the disease information of each individual appearing in $\tilde{X}$ with a probability of 1.0. Without loss of generality, let Table. 3 be the perturbed database $\tilde{X}$. For Alice, the adversary can determine that Alice has SARS from $\tilde{X}$. In this case, the malicious adversary can spread the disease information of Alice and claim that Alice does have $SARS$. Specifically, Alice and the data publisher cannot deny the disease information that the malicious adversary claims, because such deny requires extra information of the original database which would violate the privacy guarantee provided by DP as mentioned in *Example 1*. Therefore, *Alice* has to endure the reputation of $SARS$ even though Alice does not have $SARS$ in Table. 2. Additionally, since $SARS$ is an infectious disease, people around her may stay away from her, leading to negative social consequences due to inadequate privacy handling.

One may argue that stigmatization should not be considered as a privacy threat. However, the cyberbullying without any reason has become a high-frequency event in the current world, let alone such a reason that cannot be denied by the entities involved [9].

5 The Privacy Type 2 and η-Inference

In this section, we first propose the η-inference to guarantee **Privacy type** 2. Additionally, we introduce the corresponding properties and their significance.

5.1 The η-Inference Definition

We argue that the information is independent of the published database $\tilde{X}$ should not be considered when guaranteeing **Privacy type** 2. In this paper, we propose a new privacy definition called η-inference to guarantee **Privacy type** 2, which is defined as follows:

Definition 2 (η-inference)

$$\forall a \in A, \tilde{x} \in \tilde{X}, \boldsymbol{\rho}_{\tilde{x}} \in \boldsymbol{P}_{\tilde{x}},$$
$$\max(\boldsymbol{\rho}_{\tilde{x}}) \cdot Pr[\tilde{x} \leftarrow \boldsymbol{\rho}_{\tilde{x}} | a] \leq \eta, \qquad (3)$$

where a is a piece of background knowledge in the domain A and η is the threshold for the probability. The significance of η-inference is that based on any background knowledge $a \in A$, the probability of the malicious adversary to obtain the information in $\tilde{X}$ is limited by η.

The drawback of the η-inference model is that it requires an extra construction for the background knowledge domain. As discussed in [2], inference on statistics is not considered a privacy violation. Specifically, based on the background knowledge that "Mr. S is a smoker" in [2], an adversary can infer that "Mr. S has an elevated cancer risk" even if differential privacy is adopted, when the output database indicates a strong correlation between smoking and cancer. Therefore, background knowledge about specific individuals should not be permitted; otherwise, privacy cannot be maintained. In this paper, we construct the background knowledge by following the methods in [29]

Theorem 2. *To provide **Privacy Type** 2 privacy information related to $\tilde{X}$ is equivalent to provide **Privacy Type** 2 privacy information in $\tilde{X}$.*

Proof. For **Privacy Type** 2, if the probability of the malicious adversary to obtain the privacy information in $\tilde{X}$ is not limited, **Privacy Type** 2 cannot be guaranteed. Therefore, we only to prove that if the probability of the malicious adversary to obtain the privacy information in $\tilde{X}$ is limited, the probability of the malicious adversary to obtain the privacy information related to $\tilde{X}$ is also limited at least to the same extent.

According to the condition, the following equation should hold:

$$\forall a \in A, \tilde{x} \in \tilde{X}, \rho_{\tilde{x}} \in P_{\tilde{x}}$$
$$\max(\rho_{\tilde{x}}) \cdot Pr[\tilde{x} \leftarrow \rho_{\tilde{x}}|a] \leq \eta,$$

Specifically, the privacy information related to $\tilde{X}$ can be separated into two parts: X and $\tilde{X} - X$. The first part contains the privacy information of the original database X and the second part contains the extra privacy information added by the privacy protection mechanism.

It is worth noting that privacy information in $\tilde{X} - X$ and $\tilde{X} \cap X$ is already limited due to the condition. For $X - \tilde{X}$, the malicious adversary cannot obtain the privacy information in $X - \tilde{X}$ because it is not meant for publishing. Therefore, the malicious adversary has to infer the privacy information in $X - \tilde{X}$ based on $\tilde{X}$.

In this case, we assume there is a max transfer probability $Pr_{max}[X|\tilde{X}]$ from each piece of information in $\tilde{X}$ to each piece of information in X. Therefore, the maximum probability of the malicious adversary to obtain the privacy information in $X - \tilde{X}$ should be

$$\max_{x \in X, \rho_x \in P_x} \max_{\tilde{x} \in \tilde{X}, \rho_{\tilde{x}} \in P_{\tilde{x}}, a \in A} \max(\rho_x) \cdot Pr[x \leftarrow \rho_x|\tilde{x} \leftarrow \rho_{\tilde{x}}] \cdot Pr[\tilde{x} \leftarrow \rho_{\tilde{x}}|a]$$
$$= \max_{x \in X, \rho_x \in P_x} \max_{\tilde{x} \in \tilde{X}, \rho_{\tilde{x}} \in P_{\tilde{x}}, a \in A} Pr[x \leftarrow \rho_x|\tilde{x} \leftarrow \rho_{\tilde{x}}] \cdot \max(\rho_{\tilde{x}}) \cdot Pr[\tilde{x} \leftarrow \rho_{\tilde{x}}|a]$$
$$\leq Pr_{max}[X_p|\tilde{X}_p] \cdot \eta \leq \eta,$$

which is also limited by η. Therefore, the theorem is proved.

5.2 The Property of Privacy Type 2 and η-Inference

There is a data incrementally invariant for **Privacy type** 2 with η-inference, which especially suit for the privacy protection in dynamic publishing. It is defined as,

Definition 3 (Data incrementally invariant) *For a privacy definition PD, PD is data incrementally invariant if for arbitrary two databases X_1 and X_2 that are both protected by PD with the parameter η, the merged database $X_3 = X_1 \cup X_2$ is also protected by PD with η.*

Theorem 3. *η-inference is incrementally invariant on the published database.*

Proof. Let X_1 and X_2 be the databases protected by η-inference with parameter η. In this case, we have,

$$\forall a \in A, x_1 \in X_1, \boldsymbol{\rho}_{x_1} \in \boldsymbol{P}_{x_1}$$
$$\max(\boldsymbol{\rho}_{x_1}) \cdot Pr[x_1 \leftarrow \boldsymbol{\rho}_{x_1}|a] \leq \eta,$$

and

$$\forall a \in A, x_2 \in X_2, \boldsymbol{\rho}_{x_2} \in \boldsymbol{P}_{x_2}$$
$$\max(\boldsymbol{\rho}_{x_2}) \cdot Pr[x_2 \leftarrow \boldsymbol{\rho}_{x_2}|a] \leq \eta.$$

In this case, for the database $X_3 = X_1 \cup X_2$ we have,

$$\forall a \in A, x_3 \in X_3, \boldsymbol{\rho}_{x_3} \in \boldsymbol{P}_{x_3}$$
$$\max(\boldsymbol{\rho}_{x_3}) \cdot Pr[x_3 \leftarrow \boldsymbol{\rho}_{x_3}|a] \leq \eta.$$

Therefore, it is proved.

Based on **Definition 3** and **Theorem 3**, if all the databases for publishing are protected by **Privacy type** 2 with η-inference, it will not introduce extra privacy risk if we publish these databases separately and dynamically.

5.3 Comparing η-Inference with Differential Privacy

In this paper, we argue that the privacy should be handled to **Privacy type** 2 with η-inference as it can be used to deal with the limitations of DP mentioned in Sec 4.

The Misusing of the Sequential Composition. For DP, the core issue with the misusing of the sequential composition lies in its focus on the privacy protection mechanism from the original database X to the published database $\tilde{X}$. The mechanism limits the amount of privacy information that can be inferred from X based on $\tilde{X}$. However, when data publishers aim to determine which of a series of generated published databases performs better, they need to use the original database as a reference for comparison. This introduces additional information from X, which undermines the strict guarantees of DP.

For η-inference, the privacy is guaranteed by protecting the privacy information in $\tilde{X}$. In this case, even though the data publisher may also refer to the original database X to determine which of a series of generated published databases performs better. As long as the data publisher does not make extra changes to the published database $\tilde{X}$, the privacy guarantee cannot be weakened.

Presence of Dishonest Entities. The key of the presence of dishonest entities is that DP cannot be verified at the end side. For η-inference, any user who can access the published database can perform a verification. According to **Theorem 2**, one can verify whether the published database $\tilde{X}$ is protected with η-inference by checking the inference confidence of the privacy information contained in $\tilde{X}$ based on their own background knowledge. Therefore, scenarios where the claimed privacy level does not match the actual privacy level can be detected. In this way, dishonest data publishers can be identified and defended against. Similarly, dishonest local users can also be defended against in the same manner.

The Stigmatization. For DP, it does not ensure the privacy of the published database $\tilde{X}$. Therefore, a malicious adversary can take advantage of the imbalanced distribution of the output database and attempt tp make some false information seem true. For η-inference, the adversary cannot accurately obtain privacy information based on the published database, effectively preventing stigmatization.

The privacy of DP can be inherited by η-inference Considering there are two processes. The first process transforms the original database X into X', which is implemented by a DP mechanism. The second process transforms X' into the published database $\tilde{X}$ to ensure η-inference. In this case, the entire process from X to $\tilde{X}$ is protected by both DP and η-inference. However, if we adopt η-inference in the first process and adopt DP in the second process. The privacy guaranteed by η-inference cannot be maintained if DP is applied after η-inference.

5.4 To Adopt η-Inference Model in Practice

According to **Definition 3** and **Theorem 3**, protecting an original database X with the η-inference model can be achieved by protecting arbitrary subsets of X. In this case, we introduce two steps to implement η-inference in practice.

(1) **Step 1** First, the data publisher can adopt the η-inference to the original database X. Then, the database can be divided into two subsets, one that already satisfies η-inference and the other that does not.
(2) **Step 2** The subset which does not satisfy η-inference can be generalized to decrease the inference confidence of the adversary. In this paper, we adopt the method in [29] to generalize the database to satisfy η-inference.

Consequently, additional anonymization for the subset that already satisfies η-inference can be avoided, resulting in better utility.

6 Experiments

6.1 Setup

Datasets. We choose four benchmark datasets which are also adopted in [3, 24, 31]. They are **Adult** dataset with 45222 tuples and 15 attributes, **Nltcs** dataset with 21572 tuples and 16 attributes, **Acs** dataset with 47461 tuples and 23 tuples, and **BR2000** dataset with 28000 tuples and 14 attributes.

Privacy Metric. In this paper, we mainly focus on the privacy performance, where the disclosure risk is adopted.

- **Disclosure risk:** It is defined as the leakage probability under the inference for a certain kind of privacy information, where the values of sensitive attributes are considered as privacy information, and sensitive attributes are chosen at random in this paper.

Methods. We choose two open-source methods P3GM [24] and PrivMRF [3] in our experiments. They generate databases satisfying DP by synthesizing. In our design, we aim to exam the disclosure risk of the published database which is protected by DP.

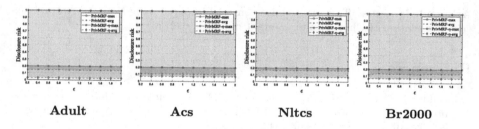

Adult **Acs** **Nltcs** **Br2000**

Fig. 1. The disclosure risk for PrivMRF

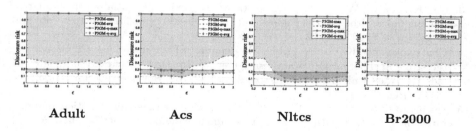

Adult **Acs** **Nltcs** **Br2000**

Fig. 2. The disclosure risk for P3GM

6.2 Results and Analysis

The Disclosure Risk Under DP Synthesizing Methods. As shown in Fig. 1 and Fig. 2, the maximum disclosure risk for the published databases based on PrivMRF and P3GM is both 1.0. Specifically, this disclosure risk also reflects the success rate of an adversary's stigmatization attempt. This indicates that Differential Privacy (DP) is not fully effective in preventing stigmatization. Once a stigmatization threat arises, there will inevitably be victims.

In contrast, by applying η-inference with $\eta = 0.2$ to generalize the results of PrivMRF and P3GM, we successfully reduce the maximum leakage risk across the four datasets to 0.2. This demonstrates that η-inference can effectively mitigate the issue of stigmatization.

7 Conclusion

In this paper, we explored the limitations and challenges associated with differential privacy, particularly in the context of data publishing. To address these issues, we introduced the η-inference model, which ensures information limited privacy. Our experiments on four real datasets demonstrate that DP is indeed vulnerable to the threat of stigmatization, while η-inference effectively alleviates this problem.

Acknowledgments. This research is sponsored in part by the National Natural Science Foundation of China (contract/grant numbers: 62272084).

References

1. Bindschaedler, V., Shokri, R., Gunter, C.A.: Plausible deniability for privacy-preserving data synthesis. Proc. VLDB Endow. **10**(5), 481–492 (jan 2017). https://doi.org/10.14778/3055540.3055542
2. Bun, M., Desfontaines, D., Dwork, C., Naor, M., Nissim, K., Roth, A., Smith, A., Steinke, T., Ullman, J., Vadhan, S.: Statistical inference is not a privacy violation. Differential-Privacy.org (06 2021), https://differentialprivacy.org/inference-is-not-a-privacy-violation/
3. Cai, K., Lei, X., Wei, J., Xiao, X.: Data synthesis via differentially private markov random fields. Proc. VLDB Endow. **14**(11), 2190–2202 (jul 2021). https://doi.org/10.14778/3476249.3476272
4. Chatzikokolakis, K., Andrés, M.E., Bordenabe, N.E., Palamidessi, C.: Broadening the scope of differential privacy using metrics. In: De Cristofaro, E., Wright, M. (eds.) Privacy Enhancing Technologies, pp. 82–102. Springer, Berlin Heidelberg, Berlin, Heidelberg (2013)
5. Duchi, J.C., Jordan, M.I., Wainwright, M.J.: Local privacy and statistical minimax rates. In: 2013 IEEE 54th Annual Symposium on Foundations of Computer Science. pp. 429–438 (2013). https://doi.org/10.1109/FOCS.2013.53
6. Dwork, C.: Differential privacy. In: International Colloquium on Automata, Languages, and Programming. pp. 1–12 (2006)

7. Ekenstedt, E., Ong, L., Liu, Y., Johnson, S., Yeoh, P.L., Kliewer, J.: When differential privacy implies syntactic privacy. IEEE Trans. Inf. Forensics Secur. **17**, 2110–2124 (2022). https://doi.org/10.1109/TIFS.2022.3177953
8. He, X., Machanavajjhala, A., Ding, B.: Blowfish privacy: Tuning privacy-utility trade-offs using policies. In: Proceedings of the 2014 ACM SIGMOD International Conference on Management of Data. pp. 1447–1458. SIGMOD '14, Association for Computing Machinery, New York, NY, USA (2014). https://doi.org/10.1145/2588555.2588581
9. Idrizi, E., Hamiti, M.: Classification of text, image and audio messages used for cyberbulling on social medias. In: 2023 46th MIPRO ICT and Electronics Convention (MIPRO). pp. 797–802 (2023). https://doi.org/10.23919/MIPRO57284.2023.10159835
10. Jiang, B., Li, M., Tandon, R.: Local information privacy and its application to privacy-preserving data aggregation. IEEE Trans. Dependable Secure Comput. **19**(3), 1918–1935 (2022). https://doi.org/10.1109/TDSC.2020.3041733
11. Jin, F., Hua, W., Francia, M., Chao, P., Orlowska, M., Zhou, X.: A survey and experimental study on privacy-preserving trajectory data publishing. IEEE Transactions on Knowledge and Data Engineering pp. 1–1 (2022). https://doi.org/10.1109/TKDE.2022.3174204
12. Kenny, C.T., Kuriwaki, S., McCartan, C., Rosenman, E.T.R., Simko, T., Imai, K.: The use of differential privacy for census data and its impact on redistricting: The case of the 2020 u.s. census. Science Advances **7**(41), eabk3283 (2021). https://doi.org/10.1126/sciadv.abk3283
13. Kifer, D., Machanavajjhala, A.: A rigorous and customizable framework for privacy. In: Proceedings of the 31st ACM SIGMOD-SIGACT-SIGAI Symposium on Principles of Database Systems. pp. 77–88. PODS '12, Association for Computing Machinery, New York, NY, USA (2012). https://doi.org/10.1145/2213556.2213571
14. Li, N., Li, T., Venkatasubramanian, S.: t-closeness: Privacy beyond k-anonymity and l-diversity. In: 2007 IEEE 23rd International Conference on Data Engineering. pp. 106–115 (2007). https://doi.org/10.1109/ICDE.2007.367856
15. Machanavajjhala, A., Kifer, D., Gehrke, J., Venkitasubramaniam, M.: L-diversity: Privacy beyond k-anonymity. ACM Trans. Knowl. Discov. Data **1**(1), 3–es (mar 2007). https://doi.org/10.1145/1217299.1217302
16. Majeed, A., Lee, S.: Anonymization techniques for privacy preserving data publishing: A comprehensive survey. IEEE Access **9**, 8512–8545 (2021). https://doi.org/10.1109/ACCESS.2020.3045700
17. McSherry, F., Talwar, K.: Mechanism design via differential privacy. In: 48th Annual IEEE Symposium on Foundations of Computer Science (FOCS'07). pp. 94–103 (2007). https://doi.org/10.1109/FOCS.2007.66
18. Mironov, I.: Rényi differential privacy. In: 2017 IEEE 30th Computer Security Foundations Symposium (CSF). pp. 263–275 (2017). https://doi.org/10.1109/CSF.2017.11
19. Murakami, T., Kawamoto, Y.: Utility-Optimized local differential privacy mechanisms for distribution estimation. In: 28th USENIX Security Symposium (USENIX Security 19). pp. 1877–1894. USENIX Association, Santa Clara, CA (Aug 2019), https://www.usenix.org/conference/usenixsecurity19/presentation/murakami
20. Nergiz, M.E., Atzori, M., Clifton, C.: Hiding the presence of individuals from shared databases. In: Proceedings of the 2007 ACM SIGMOD International Conference on Management of Data. pp. 665–676. SIGMOD '07, Association for Computing Machinery, New York, NY, USA (2007). https://doi.org/10.1145/1247480.1247554

21. du Pin Calmon, F., Fawaz, N.: Privacy against statistical inference. In: 2012 50th Annual Allerton Conference on Communication, Control, and Computing (Allerton). pp. 1401–1408 (2012). https://doi.org/10.1109/Allerton.2012.6483382
22. Rastogi, V., Suciu, D., Hong, S.: The boundary between privacy and utility in data publishing. In: Proceedings of the 33rd International Conference on Very Large Data Bases. pp. 531–542. VLDB '07, VLDB Endowment (2007)
23. Sweeney, L.: K-anonymity: A model for protecting privacy. Int. J. Uncertain. Fuzziness Knowl.-Based Syst. **10**(5), 557–570 (oct 2002). https://doi.org/10.1142/S0218488502001648
24. Takagi, S., Takahashi, T., Cao, Y., Yoshikawa, M.: P3gm: Private high-dimensional data release via privacy preserving phased generative model. In: 2021 IEEE 37th International Conference on Data Engineering (ICDE). pp. 169–180 (2021). https://doi.org/10.1109/ICDE51399.2021.00022
25. Wong, R.C.W., Li, J., Fu, A.W.C., Wang, K.: (α, k)-anonymity: An enhanced k-anonymity model for privacy preserving data publishing. In: Proceedings of the 12th ACM SIGKDD International Conference on Knowledge Discovery and Data Mining. pp. 754–759. Association for Computing Machinery, New York, NY, USA (2006). https://doi.org/10.1145/1150402.1150499
26. Xu, C., Ren, J., Zhang, Y., Qin, Z., Ren, K.: Dppro: Differentially private high-dimensional data release via random projection. IEEE Trans. Inf. Forensics Secur. **12**(12), 3081–3093 (2017). https://doi.org/10.1109/TIFS.2017.2737966
27. Yang, B., Sato, I., Nakagawa, H.: Bayesian differential privacy on correlated data. In: Proceedings of the 2015 ACM SIGMOD International Conference on Management of Data. p. 747–762. SIGMOD '15, Association for Computing Machinery, New York, NY, USA (2015). https://doi.org/10.1145/2723372.2747643
28. Yang, X., Wang, T., Ren, X., Yu, W.: Survey on improving data utility in differentially private sequential data publishing. IEEE Transactions on Big Data **7**(4), 729–749 (2021). https://doi.org/10.1109/TBDATA.2017.2715334
29. Yao, L., Wang, X., Hu, H., Wu, G.: A utility-aware anonymization model for multiple sensitive attributes based on association concealment. IEEE Transactions on Dependable and Secure Computing pp. 1–12 (2023). https://doi.org/10.1109/TDSC.2023.3299641
30. Zhang, M., Li, X., Ren, Y., Luo, B., Miao, Y., Liu, X., Deng, R.H.: Privacy-preserved data trading via verifiable data disturbance. IEEE Transactions on Dependable and Secure Computing pp. 1–14 (2023). https://doi.org/10.1109/TDSC.2023.3323669
31. Zhang, Z., Wang, T., Li, N., Honorio, J., Backes, M., He, S., Chen, J., Zhang, Y.: PrivSyn: Differentially private data synthesis. In: 30th USENIX Security Symposium (USENIX Security 21). pp. 929–946. USENIX Association (Aug 2021), https://www.usenix.org/conference/usenixsecurity21/presentation/zhang-zhikun
32. Zhu, T., Li, G., Zhou, W., Yu, P.S.: Differentially private data publishing and analysis: A survey. IEEE Trans. Knowl. Data Eng. **29**(8), 1619–1638 (2017). https://doi.org/10.1109/TKDE.2017.2697856

Secure Why-Not Spatial Keyword Top-k Queries in Cloud Environments

Yiping Teng[✉], Miao Li, Shiqing Wang, Huan Wang, Chuanyu Zong, and Chunlong Fan

School of Computer, Shenyang Aerospace University, Shenyang, China
{typ,zongcy,FanCHL}@sau.edu.cn,
{limiao1,wangshiqing1,wanghuan}@stu.sau.edu.cn

Abstract. With the rapid growing volume of spatial data, answering why-not questions on spatial keyword top-k queries, aiming at refining spatial keyword queries to include the missing objects in query results with minimal costs, has attracted much attention in the field of spatial databases. To alleviate the burden of local storage and computation, when service providers outsource the why-not query services to public cloud, it may raise privacy concerns. To address these issues, in this paper we first present a basic secure why-not spatial keyword top-k query (BSWoSKQ) scheme, featuring a secure weight vector generation method to obtain the best approximate refined query with minimal costs. Furthermore, to improve the query efficiency, we propose an optimized scheme named SWoSKQ that employs a new secure index structure, i.e., SSR-tree, and efficient pruning methods based on such secure index. Comprehensive analysis demonstrates the security and computational complexity of our approach, and extensive experiments on real and synthetic datasets validate the query performance of the proposed methods.

Keywords: Why-not question · Spatial keyword query · Query refinement · Data encryption

1 Introduction

With the rapid development of global positioning technologies and smart mobile devices, location-based services have become more prominent. The spatial keyword top-k query is a crucial technique that utilizes the user's location and keywords to retrieve the top-k objects. After issuing a spatial keyword query and receiving the results, the user may notice the absence of expected objects. This suggests that other valuable objects, unknown objects could also be missing and the user has reason to question the overall effectiveness and completeness of the query results through a why-not query. For example, Alice is in San Francisco looking for a nearby bakery. She runs a top-5 spatial query with the keyword "bakery", but the results do not show Sweet Treats. Alice may ask a why-not question, i.e., *"why is Sweet Treats **not** in the results?"* and wonders how to make it appear. In this situation, providing suggestions for modifying the query to include the desired objects in the results is crucial for a better query experience and utility.

As the proliferation of spatial data, many data service providers are outsourcing why-not queries to cloud servers to alleviate the burden of local storage and computation. However, despite the benefits, public cloud platforms are not completely reliable. They may potentially access users' sensitive data (e.g., locations, keywords, and desired objects), leading to serious privacy breaches. Hence, it holds considerable importance to investigate the privacy-preserving challenges associated with addressing why-not questions in spatial keyword top-k queries.

In recent years, several secure methods [17,18] have been proposed for addressing secure why-not queries, as the most related work of this paper. However, these methods exhibit long query times and fail to securely address the why-not question by adjusting the initial query parameters that better reflect user preferences. In addition, the secure techniques presented in existing research on secure spatial (keyword) queries [9,13,19–21,23] can hardly be fully utilized to answer the why-not questions [2–4,10,25,26]. Therefore, effective strategies are essential to address the secure why-not spatial keyword top-k query problem.

To achieve secure why-not spatial keyword top-k query processing, in this paper, we address two main challenges. The first challenge is how to obtain the best approximate refinement weight in the secure retrieval of refined queries. To this end, we first propose a straightforward method (BSWoSKQ) to securely map the objects and weight vectors onto a two-dimensional plane, allowing for a geographical interpretation of candidate weight vectors and k. The candidate query with the smallest penalty is then selected as the optimal query refinement. To solve the efficiency problem of the basic approach, we propose a novel secure index SSR-tree that satisfies the secure computation of spatial proximity and textual similarity in the optimized scheme (SWoSKQ). By traversing SSR-tree, unnecessary object accesses can be reduced for quickly obtaining the ranking of candidate weights, and candidates that are unlikely to be the best-refined queries can be eliminated by comparing penalty values. We analyze the computational complexity and security of our approaches and conduct extensive experiments on real and synthetic datasets. The experimental results demonstrate the effectiveness of the proposed methods.

We summarize the contribution as follows.

- We first propose a straightforward BSWoSKQ method, which achieves the candidate weight vectors selecting through the intersection of line segments from incomparable and missing points and secures the penalty computation to find the optimal refined queries.
- We propose a new secure indexing structure SSR-tree that satisfies the secure computation of spatial proximity and textual similarity, which obtains the ranking of candidate weights without direct access to data points.
- We further propose the SWoSKQ scheme, which prunes candidate weights using penalty value comparisons, followed by traversing the SSR-tree to quickly obtain the candidate k.

- We analyze the computational complexity and security of our approach and conduct extensive experiments on real and synthetic datasets. The results demonstrate the effectiveness of the proposed method.

The rest of the paper is organized as follows. Section 2 discusses related work. Section 3 presents the problem definition and preliminaries. Section 4 explains two solutions to secure why-not spatial keyword queries, i.e., BSWoSKQ and SWoSKQ. The security and computational complexity of our approach are analyzed in Sect. 5, and the experimental results are in Sect. 6. Section 7 concludes the paper.

2 Related Work

The previous work on the secure why-not problem has made significant progress. In [18], a safe trade-off space generation method is proposed to obtain the best approximate refined query, which is optimized by pruning conditions and early termination conditions to improve query efficiency. [17] is the most relevant to our proposed research direction. They introduced two novel protocols for secure computation and a method to retrieve refined queries with minimal penalty by determining the best approximate refinement direction. However, none of the above studies explored how to efficiently and safely modify the weight vectors and k to obtain the best refined queries. In the study of secure why-not spatial keyword top-k query, we survey the related work from the following two perspectives: answering why-not questions, secure spatial/spatial keyword query.

2.1 Answering Why-Not Questions

To address the availability issue of database query results, the why-not question was initially introduced by Chapman et al. [1]. There are four main existing approaches to solving the why-not question, where query refinement is commonly employed [15]. Chen et al. [3,4] addressed the why-not question for top-k spatial keyword queries by adjusting parameters in the initial query. In addition, Chen et al. [2] proposed effective refinement techniques to recover missing objects by minimally modifying users' direction-aware queries. Zheng et al. [26] solved the why-not question for grouped queries by filtering the search space to exclude unwanted queries and then obtaining candidate weights for promising queries. Zhang et al. [25] proposed a new index called Shadow based on the probability value of a moving object appearing in a special spatial region over a period of time to deal with the why-not question in top-k SKQ queries for moving objects. Li et al. [10] designed a DAPC indexing structure, answering the why-not question about direction-aware augmented spatial keyword top-k queries. However, none of these studies addressed the privacy aspects of the why-not question.

2.2 Secure Spatial/Spatial Keyword Query

Recently, several methods for secure spatial queries and spatial keyword queries have emerged. We first briefly discuss the existing work for spatial queries. Tong et al. [20] proposed the Hu-Fu system to decompose the security handling of spatial queries into as many plaintext operations as possible and as few security operations as possible. Li et al. [9] proposed ContactGuard, an efficient framework that enhances Secure Multiparty Computation by leveraging Geo-Indistinguishability to optimize location. The spatial keyword search allows users to submit more detailed queries. Wang et al. [21] defined the concept of spatial keyword structured encryption and proposed several specific SKSE constructs with various efficiency security trade-offs. In Zhang et al. [23], fr-tree indexing was proposed to efficiently perform spatial keyword similarity queries. Tong et al. [19] developed a comparable product encoding strategy to retrieve objects within the query range that have the highest textual similarity. Gong et al. [13] introduced a straightforward scheme that integrates the Geohash algorithm with EASPE and devised a GR-tree and a pruning strategy to enhance query efficiency. However, few of them can address the problem of secure why-not spatial keyword top-k queries.

3 Problem Definition and Preliminaries

3.1 Problem Definition

Given a spatio-textual dataset $\mathcal{D}$, each object $o \in \mathcal{D}$ is represented as a tuple $(o.loc, o.doc)$, where $o.loc$ is a multi-dimensional location, and $o.doc$ is a set of keywords. A spatial keyword top-k query q consists of four parameters $(loc, doc, k, \boldsymbol{w})$ and returns the top-k objects based on a scoring function that accounts for both spatial proximity and textual similarity. In this paper, we assume the scoring function is defined as follows:

$$ST(o, q, \boldsymbol{w}) = w_s DSim(o, q) + w_t TSim(o, q)$$
$$= w_s \left(1 - \frac{\text{dist}(o.loc, q.loc)^2}{\text{dist}_{max}^2}\right) + w_t \frac{|o.doc \cap q.doc|}{|o.doc \cup q.doc|} \quad (1)$$

In Eq. 1, $\boldsymbol{w} = \langle w_s, w_t \rangle$ is a weighting vector is used to represent the relative preference between spatial proximity and textual similarity. $DSim(o, q)$ denotes spatial proximity, where $\text{dist}(o.loc, q.loc)$ is the Euclidean distance and dist_{max} is the maximum Euclidean distance. $TSim(o, q)$ denotes the textual similarity using the Jaccard similarity [16].

Definition 1: **(Spatial Keyword Top-k Query)** [3] . *Given a spatial dataset $\mathcal{D}$ and a query q, a spatial keyword top-k query q returns k objects in D that maximize the scoring function according to Eq. 1, where $\forall o \in \mathcal{R}(\forall o^{'} \in \mathcal{D} - \mathcal{R}, (ST(o, q, \boldsymbol{w}) \geq ST(o^{'}, q, \boldsymbol{w})))$*

When a user issues a spatial keyword top-k query $q = (loc, doc, k_0, \boldsymbol{w})$, the query result may lack expected objects. The user might propose a why-not query with a set of missing objects $M = \{m_1, m_2, ..., m_k\}$. To address this, we generate a refined query $q^{'} = (loc, doc, k^{'}, \boldsymbol{w}^{'})$ by adjusting k and $\boldsymbol{w}$ to ensure that all objects in the set M are included in the result. To evaluate the quality of the refined querys, we employ a penalty model [3,7,8], which uses $\Delta k = max(0, k' - k_0)$ and $\Delta w = \|w' - w_0\|_2^2$) to measure the degree of adjustment relative to the original query. Where λ denotes the user's preference for modifying $q.k$ and $q.w$, and $R(o, q, w_0)$ denotes the initial ranking of the missing objects. Based on this, the definition of penalty is as follows:

$$Penalty(k', w') = \lambda \frac{\Delta k}{R(o, q, w_0) - k_0} + (1 - \lambda) \frac{\Delta w^2}{1 + w_{s_0}^2 + w_{t_0}^2} \quad (2)$$

Definition 2: (**Why-Not Spatial Keyword Top-k Query**) [3]. *Given an object set $\mathcal{D}$, a missing object set $M \subset \mathcal{D}$ and an original spatial keyword query $q = (loc, doc, k_0, w_0)$, the why-not spatial keyword top-k query returns a refined query $q' = (loc, doc, k', w')$ with the minimum penalty value according to Eq. 2 and the result of which contains all objects in M.*

Definition 3: (**Secure Why-Not Spatial Keyword Top-k Query**). *Given an encrypted object set $E_{pk}(\mathcal{D})$, an encrypted missing object set and encrypted spatial keyword top-k query $\{E_{pk}(M), E_{pk}(q)\}$ as the secure why-not request, a secure why-not spatial keyword top-k query is to find a refined top-k query $q' = (E_{pk}(loc), E_{pk}(doc), E_{pk}(k'), E_{pk}(w'))$ in ciphertext, of which the result includes $E_{pk}(M)$ with the minimum penalty value calculated under encryption. Here pk is the public key of asymmetric cryptography.*

3.2 System Framework

In this paper, we consider the dual cloud server based spatial location data outsourcing scenario that has been applied in many related studies [5,6,11,12,24]. The system framework of our schemes mainly consists of three entities: user ($\mathcal{U}$), data owner ($\mathcal{DO}$) and cloud servers($\mathcal{C}_1, \mathcal{C}_2$), as shown in Fig. 1.

Data Owner. $\mathcal{DO}$ first generates the key $K = \langle pk, sk \rangle$, where pk is public key and the sk is private key. Before outsourcing, $\mathcal{DO}$ encrypts the dataset and the constructed index with pk as $E_{pk}(\mathcal{D}), E_{pk}(\mathcal{T})$, and then uploads $pk, E_{pk}(\mathcal{D})$ and $E_{pk}(\mathcal{T})$ to $\mathcal{C}_1$, sends pk and sk to $\mathcal{C}_2$, and sends pk to $\mathcal{U}$.

User. $\mathcal{U}$ encrypts the initial spatial keyword top-k query and the set of missing objects using pk as a secure why-not request $\{E_{pk}(q), E_{pk}(M)\}$, and then submits it to $\mathcal{C}_1$.

Cloud Server. When receiving $\{E_{pk}(q), E_{pk}(M)\}, E_{pk}(\mathcal{D})$ and $E_{pk}(\mathcal{T}), \mathcal{C}_1$ executes the secure why-not spatial keyword top-k query through secure protocols with the help of $\mathcal{C}_2$ and returns refined query $E_{pk}(q')$ to $\mathcal{U}$.

3.3 Security Model

In our study, we adopt the *semi-honest* adversary model, and assume the cloud severs follow *"honest but curious"* model, which has been used in related studies [6,11]. However, cloud servers are still curiously capture and analyze meaningful information in query requests, query processing, and query results. The model

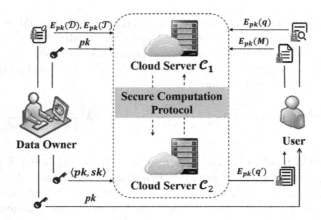

Fig. 1. System Framework

implicitly assumes that the two cloud servers $(\mathcal{C}_1, \mathcal{C}_2)$ do not collude. To ensure the privacy of query services, the privacy requirements are stated as follows.

Data Privacy. $\mathcal{C}_1$ and $\mathcal{C}_2$ know nothing about any plaintext of $\mathcal{D}$. Except for q', $\mathcal{U}$ knows nothing about $\mathcal{D}$.

Query Privacy. $\mathcal{DO}$, $\mathcal{C}_1$ and $\mathcal{C}_2$ know nothing about the plaintext of q during the query processing.

Result Privacy. $\mathcal{DO}$, $\mathcal{C}_1$ and $\mathcal{C}_2$ know nothing about q'.

Access Patterns Privacy. $\mathcal{C}_1$ and $\mathcal{C}_2$ know nothing about the access patterns from intermediate results.

Notice that, we do not focus on the issues related to access control and channel attacks in our work. We assume that $\mathcal{DO}$ can be trusted, $\mathcal{U}$ is well-authorized, and the transmission channels are secure.

3.4 Preliminaries

Paillier Cryptosystem [14]. To support ranking updates and penalty value calculations for refined queries with security guarantee, we adopt the semantic secure paillier cryptosystem. The paillier cryptosystem supports homomorphic addition and multiplication operations on encrypted data. Specifically, given plaintext m_0 and m_1, their homomorphic properties are described as follows:

- *Homomorphic Addition*

$$D_{sk}(E_{pk}(m_0) * E_{pk}(m_1) \mod N^2) = (m_0 + m_1) \mod N$$

- *Homomorphic Multiplication*

$$D_{sk}(E_{pk}(m_0)^{m_1} \mod N^2) = m_0 * m_1 \mod N$$

Secure Computation Protocols. We introduce several secure computing protocols for basic computations on encrypted data, using them to support our secure search method. All protocols adopt a two-party semi honest model ($\mathcal{C}_1$ and $\mathcal{C}_2$). Due to space limitations, we will only briefly introduce the basic concepts of these protocols as follows.

- *Secure Multiplication* (SM) [6]: $\text{SM}(E_{pk}(m_0), E_{pk}(m_1)) \leftarrow E_{pk}(m_0 * m_1)$
- *Secure Division* (SDC) [5]: $\text{SDC}(E_{pk}(m_0), E_{pk}(m_1)) \leftarrow E_{pk}(m_0/m_1)$
- *Secure Equal* (SEQ) [24]: $\text{SEQ}(E_{pk}(m_0), E_{pk}(m_1)) \leftarrow E_{pk}(Bool(m_1 == m_2)$
- *Secure Less* (SLESS) [24]: $\text{SLESS}(E_{pk}(m_0), E_{pk}(m_1)) \leftarrow E_{pk}(Bool(m_1 < m_2)$
- *Secure Max* (SMAX) [24] : $\text{Smax}(E_{pk}(m_0), E_{pk}(m_1)) \leftarrow E_{pk}(max(m_1, m_2))$
- *Secure Squared Euclidean Distance* (SSED) [11] : Assume that $m_0(x_0, y_0)$ and $m_1(x_1, y_1)$ are points.

$$\text{SSED}(E_{pk}(m_0), E_{pk}(m_1)) \leftarrow E_{pk}((x_0-x_1)^2 + (y_0-y_1)^2)$$

4 Approach

In this section, we first propose a straightforward method named Basic Secure Why-not Spatial Keyword Query (BSWoSKQ) to directly identify candidate refinement queries. After that, we design a novel secure index structure SSR-tree and propose a more efficient Secure Why-not Spatial Keyword Query (SWoSKQ) scheme to improve the query efficiency. For simplicity, we define relevant notations in Table 1.

4.1 Basic Secure Why-Not Spatial Keyword Top-k Query

To identify candidate refined queries, BSWoSKQ represent data objects as line segments on a 2D plane and define promotion and demotion points to obtain refined weights. Finally, the optimal refinement query with the lowest penalty value is obtained.

Table 1. Summary of Notations

Notation	Definition
l_w	The line of weighting vector $\boldsymbol{w}$.
$[l_{w_1}, l_{w_2}]$	The angle interval from $\boldsymbol{w}_1$ to $\boldsymbol{w}_2$.
S_o	The fixed-score segment of object o.
p_o	The intersection point of S_o and l_w.
$Pro(De)$	The set of promotion(demotion) points for m.
$n_p(p_1, p_2)$	The number of Pro_i between p_1 and p_2.
$n_d(p_1, p_2)$	The number of De_i between p_1 and p_2.

We use the transformation introduced by Chester etal [14], which projects the objects and weight vectors onto a "$ws - wt$" plane, visualizing the rankings of the objects as the weight vectors change. After the user specifies the initial secure spatial keyword top-k query. Then, C_1 calculates object's spatial proximity $E_{pk}(DSim_i)$, textual similarity $E_{pk}(TSim_i)$ and score $E_{pk}(ST_i)$ based on Eq. 3, Eq. 4 and Eq. 5. we generate a fixed score segment S_o with endpoints $Ss_o(\frac{E_{pk}(C)}{E_{pk}(DSim_i)}, E_{pk}(0))$ and $St_o(E_{pk}(0), \frac{E_{pk}(C)}{E_{pk}(TSim_i)})$ for each data object, where $E_{pk}(C)$ is a randomly selected positive real number.

$$E_{pk}(DSim_i) = E_{pk}(1) * SDC(SSED(E_{pk}(o_i.loc), E_{pk}(q.loc)), E_{pk}(maxD^2))^{N-1} \quad (3)$$

$$E_{pk}(TSim_i) = SDC(E_{pk}(0) * SEQ(E_{pk}(o_i.doc), E_{pk}(q.doc)), E_{pk}(|o_i.doc|*|q.doc|) \\ * (E_{pk}(0) * SEQ(E_{pk}(o_i.doc), E_{pk}(q.doc)))^{N-1}) \quad (4)$$

$$E_{pk}(ST_i) = SM(E_{pk}(w_s), E_{pk}(DSim_i)) * SM(E_{pk}(w_t), E_{pk}(TSim_i)) \quad (5)$$

When the weight vector changes, if the fixed score segments of two objects do not intersect, their rankings remain unaffected. However, if their fixed score segments intersect at point p, when the rotation of the weight vector to the point p and the ranking of o is incremented, p is considered a promotion point. If the ranking of o is decremented, p is considered a demotion point. Promotion points are the intersections of the fixed score segments of the high-ranked incomparable points and the missing points, while demotion points are generated by low ranked incomparable points. Therefore, the weighting vector of the best refined query for a missing object o must be either the original weighting vector or must go through a promoted point of S_o.

Based on the promotion and demotion points, we can derive an alternative method for computing the ranking of a given o from an arbitrary weight vector. Let the ranking of o under the original weight vector $E_{pk}(w_0)$ be $E_{pk}(r_0)$, and let the modified weight vector be $E_{pk}(w_1)$. Firstly compute the number of promotion and demotion points in $[l_{w_0}, l_{w_1}]$, denoted as $E_{pk}(n_p)$ and $E_{pk}(n_d)$. Therefore, the ranking of o under $E_{pk}(w_1)$ is calculated based on Eq. 6. To obtain the best refined query by calculating and comparing penalty values for all candidate queries. User's preferences $E_{pk}(\lambda)$ and $E_{pk}(q_0)$ given by $\mathcal{U}$, C_1 first calculates $E_{pk}(\eta_k) = SDC(SMAX(E_{pk}(0), E_{pk}(k^{'}) * E_{pk}(k_0)^{N-1}), E_{pk}(r_0 * k_0^{N-1}))$ and $E_{pk}(\eta_w) = SDC(SSED(E_{pk}(w^{'}), E_{pk}(w_0)), E_{pk}(1) * E_{pk}(\sum w_0[i]^2))$, then calculate the penalty for $E_{pk}(q^{'})$ based on Eq. 7.

$$E_{pk}(r_i) = E_{pk}(r_0) * E_{pk}(n_p)^{N-1} * E_{pk}(n_d) \quad (6)$$

$$\text{Penalty}(E_{pk}(k'), E_{pk}(w')) = \text{SM}(E_{pk}(\lambda), E_{pk}(\eta_k)) * \text{SM}(E_{pk}(1-\lambda), E_{pk}(\eta_w)) \quad (7)$$

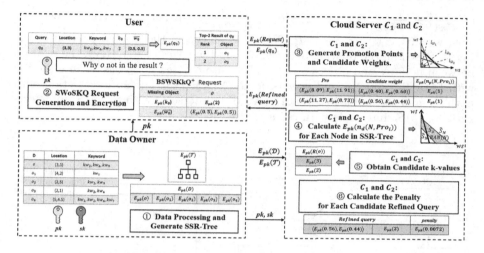

Fig. 2. An example of SWoSKQ processing.

In this basic scheme, the method of generating weight vectors avoids the exhaustive enumeration of an infinite number of candidate weight vectors. The algorithm still has a high computational cost because the number of degraded points can be large. To enhance efficiency, we propose a novel secure index structure called SSR-tree. By traversing the SSR-tree, we can quickly obtain the ranking of candidate weights without directly visiting the data points.

4.2 SSR-Tree: Secure Set R-Tree

In this paper, we propose a secure index called SSR-tree that can estimate spatial proximity and textual similarity bounds. Each leaf stores entries of the form $(o.cnt, o.mbr, o.k)$, where $o.cnt$ is the number of data objects in the leaf node, $o.mbr$ is the minimum bounding rectangle of data objects and $o.k$ is the keyword of data objects. Each non-leaf node N stores entries of the form $(N.pc, N.cnt, N.mbr, N.Union, N.Intersection)$ where $N.pc$ is a pointer to the subtree, $N.cnt$ denotes the number of data objects in the subtree of the node N, $N.mbr$ is the minimum bounding rectangle of the child node, $N.Union$ denotes the union of the keyword sets of all objects indexed in the subtree rooted at this node, and $N.Intersection$ denotes the intersection of these keyword sets. $\mathcal{DO}$ encrypts all the information stored constructs the SSR-tree. Figure 3 shows an example of SSR-tree, where the keyword sets associated with each leaf and non-leaf node are shown. The union(intersection) set of a non-leaf node N_1 is the union(intersection) of keyword sets associated with o_1, o_2 and o_3.

Based on the SSR-tree, we can get spatial proximity and text similarity boundary segments between each node N and $E_{pk}(q)$, where $\hat{S}_N$ is the segment with endpoints $(\frac{E_{pk}(C)}{E_{pk}(\check{D}Sim(N,q))}, E_{pk}(0))$, $(E_{pk}(0), \frac{E_{pk}(C)}{E_{pk}(\check{T}Sim(N,q))})$, and $\check{S}_N$ is $(\frac{E_{pk}(C)}{E_{pk}(\hat{D}Sim(N,q))}, E_{pk}(0))$, $(E_{pk}(0), \frac{E_{pk}(C)}{E_{pk}(\hat{T}Sim(N,q))})$. The region consisting of ws-axis, wt-axis, $\hat{S}_N$ and $\check{S}_N$ is $RAN(N)$, as shown in the shaded portion of the Fig. 3. Based on the relation between S_o and the edges of $RAN(N)$, the relation between node N and object o is classified into 5 cases. 1)S_o has no intersection with both $\check{S}_N$ and $\hat{S}_N$ and is outside $RAN(N)$; 2)S_o has no intersection with both $\check{S}_N$ and $\hat{S}_N$ and is inside $RAN(N)$; 3)S_o intersects with $\check{S}_N$; 4)S_o intersects with $\hat{S}_N$; 5)S_o intersects with $\check{S}_N$ and $\hat{S}_N$.

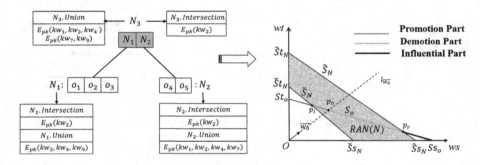

Fig. 3. An Example of SSR-tree and $RAN(N)$

4.3 Pruning Optimization Based on SSR-Tree

In the optimization algorithm, we prune the candidate weights according to the properties of Eq. 6 and Eq. 7. First, $E_{pk}(n_p)$ for each promotion point is calculated, and the temporary missing point rankings and penalty values are obtained. Then, compare these penalty values with the penalty values initialized by $E_{pk}(r_0)$ and $E_{pk}(w_0)$. If they are larger than this value, the candidate weights generated by the promotion points can be pruned.

To quickly obtain rankings under candidate weights, we first estimate the number of demotion points in a node $E_{pk}(n_d(N))$, with its minimum value $E_{pk}(\check{n}_d(N))$. Since node N contains four types of objects, i.e. $Dom(o)$ is the dominators of o, $DomBy(o)$ is the dominated points of o, $IH(o)$ is the high ranked incomparable points of o and $IL(o)$ is the low ranked. Because the sum of $E_{pk}(|N.IH(o)|)$ and $E_{pk}(|N.IL(o)|)$ is the same as the sum of $E_{pk}(n_p(N))$ and $E_{pk}(n_d(N))$. Thus, $E_{pk}(n_d(N))$ generated by objects in the subtree of N on S_o can be expressed as:

$$E_{pk}(n_d(N)) = E_{pk}(N.cnt) - E_{pk}(|N.DomBy(o)|) \\ - E_{pk}(|N.Dom(o)|) - E_{pk}(n_p(N)) \quad (8)$$

Based on Eq. 8 and the relationship between nodes N and S_o, we estimate $E_{pk}(d(N))$ for each relation. For relation 1, $E_{pk}(N.IL(o)) = \emptyset$. Therefore, they can be pruned. For relation 2 and 3, $E_{pk}(d(N)) = E_{pk}(0)$. For relations 4, $E_{pk}(N.DomBy(o)) = \emptyset$ and hence $E_{pk}(d(N))$ =SMAX$\{E_{pk}(0), E_{pk}(N.cnt - n_p(N, Pro_i) - |Dom(o)|)\}$. For relations 5, nodes contain objects that belong to $IH(o)$ or $IL(o)$, $E_{pk}(d(N))$ =SMAX$\{E_{pk}(0), E_{pk}(N.cnt - n_p(N, Pro_i))\}$.

To determine the number of demotion points between promotion points and node N called $E_{pk}(n_d(N, Pro_i))$, the position of $E_{pk}(Pro_i)$ must first be identified. S_o is divided into four disjoint parts using l_{w_o} and $RAN(N)$, as shown in Fig. 3, an example of relation 5. Here, $[E_{pk}(p_l), E_{pk}(p_o))$ is the promotion part, $(E_{pk}(p_o), E_{pk}(p_r))$ is the demotion part, and $[E_{pk}(p_r), E_{pk}(Ss_o))$ is the influential part. The point on the promotion part must be an promotion point, the same applies to the demotion part. If $E_{pk}(Pro_i)$ is located in the influential part of N, $E_{pk}(n_d(N, Pro_i)) = E_{pk}(n_d(N))$. We can estimate $E_{pk}(n_d(N, Pro_i))$ as Eq. 9. If $E_{pk}(Pro_i)$ lies in the demotion part, promotion part or the remaining part of N, $E_{pk}(n_d(N, Pro_i)) = E_{pk}(0)$.

$$E_{pk}(n_d(N,Pro_i)) = \begin{cases} SMAX\{E_{pk}(0), E_{pk}(N.cnt - n_p^+(N, Pro_i) - |Dom(o)|)\} & \text{Relation 4} \\ SMAX\{E_{pk}(0), E_{pk}(N.cnt - n_p^+(N, Pro_i))\} & \text{Relation 5} \end{cases} \quad (9)$$

Therefore, the ranking of the missing points under the candidate weights can be obtained $E_{pk}(r_i) = E_{pk}(r_0) * E_{pk}(n_p(N, Pro_i))^{N-1} * E_{pk}(n_d(N, Pro_i))$.

4.4 Secure Why-Not Spatial Keyword Top-k Query

Algorithm 1: SWoSKQ Processing

Input: $E_{pk}(\mathcal{D})$, $E_{pk}(\mathcal{T})$, $E_{pk}(m)$ and $E_{pk}(q_0)$;
Output: $q'(k', w')$;
$\mathcal{C}_1$ and $\mathcal{C}_2$:
1 $E_{pk}(r_0), E_{pk}(p_o), E_{pk}(S_o)$ and $E_{pk}(Pro)$ is calculated through Security Protocols;
2 Initialize $E_{pk}(Pen_{min}), E_{pk}(k_{min})$ and $E_{pk}(w_{min})$;
3 **for** $i = 1$ to $|E_{pk}(E_{pk}(Pro))|$ **do**
4 $\quad E_{pk}(n_p(p_o, Pro_i)) \leftarrow$ Calculate the number of promotion points;
5 $\quad E_{pk}(r_i) \leftarrow E_{pk}(r_0) * E_{pk}(n_p(p_o, Pro_i))^{N-1}$;
6 $\quad$ Compute $E_{pk}(Pen_{min}), E_{pk}(k_{min})$ and $E_{pk}(w_{min})$ using $E_{pk}(r_i)$, $E_{pk}(w_i)$;
7 $\quad$ **if** SLESS($E_{pk}(Pen_i), E_{pk}(Pen_{min})$) **then**
8 $\quad\quad$ Prune $E_{pk}(Pro_i)$ from $E_{pk}(Pro)$;

9 Create an empty queue Q and insert $E_{pk}(\mathcal{T}.root)$ into queue Q;
10 **while** Q *is not empty* **do**
11 $\quad N \leftarrow Dequeue(Q)$;
12 $\quad$ Compute N's promotion, demotion and influential part;
13 $\quad$ **if** N *lacks demotion part or no existing* $E_{pk}(Pro_i)$ *on demotion part* **then**
14 $\quad\quad$ continue
15 $\quad$ **for** *each existing* $E_{pk}(Pro_i)$ *on N's demotion part* **do**
16 $\quad\quad E_{pk}(n_d) \leftarrow E_{pk}(0)$;
17 $\quad\quad$ **for** *each child c of N* **do**
18 $\quad\quad\quad E_{pk}(n_d) \leftarrow E_{pk}(n_d) * E_{pk}(n_d(c, Pro_i))$;
19 $\quad\quad E_{pk}(r_i) \leftarrow E_{pk}(r_i) * E_{pk}(n_d) * E_{pk}(n_d(N, Pro_i))^{N-1}$;
20 $\quad\quad$ Compute $E_{pk}(Pen_i)$ using $E_{pk}(r_i)$ and $E_{pk}(w_i)$;
21 $\quad$ Insert each child c of N into Q;

22 **for** $i = 1$ to $|E_{pk}(Pro)|$ **do**
23 $\quad$ **if** SLESS($E_{pk}(Pen_i), E_{pk}(Pen_{min})$) **then**
24 $\quad\quad E_{pk}(k_{min}) \leftarrow E_{pk}(r_i)$, $E_{pk}(w_{min}) \leftarrow E_{pk}(w_i)$;

$\mathcal{C}_1$:
25 Generate random noise $\varepsilon_1, \varepsilon_2$ and Send $\varepsilon_1, \varepsilon_2$ to $\mathcal{U}$;
26 $E_{pk}(k_{med}) \leftarrow E_{pk}(k') * E_{pk}(\varepsilon_1)$, $E_{pk}(w_{med}) \leftarrow E_{pk}(w_{min}) * E_{pk}(\varepsilon_2)$;
27 Send $E_{pk}(k_{med}), E_{pk}(w_{med})$ to $\mathcal{C}_2$;
$\mathcal{C}_2$:
28 $k_{med} \leftarrow D_{sk}(k_{med})$, $w_{med} \leftarrow D_{sk}(w_{med})$, and send k_{med} and w_{med} to $\mathcal{U}$;
$\mathcal{U}$:
29 Receive $k_{med}, w_{med}, \varepsilon_1$ and ε_2, $k' \leftarrow k_{med} - \varepsilon_1$, $w' \leftarrow w_{med} - \varepsilon_2$;
30 $q'(k', w')$ as the refined query;

We propose the SWoSKQ scheme based on SSR-tree and an efficient pruning optimisation strategy. The details of the SWoSKQ scheme process are shown in Algorithm 1. $\mathcal{C}_1$ and $\mathcal{C}_2$ first calculate the ranking $E_{pk}(r_0)$, the fixed score segment $E_{pk}(S_o)$, all promotion points $E_{pk}(Pro)$ and the intersection $E_{pk}(p_o)$ of $E_{pk}(l_{w_0})$ with $E_{pk}(S_o)$ (Line 1). Initialise minimum penalty value, k and weight using $E_{pk}(w_0)$ and the missing point ranking $E_{pk}(r_0)$ (Line 2). By sorting the promotion points, it is straightforward to obtain $E_{pk}(n_p(p_o, Pro_i))$ for each promotion point. Calculate the penalty value $E_{pk}(Pen_i)$. Prune the promotion points by comparing the penalty values (Lines 3–8). Next, we traverse the SSR-tree by starting from the root (Lines 9–21). Compute the parts of each node and determine if $E_{pk}(n_d(N, Pro_i))$ can still be tightened based on the location of the promotion point, i.e. whether the promotion point is located in the demotion part. If a node cannot tighten the bounds for any promotion point, we prune it (Lines 13–14). Otherwise, it continue to traverse its children to compute the demotion point minimum value (Lines 15–21). By pruning promotion points and nodes in the tree, we obtain the refined query with minimum penalty value (Lines 22–24). $\mathcal{C}_1$ adds the randomly generated noise ε_1 and ε_2 to $E_{pk}(k')$ and $E_{pk}(\boldsymbol{w}')$. Next, $\mathcal{C}_1$ sends ε_1 and ε_2 to $\mathcal{U}$ and sends $E_{pk}(k_{med})$ and $E_{pk}(\boldsymbol{w}_{med})$ to $\mathcal{C}_2$. $\mathcal{C}_2$ decrypts $E_{pk}(k_{med})$ and $E_{pk}(\boldsymbol{w}_{med})$ and sends k_{med} and $\boldsymbol{w}_{med}$ to $\mathcal{U}$. $\mathcal{U}$ removes the noise to obtain the refined query $q'(k', \boldsymbol{w}')$ (Lines 25–29).

Figure 2 shows an example of SWoSKQ processing. Firstly, $\mathcal{DO}$ processes and encrypts dataset and SSR-tree. Based on the top-2 result, $\mathcal{U}$ generates and encrypts a SWoSKQ request, including the encrypted missing object $E_{pk}(o)$, $k_0(E_{pk}(2))$, $\boldsymbol{w}_0\langle E_{pk}(0.5), E_{pk}(0.5)\rangle$. After steps 3, 4 and 5, the optimal refined query $q'((E_{pk}(2)), \langle E_{pk}(0.56), E_{pk}(0.44)\rangle)$ with the smallest penalty value obtained by calculating and comparing the penalty values of all candidate weights is obtained as the result of SWoSKQ.

5 Analysis

5.1 Computational Complexity Analysis

BSWoSKQ. For the complexity, the BSWoSKQ approach requires $\mathcal{O}(|E_{pk}(H)|^2 + |E_{pk}(L)|^2)$ encryptions and decryptions, where $|E_{pk}(H)|$ is the number of data points ranked higher and $|E_{pk}(L)|$ is the number of data points ranked lower.

SWoSKQ. It involves secure rank calculation leading to $\mathcal{O}(n)$ encryptions and decryptions in Line 1. It also executes $3|E_{pk}(Pro)|$ times SM, SDC and $|E_{pk}(Pro)|$ times SLESS protocols and secure penalty computation leading to $11|E_{pk}(Pro)|$ encryptions and $21|E_{pk}(Pro)|$ decryptions in Lines 3–8. It also requires $4|N| \cdot |E_{pk}(Pro)|$ encryptions and $8|N| \cdot |E_{pk}(Pro)|$ decryptions in Line 19. It requires $|E_{pk}(Pro)|$ encryptions and decryptions in Line 23. Finally, it requires 3 decryptions in Line 28. The overall complexity of the SWoSKQ approach is $\mathcal{O}(|E_{pk}(Pro)| \cdot |N|)$ encryptions and decryptions, where $|E_{pk}(Pro)|$ is the number of promotion points and $|N|$ is the maximum number of nodes in the SSR-tree.

5.2 Security Analysis

We utilize the formal definitions and theorems provided by [5,22] to analyze the security of our proposed approach as follows.

Security of Approach. In the BSWoSKQ approach, the computation of the incomparable points and the candidate weight vectors are computed in ciphertext, so neither $\mathcal{C}_1$ and $\mathcal{C}_2$ will know any information about the best refined query. Therefore, the BSWoSKQ approach is secure.

In the SWoSKQ approach, the primary difference is that it prunes the candidate weights and quickly obtains the ranking of the lost points under the candidate weights by traversing the SSR-tree. Skipping certain weight vectors during pruning only reduces the number of candidates, without revealing any information about the chosen ranks or weighting vectors for the best refined query. Since $k^{'}$ and $w^{'}$ are perturbed by ε_1 and ε_2, which are randomly generated by $\mathcal{C}1$, no information about the refined queries is revealed to $\mathcal{C}2$. Therefore, the SWoSKQ approach remains secure.

Privacy Requirements. We ensure that our proposed approaches meet the privacy requirements outlined in Section 3. Firstly, since the inputs $E_{pk}(\mathcal{D})$ and $E_{pk}(\mathcal{T})$ is encrypted by $\mathcal{DO}$, and $E_{pk}(q)$ is encrypted by $\mathcal{U}$, both data and query privacy are protected from $\mathcal{C}_1$ and $\mathcal{C}_2$ by the Paillier cryptosystem. Next, Since the results of the refined queries are in ciphertext stored in $\mathcal{C}_1$, and since the results of refined queries are randomized by noise ε_1 and ε_2. $\mathcal{C}_1$ and $\mathcal{C}_2$ cannot learn anything about the plaintext of the results. This guarantees the privacy of the results. In each iteration of BSWoSKQ and SWoSKQ processing, neither $\mathcal{C}1$ nor $\mathcal{C}2$ can find which candidate query represents the optimal refined query with the lowest penalty value. Therefore, the access pattern is ensured.

6 Experiment Evaluation

6.1 Experimental Setups

Algorithms for Comparison. To evaluate the performance of our proposed algorithms BSWoSKQ and SWoSKQ, We compare them with the existing SDAWNkQ algorithm [11]. In comparison to SDAWNkQ, our scheme introduces new functionalities and enhances the efficiency of retrieval.

Setup. We implement all the approaches above in Python3.9. All experiments are conducted on a *Tower Server* with 256G *RAM, Intel Xeon CPU 2.10Ghz*, running *Ubuntu 20.04*. Note that $\mathcal{C}_1$ and $\mathcal{C}_2$ are running on the same workstation.

Datasets. We use two real datasets POI[1] and SINA[2]. For POI, multiple real points of interest categories are aggregated into textual content, retaining their original locations. For SINA, the original texts are tokenized, preserving high-frequency keywords. Furthermore, we construct two datasets, TF and FT, from SINA. TF comprises real locations paired with randomly generated texts, whereas FT consists of randomly generated locations matched with real texts. Detailed dataset information is provided in Table 2.

Parameter Setting. We evaluate the performance of our approach by varying various system parameters. For the key size, we respectively choose 256, 512, and

---

[1] http://www.poilist.cn/.
[2] https://hub.hku.hk/.

Table 2. Spatio-textual dataests

Dataset	POI	SINA	TF	FT
# of Objects	5,000	10,000	5,000	5,000
Total # of keywords	10,434	58,814	49,822	49,691
Avg. # of keywords	2	6	10	10

1024. For the initial rank r_0, we choose the 10-th, 15-th, 20-th, 30-th and 45-th ranked objects, with k of the top-k query set to 5, 10, 15, 25 and 40. Regarding the user's preference w, we set w_s to 0.1, 0.3, 0.5, 0.7 and 0.9. For the parameter λ, we choose five different values as 0.1, 0.3, 0.5, 0.7 and 0.9.

6.2 Experimental Evaluations

Evaluation on Encryption Costs. We conducted experiments to evaluate the time and storage costs of encrypting spatial datasets with different key sizes using our approach. In Fig. 4, the encryption time and storage costs clearly increase with the number of data objects and the dimension of data vectors. In addition, the encryption time and storage cost also increase with the increase of key size.

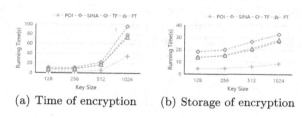

(a) Time of encryption (b) Storage of encryption

Fig. 4. Evaluation on Encryption Costs

Performance Evaluation on $\mathcal{K}$. Figure 5 illustrates the runtime variation with different values of $\mathcal{K}$ across four datasets. It's evident that as $\mathcal{K}$ increases, the time costs of all three schemes rise. This is mainly due to the longer key length, resulting in higher computational time for data encryption, query modification, penalty value calculation and comparison.

Performance Evaluation on k_0. In this set of experiments, we investigate the impact of varying k_0 in the initial query. The ranking of the missing objects varies with k_0, i.e. $r_0 = 10k_0 + 1$. As shown in Fig. 6, the running time of our algorithm increases with k_0. In terms of total query time, SDAWNkQ is significantly worse than BSWoSKQ and SWoSKQ.

Performance Evaluation on w_0. We examine the impact of varying the weight vector w_0 in the initial query. As depicted in Fig. 7, changing w_0 has little effect on all three algorithms. However, the overall running time of SDAWNkQ is

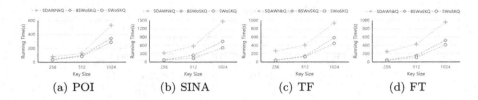

Fig. 5. Running time under impact of $\mathcal{K}$

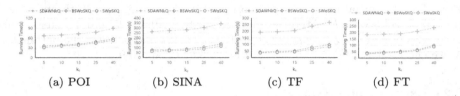

Fig. 6. Running time under impact of k_0

longer. This shows that BSWoSKQ and SWoSKQ are better when changing the initial weight vector w_0.

Performance Evaluation on λ. When changing λ, we observe from Fig. 8 that three algorithms are minimally affected. This is because in BSWoSKQ, λ is only utilized in the penalty calculation step for all refined queries. In SWoSKQ, different λ settings do not significantly affect the pruning efficiency. In SDAWNkQ, the process of obtaining candidate queries does not involve λ.

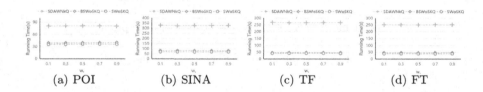

Fig. 7. Running time under impact of w_0

7 Conclusion

In this paper, we address the secure why-not spatial keyword top-k query problem. We first introduce a basic scheme(BSWoSKQ), which obtains the best approximate refined query with minimal penalty by modifying the weight vector w and k in the initial query. Additionally, we further propose an enhanced query scheme(SWoSKQ), featuring a new secure index structure SSR-tree and efficient pruning methods to improve query efficiency. We analyze the computational

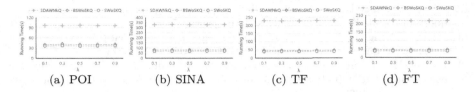

Fig. 8. Running time under impact of λ

complexity and security of our approach and conduct extensive experiments to demonstrate its performance. For future work, considering the query efficiency and data distribution issues, we will focus on the secure why-not query scheme through modifying other parameters in spatial keyword queries.

Acknowledgement. The work is partly supported by the Young and Middle-aged Science and Technology Innovation Talent Support Plan of Shenyang (RC230832), the Scientific Research Project of Education Department of Liaoning Provin-ce (JYTMS20230272, JYTMS20230270) and the Natural Science Foundation of Liaoning Province (Nos. 2022-MS-303).

References

1. Chapman, A., Jagadish, H.V.: Why not? In: SIGMOD Conference. pp. 523–534. ACM (2009)
2. Chen, L., Li, Y., Xu, J., Jensen, C.S.: Towards why-not spatial keyword top-k queries: A direction-aware approach. TKDE **30**(4), 796–809 (2018)
3. Chen, L., Lin, X., Hu, H., Jensen, C.S., Xu, J.: Answering why-not questions on spatial keyword top-k queries. In: ICDE. pp. 279–290 (2015)
4. Chen, L., Xu, J., Lin, X., Jensen, C.S., Hu, H.: Answering why-not spatial keyword top-k queries via keyword adaption. In: ICDE. pp. 697–708 (2016)
5. Cui, N., Yang, X., Wang, B., et al.: SVkNN: Efficient secure and verifiable k-nearest neighbor query on the cloud platform*. In: IEEE ICDE. pp. 253–264 (2020)
6. Elmehdwi, Y., Samanthula, B.K., Jiang, W.: Secure k-nearest neighbor query over encrypted data in outsourced environments. In: IEEE ICDE. pp. 664–675 (2014)
7. Gao, Y., Liu, Q., Chen, G., Zheng, B., Zhou, L.: Answering why-not questions on reverse top-k queries. Proc. VLDB Endow. **8**(7), 738–749 (2015)
8. He, Z., Lo, E.: Answering why-not questions on top-k queries. TKDE **26**(6), 1300–1315 (2014)
9. Li, M., Zeng, Y., Zheng, L., Chen, L., Li, Q.: Accurate and efficient trajectory-based contact tracing with secure computation and geo-indistinguishability. In: DASFAA I. Lecture Notes in Computer Science, vol. 13943, pp. 300–316. Springer (2023)
10. Li, Y., Zhang, W., Gao, Y., Li, Q., Shu, L., Luo, C.: DAPC: answering why-not questions on top-k direction-aware ASK queries in polar coordinates. IEEE TITS **24**(5), 4932–4947 (2023)
11. Liu, A., Zheng, K., Li, L., Liu, G., Zhao, L., Zhou, X.: Efficient secure similarity computation on encrypted trajectory data. In: ICDE. pp. 66–77. IEEE Computer Society (2015)

12. Miao, Y., Song, L., Li, X., et al.: Privacy-preserving arbitrary geometric range query in mobile internet of vehicles. TMC pp. 1–15 (2023)
13. Miao, Y., Yang, Y., Li, X., Wei, L., Liu, Z., Deng, R.H.: Efficient privacy-preserving spatial data query in cloud computing. TKDE **36**(1), 122–136 (2024)
14. Paillier, P.: Public-key cryptosystems based on composite degree residuosity classes. In: EUROCRYPT. Lecture Notes in Computer Science, vol. 1592, pp. 223–238. Springer (1999)
15. Qing, L., Yunjun, G.: Survey of database usability for query results. J. Comput. Res. Develop. **54**(6), 1198–1212 (2017)
16. Schütze, H., Manning, C.D., Raghavan, P.: Introduction to information retrieval, vol. 39. Cambridge University Press Cambridge (2008)
17. Teng, Y., Wang, H., Qi, J., Li, M., Fan, C., Xu, L.: Privacy-preserving direction-aware why-not spatial keyword top-k queries on cloud platform. In: ICPADS. pp. 1107–1114. IEEE (2023)
18. Teng, Y., Zhao, W., Zong, C., Xu, L., Fan, C., Wang, H.: Answering why-not questions on top k queries with privacy protection. In: TrustCom. pp. 476–485. IEEE (2021)
19. Tong, Q., Miao, Y., Li, H., Liu, X., Deng, R.H.: Privacy-preserving ranked spatial keyword query in mobile cloud-assisted fog computing. TMC **22**(6), 3604–3618 (2023)
20. Tong, Y., Pan, X., Zeng, Y., Shi, Y., Xue, C., Zhou, Z., Zhang, X., Chen, L., Xu, Y., Xu, K., Lv, W.: Hu-fu: Efficient and secure spatial queries over data federation. Proc. VLDB Endow. **15**(6), 1159–1172 (2022)
21. Wang, X., Ma, J., Li, F., Liu, X., Miao, Y., Deng, R.H.: Enabling efficient spatial keyword queries on encrypted data with strong security guarantees. IEEE TIFS. **16**, 4909–4923 (2021)
22. Yao, A.C.: How to generate and exchange secrets (extended abstract). In: FOCS. pp. 162–167. IEEE Computer Society (1986)
23. Zhang, S., Ray, S., Lu, R., Guan, Y., Zheng, Y., Shao, J.: Efficient and privacy-preserving spatial keyword similarity query over encrypted data. IEEE TDSC **20**(5), 3770–3786 (2023)
24. Zhang, S., Ray, S., Lu, R., Zheng, Y., Guan, Y., Shao, J.: Achieving efficient and privacy-preserving dynamic skyline query in online medical diagnosis. IEEE Internet Things J. **9**(12), 9973–9986 (2022)
25. Zhang, W., Li, Y., Shu, L., Luo, C., Li, J.: Shadow: Answering Why-Not Questions on Top-K Spatial Keyword Queries over Moving Objects. In: Jensen, C.S., Lim, E.-P., Yang, D.-N., Lee, W.-C., Tseng, V.S., Kalogeraki, V., Huang, J.-W., Shen, C.-Y. (eds.) DASFAA 2021. LNCS, vol. 12682, pp. 738–760. Springer, Cham (2021). https://doi.org/10.1007/978-3-030-73197-7_51
26. Zheng, B., Zheng, K., Jensen, C.S., Hung, N.Q.V., Su, H., Li, G., Zhou, X.: Answering why-not group spatial keyword queries. TKDE **32**(1), 26–39 (2020)

FreqAT: An Adversarial Training Based on Adaptive Frequency-Domain Transform

Denghui Zhang[1,2], Yanming Liang[1], Qiangbo Huang[1], Xvxin Huang[1], Peixin Liao[1], Ming Yang[2], and Liyi Zeng[3(✉)]

[1] Cyberspace Institute of Advanced Technology, Guangzhou University, GuangZhou, China
{yanmingliang,huangqiangbo,huangxvxin,liaopeixin}@e.gzhu.edu.cn
[2] Qilu University of Technology, Shandong, China
yangm@sdas.org
[3] Pengcheng Laboratory, ShenZhen, China
zengly@pcl.ac.cn

Abstract. With the advancement and popularity of deep learning (DL), the complexity and black-box of its decision-making process have posed novel vulnerabilities. Adversarial attacks (AATs), which craft clean samples to adversarial examples (AEs), compromise the reliability of DL-based models. By failing adversarial perturbation, transformation-based adversarial defenses have emerged as effective strategies against AEs. While these transformations can filter out AEs, they fail to fundamentally enhance the robustness of models and ignore the interference of perturbations on the visual fidelity of images. Instead of spatial-domain, frequency-domain transformations provide an imperceptible transformation and defense strategy. This paper introduces a novel adaptive frequency-domain-based adversarial training (AT) mechanism (called FreqAT). FreqAT first leverages frequency-domain transformations to improve the robustness of models and reduce visual interference on clean images. It further enhances the generalization of models across architectures and datasets by an adaptive transformation. Experimental results show that by combining frequency-domain transformations and AT, we can enhance model robustness while mitigating the quality degradation of clean samples.

Keywords: Attack Defense · Adversarial Attack · Frequency-domain Transformation · Adversarial Training

1 Introduction

The rapid accumulation of data and advancements in hardware technologies have fueled the exponential growth of DL [3,26]. DL-powered intelligent systems are now being deployed at an unprecedented scale across domains including speech recognition, natural language processing (NLP), and computer vision (CV), even surpassing human performance in tasks like image classification and the game of Go [12,23]. The widespread adoption of efficient DL libraries [13] and the continuous refinement of convolutional neural network (CNN) architectures have accelerated their deployment in safety-critical scenarios including autonomous driving [6], real-time monitoring, and malware detection [18].

The inherent vulnerability of DL presents a novel threat to its deployment, where subtle perturbations added to inputs can craft AEs in the inference stage, leading to misguided model predictions with high confidence [31]. The growing complexity of AATs has driven the creation of defense mechanisms. These strategies can be grouped into gradient obfuscation, input transformation, AE detection, and robust optimization [29]. ATs stand out among these approaches. AT involves injecting AEs into the training process. During each iteration, the model is equipped with AEs to handle malicious attacks and noise interference, enhancing its robustness.

Transformation-based adversarial defense has also gained attention for its effectiveness and robustness [17]. It filters out malicious perturbation by preprocessing both clean and adversarial samples. Existing defenses focus on improving success rates but may overlook how transformations affect image quality and fidelity. Recent studies have looked into that CNNs are more sensitive to features in the frequency domain [7]. It is crucial for transformation-based defenses to prioritize imperceptibility, enhancing model robustness and accuracy.

In this paper, we propose a novel AT mechanism (FreqAT). Compared with directly disturbing pixels, transformation in the frequency domain can reduce visual interference to clean images. While frequency-domain transformation can filter out AEs, it fails to fundamentally enhance the robustness of the model, which may still be vulnerable to novel attacks. Therefore, we further mitigate its vulnerability through AT. The contributions of this paper are as follows:

- We propose an adversarial training strategy based on frequency-domain transformation to improve the robustness of the model.
- We further introduce an adaptive transformation-based AT to enhance the generalization of models across different architectures and datasets.
- We undertake extensive experiments to demonstrate the effectiveness and benefits of our proposed framework.

The remainder of this paper is organized as follows: In Sect. 2, a comprehensive review of existing literature is carried out, covering adversarial attacks, adversarial defenses, and studies on frequency-domain robustness. In Sect. 3, the FreqAT approach is presented in detail, aiming to enhance the generalization and robustness of models. Section 4 demonstrates the performance of FreqAT on different datasets and attacks, in addition to its interaction with various AT methods. Differences among various image transformations are also explored. In Sect. 5, the paper is summarized and future work is outlined.

2 Related Works

2.1 Adversarial Attacks

As deep learning evolves, its vulnerabilities and security risks are revealed. Goodfellow et al. [10] first introduce the Fast Gradient Sign Method (FGSM), which generates adversarial perturbations through the back-propagated gradient. FGSM adds optimal perturbations to clean images, turning them into AEs that mislead classification models. Kurakin et al. [15] propose an iterative attack method known as the Basic Iterative Method (BIM). Madry et al. [20] propose the Projected Gradient Descent (PGD), generating stronger AEs through multiple gradient updates. Moosavi et al. [22] propose DeepFool to identify the closest decision boundary of classification models in any given pixel space for sample forgery. Carlini et al. [4] propose the C&W method, estimating the size of the adversarial perturbations through three norm metrics (L_0, L_2, and L_∞), and designing three corresponding attacks to generate high-confidence AEs.

2.2 Adversarial Defense

The defense strategies can be grouped into gradient obfuscation, input transformation, AE detection, and robust optimization. Most AATs exploit the gradient information of targeted models. The defense mechanism of gradient obfuscation [2] thwarts AATs by concealing gradient information.

Input-transformation-based methods mitigate adversarial perturbations in input images through subtle adjustments, such as color depth restoration [28], JPEG compression [8], spatial smoothing, and total variance minimization, etc.

AEs detection methods distinguish input images into adversarial and nonadversarial examples [9]. Harder et al. [11] exploited Fourier analysis of input images to detect AATs. This method highlights the difference between adversarial and non-adversarial examples in the frequency domain instead of the usual spatial domain.

Robust optimization methods bolster the robustness of models by adversarial training, which is one of the most effective defensive schemes. AT does not rely on other models for predictions or the modification of model structures. By adopting the loss function, researchers have achieved a balance between the robustness and generalization ability of models [20,25,30]. Zhang et al. also find that the right choice of datasets could enhance a model's robustness without compromising its generalization [31].

2.3 Model Robustness in Frequency-Based Transformation

The above defense methods are similar to the AATs, directly perturbing AEs to randomize noise. The optimization goals are to maximize attack success rate, without preserving the quality of samples. Zhang et al. [34] achieve commendable results in AT by suppressing the high-frequency components of AEs, which shows that retaining crucial frequency-domain range can enhance the effectiveness of AT.

Zhang et al. [33] enhances the models' generalization and robustness by suppressing high-frequency data generated in signal neighborhoods to adversarially train the models. However, Chen et al. [5] find that models are more susceptible to attacks at high frequencies than the low. Research on the influence of different frequency domains on various datasets and models infer that AATs can occur in any frequency band, not just high frequencies [21]. Chen et al. [5] demonstrate that by quantifying and reutilizing important frequency-domain information to reconstruct images for AT, it is possible to obtain more robust models. This further substantiates that the appropriate frequency can play a pivotal role in optimizing AT.

3 Methodology

The frequency transformation of images has received increasing attention in AATs [7] and defense [5,11,33], due in part to perturbations in the non-critical frequency range result in minimal visual alteration. Research suggests that the significant frequencies from the dataset are sufficient in ensuring the generalizability of models [21]. Although significant frequencies vary for different datasets, lower-frequency information stands out as more crucial. Based upon these insights, and inspired by the positive effects of frequency suppression and data reconstruction through pivotal frequency information on AATs [5,33], we have proposed the FreqAT to enhance the robustness and generalizability of CNNs. As shown in Fig. 1.

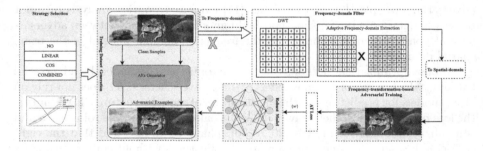

Fig. 1. Architecture of the proposed FreqAT.

3.1 Frequency-Domain Transformation

The frequency-domain transformation is broadly used to smoothen the texture and edges of an image and decrease visual noise by suppressing certain frequency components. This approach can enhance the visual quality of images without compromising their overall recognizability. Based on the observation that CNNs are sensitive towards critical frequency ranges, we can improve the robustness of models by training only on these frequencies [21,33]. Given the characteristics of different frequency-domain transformations, we adopt two different transformation methods.

The first method is the Fourier Transform, which offers an intuitive perspective to observe the various frequency components within an image. By applying designed filters in the frequency domain, we gain the ability to flexibly define and adjust the frequency cut-off threshold, consequently allowing for precise control over the range of frequencies that we wish to retain or suppress.

The Fourier Transform decomposes a function into its constituent frequencies. For signals sampled at equidistant points, this is referred to as the Discrete Fourier Transform. The two-dimensional Fourier transformation of an image $\mathbf{x}$ of size $N \times N$ is expressed as follows:

$$F(u,v) = \frac{1}{NN} \sum_{a=0}^{N-1} \sum_{b=0}^{N-1} \mathbf{x}_{a,b} e^{-j2\pi(\frac{u}{N}a + \frac{v}{N}b)} \tag{1}$$

To preserve significant frequency information of the image in the frequency domain, we need to filter out irrelevant frequencies. we select the frequency components to retain according to:

$$\hat{F}(u,v) = \mathcal{M} \odot F(u,v) \tag{2}$$

where the matrix $\mathcal{M}$ of size $N \times N$ represents the chosen separated frequency domain. For the frequency information that needs to be retained, the values in $\mathcal{M}$ should be close to 1, while for the frequency information that does not need to be preserved, the values in $\mathcal{M}$ should be close to 0. The value distribution in $\mathcal{M}$ is as follows:

$$\mathcal{M}_{u,v} = \begin{cases} 1, & 0 <= |u|, |v| <= R \\ 0, & else \end{cases} \tag{3}$$

We will turn the prepossessed frequency-domain data back to the spatial domain of normal images by the following equation:

$$\hat{\mathbf{x}} = F^{-1}(\mathcal{M} \odot \hat{F}(u,v)) \tag{4}$$

The second frequency-domain transform method selected is the wavelet transform (WT) [1]. WT addresses the decomposition problem of non-stationary signals by a set of orthogonal and rapidly decaying wavelet functions. By adjusting the scale and shift parameters, WT can capture the signal at different frequencies and times. To accommodate the discrete data in image processing, we introduce the Discrete Wavelet Transform (DWT) to decompose data into components in different frequency-domain intervals as follows:

$$W_\psi(a,b) = \frac{1}{\sqrt{a}} \sum_n \psi^* \left(\frac{n-b}{a} \right) \tag{5}$$

where a and b represent the scale and shift parameters, respectively. $\psi(\cdot)$ is the basic wavelet function, and $\psi^*(\cdot)$ is the conjugate of the $\psi(\cdot)$ function. Theoretically, the basic wavelet function in the Hilbert space can be divided into two parts: the scale function $\phi(\cdot)$which corresponds to the low-frequency part of the original function, and the wavelet function $\psi(\cdot)$ which corresponds to the high-frequency part for obtaining an orthogonal wavelet family.

The two-dimensional DWT (2D-DWT) can convert the original image into a combination of low frequency, horizontal high frequency, vertical high frequency, and diagonal high-frequency components. The 2D-DWT can produce the following four frequency bands for an image X:

$$X_{ll} = LXL^T, X_{lh} = HXL^T \\ X_{hl} = LXH^T, X_{hh} = HXH^T \quad (6)$$

where the matrix L is a circulant matrix composed of the wavelet low-pass filters $\{l_k\}_{k\in Z}$, and the matrix H is a circulant matrix formulated by wavelet high-pass filters$\{h_k\}_{k\in Z}$. The size of both matrices is $\lfloor N/2 \rfloor \times N$. We choose X_{ll} for image reconstruction an image and retain significant frequency-domain information while failing adversarial perturbations.

3.2 Adaptive Frequency-Transformation-Based Adversarial Training

AT [20,25,30] is recognized as one of the most effective means to enhance the robustness of DL models. The key idea is to artificially generate and incorporate AEs during training, thereby improving the model's robustness against AATs. Given a dataset $s = \{(x_i, y_i)\}_{i=1}^n$, where $x_i \in x$ and $y \in \{0, 1, ..., C-1\}$, the objective function for standard AT is:

$$\min_{f \in \mathcal{F}} \frac{1}{n} \sum_{i=1}^{n} \left\{ \max_{\tilde{x} \in \mathcal{B}_\epsilon[x_i]} \ell\left(f(\tilde{x}), y_i\right) \right\} \quad (7)$$

where $\tilde{x}$ denotes the adversarial data generated from x via an attack, $f(\cdot)$ stands for the score function, and l is the loss function that composes the overall loss.

AT seeks to train more robust models through potent attack mechanisms. When generating adversarial data, standard AT employs the more aggressive PGD attack to approximate the internal maximization of the Eq. 7. PGD models adversarial data as a constrained optimization problem. Given an initial point $x^{(0)} \in X$ and a step size $\alpha > 0$, PGD generates AEs as follows:

$$x^{(t+1)} = \Pi_{\mathcal{B}[x^{(0)}]} \left(x^{(t)} + \alpha \operatorname{sign} \left(\nabla_{x^{(t)}} \ell \left(f\left(x^{(t)}, y\right) \right) \right) \right), \forall t \geq 0 \quad (8)$$

where $x^{(t+1)}$ is the adversarial example at the next step, $\Pi_{\mathcal{B}[x^{(0)}]}$ denotes the projection into the set $\mathcal{B}\left[x^{(0)}\right]$.

AEs, denoted as x_{adv}, are generated with the PGD method in our method. After applying the frequency-domain transformation, we can derive $\hat{x}_{adv}$ from

x_{adv}, which are then served as inputs for subsequent ATs. The objective function of FreqAT is:

$$\min_{f \in \mathcal{F}} \frac{1}{n} \sum_{i=1}^{n} \{\max \ell (f(\hat{x}_{adv}), y_i)\} \quad (9)$$

We further propose an adaptive frequency-domain transformation method for AT. FreqAT can enhance the robustness and accuracy of models by adaptively extracting significant frequency information from the sample since finding a significant frequency range for each image is time-consuming and resource-intensive.

FreqAT takes the Fourier transform for its flexibility. To derive the optimal frequency-domain transformation method that best fits adversarial training, we have designed several transform methods, including linear, cosine, and exponential transformations, all of which have demonstrated certain effectiveness. Finally, We devise a novel AT strategy to balance robustness and accuracy. When the learning rate is high at the initial stage of training, we gradually decrease available frequency-domain information, thus enabling the model to learn from various frequency bands of images. The model takes significant frequency-domain information to learn more in the middle stage of training. Towards the last stage of training, we strive to retain as much frequency-domain information as possible to balance between generalization and robustness. To adjust the frequency-domain selection, we employ a fused cosine scheduling function [19] as follows:

$$R(t) = \begin{cases} R_1 + \frac{1}{2}(R_0 - R_1)\left(1 + \cos\left(\pi \frac{t}{0.7T}\right)\right) & \text{if } t \leq 0.7T \\ R_0 + \frac{1}{2}(R_1 - R_0)\left(1 + \cos\left(\pi \frac{t-0.7T}{0.3T}\right)\right) & \text{if } t > 0.7T \end{cases} \quad (10)$$

where T represents the total number of training rounds, t represents the current training round, and R is a parameter in the Fourier frequency-domain transformation, which varies between R_0 and R_1.

4 Experiments

To evaluate the effectiveness of our method in enhancing the generalizability and robustness, we initially compare it with conventional AT across various datasets and models. Simultaneously, we also attempt to integrate different AT strategies with our adaptive frequency-domain transformation approach. To further illuminate the outcomes of our adversarial training, we utilize the Gradient-weighted Class Activation Mapping(Grad-CAM) [24] technique to sketch the attention maps of the model. Moreover, to explore the advantages of our approach over other image transformations in enhancing the generalizability and robustness of adversarial training, we execute quality evaluations on various transformations, defense performance tests, and AT experiments with different transformations.

4.1 Experimental Setup

Datasets: We employ three classic classification datasets: CIFAR10, CIFAR100, and TinyImageNet. Both CIFAR10 [14] and CIFAR100 [14] comprise 32×32

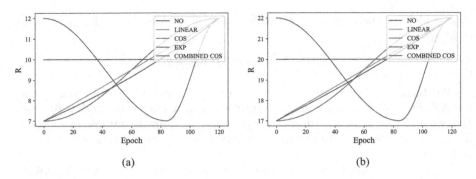

Fig. 2. Retention frequency range curve for adversarial training, (a) illustrates the variation trend for CIFAR, while (b) showcases the change pattern for TinyImageNet.

color images. The CIFAR10 dataset contains 10 categories with 6,000 images per category. The CIFAR100 dataset features 100 categories, with 600 images in each category. TinyImageNet [16] is a subset of ImageNet, comprising 64 × 64 color images and 200 categories, each with 500 images.

Evaluation Metrics: We measure the generalization of the model by the accuracy on the test set of each standard dataset. We evaluate the quality and perceptibility of robust images with metrics including LPIPS [32], SSIM, DISTS, and PSNR. We also employ Grad-CAM to reveal the attention region of trained models.

Training details: For backbone models, we employ ResNet18 [12], ResNet34, and WideResNet34-10 [29]. The ResNet18 and ResNet34 are residual models with 18 and 34 layers, respectively. WideResNet34-10 is a wide residual network model of 34 layers with a widening factor of 10. For adversarial training, we adopt PGD-10 [20], with an l_∞ norm constraint of 8/255. The robust models are enacted over 120 epochs. The weights yielding the highest test robust accuracy are selected. We choose two well-performing image defense transformation methods for comparison with our own image frequency-domain transformation method. This incorporates both independent image transformations and their combination with adversarial training. In the employed Feature Squeezing method, the color bit compression is set to 2 [28], while JPEG compressed images are reduced to 70% of their original size. To investigate the correlation between frequency discarding range and model robustness, we discard frequencies according to different schedulers. Figure 2 illustrates the trend of frequency retention during training.

4.2 Performance Evaluation

Table 1 summarizes the results of our study, where we can draw the following findings:

Frequency-Domain Transformation-Based AT can Reconcile the Conflict Between Generalization and Robustness. Across various datasets

Table 1. Results of our method on different datasets and backbones. *Std* represents the recognition accuracy of clean samples, and *Adv* represents the recognition accuracy of adversarial samples. R represents the chosen parameter in the frequency domain, and as R decreases, less frequency information is retained.

Architecture	Method	Cifar10		Cifar100		Tiny-ImageNet	
		Std	Adv	Std	Adv	Std	Adv
ResNet18	AT	82.46	53.63	58.21	30.15	45.26	15.41
	AT+ours(R=10)	83.14	58.18	58.42	37.22	45.21	36.21
	AT+ours(lin)	84.56	60.31	60.11	37.52	46.57	37.54
	AT+ours(cos)	84.92	61.44	60.23	**37.64**	46.24	36.15
	AT+ours(exp)	84.78	**61.78**	60.23	37.27	46.35	37.21
	AT+ours(com)	**86.01**	61.11	**61.56**	37.31	**47.56**	**38.05**
ResNet34	AT	84.23	55.31	58.67	30.50	46.25	16.02
	AT+ours(R=10)	83.22	57.26	59.32	37.98	46.15	37.05
	AT+ours(lin)	84.35	59.83	59.22	38.06	47.31	38.21
	AT+ours(cos)	84.01	59.23	59.68	38.16	47.16	38.44
	AT+ours(exp)	84.22	**60.28**	60.01	38.55	47.58	38.25
	AT+ours(com)	**85.41**	59.00	**61.15**	**40.14**	**48.25**	**38.66**
Wrn34-10	AT	87.41	55.40	62.35	31.66	47.23	16.57
	AT+ours(R=20)	84.16	58.24	62.04	37.22	47.21	36.11
	AT+ours(lin)	84.32	59.26	62.14	39.19	48.31	36.89
	AT+ours(cos)	85.12	59.44	62.34	39.24	48.21	37.56
	AT+ours(exp)	85.31	60.28	62.41	39.44	48.56	37.25
	AT+ours(com)	**86.27**	**63.23**	**62.55**	**41.02**	**49.69**	**39.05**

and neural network models, our method enhances both the robustness and generalization of models. This prompts us to rethink the delicate balance between generalization and robustness, which may be disrupted by the disorder of data distribution [31].

The Embedding Distance Between Adversarial and Clean Samples is Reduced. Based on the results of Frequency-domain transformation-based AT, we find that compared to only using traditional adversarial samples, utilizing datasets that have undergone frequency-domain transformation significantly enhances the model's generalization and robustness. This evidences that images processed in such a manner are more conducive to AT of the model, and can further improve its generalizability. It follows that the distance between the processed data and the clean samples is closer, which also proves advantageous for the model's adversarial training.

We schedule and carry out experiments including fixed frequency-domain, linear, cosine, exponential, and combination transformation. The results suggest that dynamic frequency-domain transformation can endow the model with

Table 2. Comparison of defense performance between our method and existing AT methods.

Method	Cifar10		Cifar100	
	Std	Adv	Std	Adv
Trades	81.54	53.31	58.44	30.20
Trades+ours($R=10$)	83.54	56.43	58.59	37.26
Trades+ours(lin)	84.40	60.14	59.27	38.48
Trades+ours(cos)	84.56	59.62	59.51	**38.71**
Trades+ours(exp)	84.88	60.31	59.65	38.24
Trades+ours(com)	**86.01**	**60.56**	**61.21**	38.34
mart	78.77	56.92	52.46	31.47
mart+ours($R=10$)	81.23	58.22	54.49	37.15
mart+ours(lin)	82.26	59.12	56.34	39.23
mart+ours(cos)	82.11	58.68	56.17	39.26
mart+ours(exp)	82.34	59.48	56.34	39.44
mart+ours(com)	**83.03**	**61.45**	**56.91**	**40.51**
AWP	78.40	54.59	52.85	31.00
AWP+ours($R=10$)	79.20	56.55	53.24	36.89
AWP+ours(lin)	79.34	**57.67**	54.32	37.24
AWP+ours(cos)	79.40	57.21	54.12	37.56
AWP+ours(exp)	79.15	57.22	54.31	38.01
AWP+ours(com)	**79.50**	57.46	**55.19**	**39.90**

enhanced generalization capabilities and robustness compared to the fixed frequency domain. We archive the best results with the combination transformation since the combination aligns better with the training process.

4.3 Integrating FreqAT to Other ATs

To further substantiate the efficacy of FreqAT, we trialed three distinct AT strategies, namely TRADES [30], MART [25], and AWP [27] on the ResNet-18 model and CIFAR-10 and CIFAR-100 datasets besides the standard AT. As illustrated in Table 2, the frequency-domain transformation is compatible with various ATs, enhancing the accuracy of both clean and AEs in the CIFAR-10 and CIFAR-100 datasets. Compared with the fixed frequency-domain method, the Adaptive Frequency-Transformation method demonstrated superior performance, with the combined transformation method achieving the best results among all methods. It highlights the adaptability and efficacy of the dynamic approach in achieving optimal outputs.

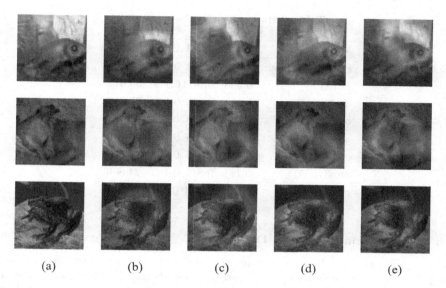

Fig. 3. Column (a) represents clean images. Column (b) represents the attention of clean samples under the normal training model. Column (c) represents the attention of AEs in the normal training model. Column (d) represents the attention of AEs in the standard AT model. Column (e) represents the attention of AEs in the frequency-domain transformation AT model.

4.4 Explainable Experiment for Defensive Model

As shown in Fig. 3, AEs have caused the focus area of the conventional model to deviate. However, through adversarial training, the model's attention can be refocused back to its original position. Normal AT cannot accurately restore the focus to the original region. Its attention is interfered with by AATs, resulting in overlaps between clean and adversarial samples. It means the robustness has not been improved. While frequency-domain-based AT is more satisfactory, where its focus area coincides with that of clean samples.

4.5 Image Quality Assessment

We evaluate the image quality after transformation for defense on the CIFAR-10 dataset. The image transformation often has negative impacts on image quality, leading to a decrease in model recognition accuracy. We compare various transformation methods with FreqAT in different spectral components. $R = 11.29$ is the optimal parameter for the adaptive frequency-domain training. As shown in Table 3, the quality of images treated with JPEG compression and our method is markedly maintained. When more frequency-domain information is eliminated, there is a significant decline in image quality. These experimental results affirm our method's proficiency in retaining more image information, which is the essential reason our technique can ensure top-tier generalization in models.

Table 3. Image Quality Assessment

Method	LPIPS	SSIM	DISTS	PSNR
Feature Squeezing	0.046	0.803	0.215	20.401
JPEG	0.006	**0.964**	0.112	30.011
ours(DWT)	0.015	0.888	0.129	27.052
ours(R=11.29)	**0.005**	0.959	**0.094**	**30.207**
ours(R=10)	0.015	0.939	0.116	28.638
ours(R=9)	0.034	0.915	0.138	27.306
ours(R=7)	0.111	0.842	0.192	24.808
ours(R=5)	0.214	0.713	0.253	22.331

4.6 Transformation-Based Defense

To ascertain whether frequency-domain transformations can minimize embedding distances between adversarial and original samples, which should belong to the same category and are closer than other categories. We conduct tests on regular models with different transformations. As illustrated in Fig. 4, all transformation methods have shown defense efficacy, with JPEG transformation standing out as the most effective. JPEG ensures high accuracy for clean samples while keeping commendable accuracy for AEs. Frequency-domain transformations demonstrate a good defense success rate for both cleaned and adversarial samples. Comparing different frequency-domain transformations, DWT yields the best results. As the frequency-domain components decrease, the defense effect increases first and then diminishes. This results from the suboptimal recognition performance of models when operating within the complete frequency domain, particularly when confronted with images that encompass only a subset of the frequency-domain components.

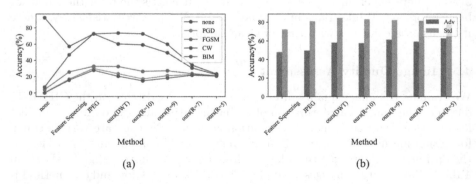

Fig. 4. (a)Results of image transformation methods used solely as a defense mechanism. (b)Results of AT using image transformation techniques.

4.7 Adversarial Training with Different Transformations

Figure 4 evaluate which transformation is better for adversarial training. Frequency-domain transformation exhibits superior performance with other transformations. The JPEG transformation, while effective in defense and preserving image quality, does not enhance the robustness of the model in adversarial training. Despite that JPEG implements frequency-domain suppression, it also applies quantization to the unsuppressed frequency domain. This results in AEs unsuitable for AT. Even though the frequency components are reduced, trained models still show good generalization, implying that our proposed FreqAT can enhance the model's performance across various contexts and datasets.

5 Conclusion

To address the limitations of existing adversarial defense strategies, this paper introduces a novel adaptive frequency-domain AT mechanism. By leveraging frequency-domain transformations and incorporating AT, we have demonstrated improvements in robustness and generalization across diverse architectures and datasets. Our proposed strategy not only reduces visual interference on clean images but also enhances the overall robustness of models against AATs. Through extensive experiments, we have validated the effectiveness and benefits of the FreqAT framework in bolstering model robustness and generalization capabilities. The integration of adaptive frequency-domain transformations in AT will fortify DL-based models against evolving AATs and improve their security in real-world applications.

In future work, our experiment is conducted under the assumption that the attacker is unaware of the defense strategy. FreqAT retains a 55% accuracy on AEs of cifar10 with ResNet18, even when attackers know the frequency-domain transformation defense. We will conduct more extensive experiments into AATs for the white-box defense. The AT introduces an additional step of frequency-domain transformation during testing, which may provide an opportunity for attackers to exploit. We will eliminate this step in the future.

Acknowledgments. This work is supported in part by the Youth Program of Humanities and Social Sciences of the MoE (23YJCZH291) and the Key Laboratory of Computing Power Network and Information Security, Ministry of Education under Grant No 2023ZD02.

References

1. Antonini, M., Barlaud, M., Mathieu, P., Daubechies, I.: Image coding using wavelet transform. IEEE Trans. Image Processing **1**, 20–5 (1992)
2. Athalye, A., Carlini, N., Wagner, D.: Obfuscated gradients give a false sense of security: Circumventing defenses to adversarial examples. In: International conference on machine learning. pp. 274–283. PMLR (2018)

3. Bansal, M., Chana, I., Clarke, S.: A survey on iot big data: Current status, 13 v's challenges, and future directions. ACM computing surveys p. 53 (2021)
4. Carlini, N., Wagner, D.: Towards evaluating the robustness of neural networks. In: 2017 ieee symposium on security and privacy (sp). pp. 39–57. Ieee (2017)
5. Chen, Y., Ren, Q., Yan, J.: Rethinking and improving robustness of convolutional neural networks: a shapley value-based approach in frequency domain. Adv. Neural. Inf. Process. Syst. **35**, 324–337 (2022)
6. Deng, B., Zhang, D., Dong, F., Zhang, J., Shafiq, M., Gu, Z.: Rust-Style Patch: A Physical and Naturalistic Camouflage Attacks on Object Detector for Remote Sensing Images. Remote Sensing **15**(4), 885 (Jan 2023). https://doi.org/10.3390/rs15040885, https://www.mdpi.com/2072-4292/15/4/885
7. Duan, R., Chen, Y., Niu, D., Yang, Y., Qin, A.K., He, Y.: Advdrop: Adversarial attack to dnns by dropping information. In: Proceedings of the IEEE/CVF International Conference on Computer Vision. pp. 7506–7515 (2021)
8. Dziugaite, G.K., Ghahramani, Z., Roy, D.M.: A study of the effect of jpg compression on adversarial images. arXiv preprint arXiv:1608.00853 (2016)
9. Gong, Z., Wang, W.: Adversarial and clean data are not twins. In: Proceedings of the sixth international workshop on exploiting artificial intelligence techniques for data management. pp. 1–5 (2023)
10. Goodfellow, I.J., Shlens, J., Szegedy, C.: Explaining and harnessing adversarial examples. arXiv preprint arXiv:1412.6572 (2014)
11. Harder, P., Pfreundt, F.J., Keuper, M., Keuper, J.: Spectraldefense: Detecting adversarial attacks on cnns in the fourier domain. In: 2021 International Joint Conference on Neural Networks (IJCNN). pp. 1–8. IEEE (2021)
12. He, K., Zhang, X., Ren, S., Sun, J.: Deep residual learning for image recognition. In: Proceedings of the IEEE conference on computer vision and pattern recognition. pp. 770–778 (2016)
13. Howard, J., Gugger, S.: Fastai: A layered api for deep learning. Information **11**(2), 108 (Feb 2020). https://doi.org/10.3390/info11020108, http://dx.doi.org/10.3390/info11020108
14. Krizhevsky, A., Hinton, G., et al.: Learning multiple layers of features from tiny images (2009)
15. Kurakin, A., Goodfellow, I., Bengio, S.: Adversarial machine learning at scale. arXiv preprint arXiv:1611.01236 (2016)
16. Le, Y., Yang, X.: Tiny imagenet visual recognition challenge. CS 231N **7**(7), 3 (2015)
17. Liao, F., Liang, M., Dong, Y., Pang, T., Hu, X., Zhu, J.: Defense against adversarial attacks using high-level representation guided denoiser. In: Proceedings of the IEEE conference on computer vision and pattern recognition. pp. 1778–1787 (2018)
18. Liao, P., Huang, X., Huang, Q., Liang, Y., Wang, Z., Zhang, D.: An explainable intrusion detection system based on feature importance. In: 2023 IEEE 12th International Conference on Cloud Networking (CloudNet). pp. 389–397. IEEE (2023)
19. Loshchilov, I., Hutter, F.: Sgdr: Stochastic gradient descent with warm restarts (2017), https://arxiv.org/abs/1608.03983
20. Madry, A., Makelov, A., Schmidt, L., Tsipras, D., Vladu, A.: Towards deep learning models resistant to adversarial attacks. arXiv preprint arXiv:1706.06083 (2017)
21. Maiya, S.R., Ehrlich, M., Agarwal, V., Lim, S.N., Goldstein, T., Shrivastava, A.: A frequency perspective of adversarial robustness. arXiv preprint arXiv:2111.00861 (2021)

22. Moosavi-Dezfooli, S.M., Fawzi, A., Frossard, P.: Deepfool: a simple and accurate method to fool deep neural networks. In: Proceedings of the IEEE conference on computer vision and pattern recognition. pp. 2574–2582 (2016)
23. Schroff, F., Kalenichenko, D., Philbin, J.: Facenet: A unified embedding for face recognition and clustering. In: 2015 IEEE Conference on Computer Vision and Pattern Recognition (CVPR). IEEE (Jun 2015). https://doi.org/10.1109/cvpr.2015.7298682, http://dx.doi.org/10.1109/cvpr.2015.7298682
24. Selvaraju, R.R., Cogswell, M., Das, A., Vedantam, R., Parikh, D., Batra, D.: Gradcam: Visual explanations from deep networks via gradient-based localization. In: Proceedings of the IEEE International Conference on Computer Vision (ICCV) (Oct 2017)
25. Wang, Y., Zou, D., Yi, J., Bailey, J., Ma, X., Gu, Q.: Improving adversarial robustness requires revisiting misclassified examples. In: International conference on learning representations (2019)
26. Woschank, M., Rauch, E., Zsifkovits, H.E.: A review of further directions for artificial intelligence, machine learning, and deep learning in smart logistics. Sustainability (2020), https://api.semanticscholar.org/CorpusID:218944310
27. Wu, D., Xia, S.T., Wang, Y.: Adversarial weight perturbation helps robust generalization. Adv. Neural. Inf. Process. Syst. **33**, 2958–2969 (2020)
28. Xu, W., Evans, D., Qi, Y.: Feature squeezing: Detecting adversarial examples in deep neural networks. arXiv preprint arXiv:1704.01155 (2017)
29. Zhang, C., Bengio, S., Hardt, M., Recht, B., Vinyals, O.: Understanding deep learning (still) requires rethinking generalization. Commun. ACM **64**(3), 107–115 (2021)
30. Zhang, H., Yu, Y., Jiao, J., Xing, E., El Ghaoui, L., Jordan, M.: Theoretically principled trade-off between robustness and accuracy. In: International conference on machine learning. pp. 7472–7482. PMLR (2019)
31. Zhang, J., Xu, X., Han, B., Niu, G., Cui, L., Sugiyama, M., Kankanhalli, M.: Attacks which do not kill training make adversarial learning stronger. In: International conference on machine learning. pp. 11278–11287. PMLR (2020)
32. Zhang, R., Isola, P., Efros, A.A., Shechtman, E., Wang, O.: The unreasonable effectiveness of deep features as a perceptual metric. In: Proceedings of the IEEE conference on computer vision and pattern recognition. pp. 586–595 (2018)
33. Zhang, S., Lin, Y., Yu, J., Zhang, J., Xuan, Q., Xu, D., Wang, J., Wang, M.: Hfad: Homomorphic filtering adversarial defense against adversarial attacks in automatic modulation classification. IEEE Transactions on Cognitive Communications and Networking (2024)
34. Zhang, Z., Jung, C., Liang, X.: Adversarial defense by suppressing high-frequency components. arXiv preprint arXiv:1908.06566 (2019)

Enhancing Network Intrusion Detection with VAE-GNN

Junyu Li[1(✉)] and Haoxi Wang[2]

[1] Harbin Engineering University, Harbin 150001, China
lijunyu0706@163.com
[2] Chinese Academy of Sciences, Beijing 100045, China

Abstract. As the network environment becomes increasingly complex, the threats it faces are becoming more severe. Intrusion detection, as a key proactive defense mechanism in network security, requires more robust and effective detection methods to address these challenges. Graph neural networks (GNNs) have shown excellent performance in anomaly detection. This paper proposes a novel intrusion detection method based on VAE-GNN, an improved graph neural network approach. Specifically, the network traffic data is first constructed into a network traffic graph. This graph is then transformed into a format suitable for processing by the graph neural network to detect anomalous traffic. To address the limitations of traditional intrusion detection methods, such as dataset imbalance and weak feature representation problems, a detection method based on VAE-GNN and the integration of feature statistical importance is proposed. Experimental results on two benchmark intrusion detection datasets demonstrate that the VAE-GNN method outperforms the original graph neural network methods, showing better detection performance and stronger noise resistance.

Keywords: Intrusion Detection System · Variational Autoencoder · Graph Neural Network · Network Security · Anomaly Detection

1 Introduction

With the continuous advancement of technology, the number of electronic devices has been growing geometrically, and the network environment has become more complex and variable. This has led to frequent network security issues, such as privacy breaches and malicious attacks, causing significant losses to the national economy [23]. Therefore, effectively preventing network attacks is an urgent problem that needs to be addressed. To solve this problem, intrusion detection systems (IDS) are typically constructed to enhance cybersecurity. These systems use proactive defense methods to effectively detect intrusions and respond in a timely manner [27], becoming one of the key means to prevent network attacks.

In this context, scholars both domestically and internationally have conducted in-depth research on intrusion detection technologies, proposing various detection algorithms such as machine learning, mathematical statistics, and neural networks [1, 16]. However, traditional machine learning methods generally emphasize feature selection and parameter training, which can be time-consuming. Data mining algorithms are sensitive to noise, and when faced with datasets with a lot of noise, they are prone to overfitting. Deep learning methods excel at extracting feature information from large, high-dimensional data, reducing the impact of noise on algorithm performance. Numerous scholars have used deep neural networks, such as convolutional neural networks (CNN), long short-term memory networks (LSTM), and recurrent neural networks (RNN), to design a series of intrusion detection algorithms.

The CNN method converts one-dimensional intrusion data into two-dimensional "image data" and then processes the intrusion data using CNN. This method considers the interactions between features within the convolution kernel of the convolutional neural network, making the feature data no longer isolated. For example, the work [13] draws on the Inception structure proposed by the Google team, designing a convolutional neural network with multiple convolution kernels of different sizes; the work [14] designs a deep convolutional neural network with different sized convolution kernels based on LeNet-5. Both of these CNN-based intrusion detection algorithms can extract the relationships between features through different sized convolution kernels, but they have the following disadvantages: *i)* CNN can only process Euclidean space data and cannot explore the complex graph structure relationships between any two feature data; *ii)* Most classical CNN models (such as VGG [18], GoogLeNet [20], and ResNet [9] etc.) have complex structures, large numbers of parameters, and high running time costs, making the introduction of classical CNN models into the field of intrusion detection need to focus on operating costs. RNN and the improved LSTM neural network algorithms based on RNN treat each sample's feature data as sequential data and input them into LSTM and RNN according to the sequence order. LSTM and RNN methods can also explore the dependencies between feature data. The works [10,26] use LSTM/RNN to mine the relationships between feature values in network intrusion data, achieving good results in the field of intrusion detection. However, since the feature data in intrusion detection datasets are input in order of their arrangement, it is difficult to accurately establish the connections between sample feature data. Additionally, these algorithms can only process Euclidean space data, making it difficult to explore the complex graph structure relationships between any two feature data.

In fact, each sample in the intrusion detection dataset contains multiple feature data, and there may be strong or weak associations between different feature data that cannot be precisely mathematically described in Euclidean space. The strong or weak associations between data features constitute the graph structure relationships in non-Euclidean space. In-depth exploration of these graph structures and applying the results to assist neural network classification decisions can effectively improve the intrusion detection performance of neural networks. Currently, a typical deep learning method for processing non-Euclidean space data is the graph neural network (GNN) [4,6,19]. However, applying GNN to intrusion detection is non-trivial as there are several challenges.

Specifically, GNN can only handle known graph structure relationships between samples and is powerless in scenarios where the graph structure relationships between internal features of samples are unknown. Furthermore, in actual network activities, normal traffic accounts for the majority, while abnormal traffic is relatively rare. This imbalance in the dataset causes classification models to focus more on the majority class during detection, thereby ignoring minority class attacks or directly overwhelming them. For IDS, detecting minority class attacks is crucial because if they are misclassified as normal traffic, it will cause greater damage to user devices. Additionally, most intrusion datasets contain high-dimensional features, which may include redundant or irrelevant features that make the differences between normal and abnormal traffic less apparent [22], thus affecting the model's ability to distinguish between normal and abnormal traffic, leading to high false alarm rates.

Contributions. We summarize the contribution of this paper as follows:

- In response to the above challenge, this paper designs a new neural network, VAE-GNN, capable of addressing the issue of imbalanced datasets and handling unknown relationships between sample data in intrusion detection.
- The VAE-GNN innovatively leverages VAE-GAN to generate new samples for facilitating the GNN model to better learn data information, resulting in a strong ability to detect intrusions in the intrusion detection system.
- We perform extensive experiments on benchmark datasets and compare our method with several baselines. Experiments validate the efficacy of our proposed method.

2 Background

2.1 Graph Neural Networks (GNNs)

Graph Neural Networks (GNNs) are a class of neural networks designed to operate on graph-structured data. Unlike traditional neural networks, which are primarily designed for Euclidean data (such as images or sequences), GNNs can handle non-Euclidean data where relationships are represented as graphs. This makes GNNs particularly suitable for tasks involving social networks, molecular structures, and communication networks.

A graph G is defined as $G = (V, E)$, where V is the set of nodes and E is the set of edges. Each node $v \in V$ can have associated features $\mathbf{x}_v$, and each edge $(u, v) \in E$ can have associated weights or features.

The core idea of GNNs is to learn a representation of each node by aggregating feature information from its neighbors. This is typically done through message passing, where each node receives messages from its neighbors and updates its state accordingly. The update process can be formulated as follows:

$$\mathbf{h}_v^{(k)} = \sigma \left(\mathbf{W}^{(k)} \cdot \text{AGGREGATE}^{(k)} \left(\left\{ \mathbf{h}_u^{(k-1)} : u \in \mathcal{N}(v) \right\} \right) \right), \tag{1}$$

where $\mathbf{h}_v^{(k)}$ is the feature vector of node v at layer k, $\mathcal{N}(v)$ is the set of neighbors of node v, $\mathbf{W}^{(k)}$ is a learnable weight matrix, σ is an activation function, and AGGREGATE is a permutation-invariant function (such as sum, mean, or max) that aggregates the features of the neighboring nodes.

One popular variant of GNNs is the Graph Convolutional Network (GCN) [12], which simplifies the aggregation function to a linear transformation followed by an average over the neighbors:

$$\mathbf{H}^{(k+1)} = \sigma \left(\tilde{\mathbf{D}}^{-\frac{1}{2}} \tilde{\mathbf{A}} \tilde{\mathbf{D}}^{-\frac{1}{2}} \mathbf{H}^{(k)} \mathbf{W}^{(k)} \right), \tag{2}$$

where $\tilde{\mathbf{A}} = \mathbf{A} + \mathbf{I}$ is the adjacency matrix with added self-connections, $\tilde{\mathbf{D}}$ is the degree matrix of $\tilde{\mathbf{A}}$, $\mathbf{H}^{(k)}$ is the matrix of node features at layer k, and $\mathbf{W}^{(k)}$ is the layer-specific trainable weight matrix.

GNNs have been successfully applied to a wide range of tasks, including node classification, link prediction, and graph classification. Their ability to capture the dependencies between nodes and edges in a graph makes them a powerful tool for analyzing and modeling complex relational data [25].

Message Passing Framework. The core idea of GNNs can be abstracted as a message-passing framework where nodes exchange information along edges and update their states accordingly:

$$\mathbf{h}_v^{(k)} = \text{UPDATE}^{(k)}\left(\mathbf{h}_v^{(k-1)}, \text{AGG}^{(k)}\left(\{\mathbf{m}_{u \to v}^{(k-1)} : u \in \mathcal{N}(v)\}\right)\right), \tag{3}$$

$$\mathbf{m}_{u \to v}^{(k)} = \text{MESSAGE}^{(k)}\left(\mathbf{h}_u^{(k-1)}, \mathbf{h}_v^{(k-1)}, \mathbf{x}_u, \mathbf{x}_v, \mathbf{x}_{u \to v}\right), \tag{4}$$

where $\mathbf{m}_{u \to v}^{(k)}$ is the message sent from node u to node v at the k-th iteration, and $\mathbf{x}_{u \to v}$ represents the features of the edge from u to v. The functions MESSAGE, AGGREGATE, and UPDATE are learned during the training process [7].

2.2 Variational Autoencoder (VAE)

Variational Autoencoders (VAEs) are a class of generative models that have gained significant attention due to their ability to learn complex data distributions. VAEs are an extension of traditional autoencoders with a probabilistic twist, allowing them to generate new data samples that are similar to the input data.

A VAE consists of two main components: an encoder and a decoder. The encoder maps the input data $\mathbf{x}$ to a latent variable $\mathbf{z}$, while the decoder reconstructs the input data from the latent variable. Unlike traditional autoencoders, VAEs assume that the latent variables are drawn from a known probability distribution, typically a multivariate Gaussian. The encoder outputs the parameters of the distribution of the latent variable, i.e., the mean μ and the logarithm of the variance $\log \sigma^2$. The latent variable $\mathbf{z}$ is sampled from this distribution:

$$\mathbf{z} \sim \mathcal{N}(\mu, \sigma^2). \tag{5}$$

The decoder then maps $\mathbf{z}$ back to the data space, producing a reconstruction $\hat{\mathbf{x}}$. The objective of the VAE is to maximize the variational lower bound (ELBO) on the marginal likelihood of the data. This involves two terms: the reconstruction loss and the regularization term.

The reconstruction loss measures how well the decoder reconstructs the input data and is typically defined as the negative log-likelihood:

$$\mathcal{L}_{\text{recon}} = -\mathbb{E}_{q(\mathbf{z}|\mathbf{x})}[\log p(\mathbf{x}|\mathbf{z})]. \tag{6}$$

The regularization term, also known as the Kullback-Leibler (KL) divergence, ensures that the distribution of the latent variables remains close to the prior distribution, usually a standard normal distribution:

$$\mathcal{L}_{\text{KL}} = D_{\text{KL}}(q(\mathbf{z}|\mathbf{x}) \| p(\mathbf{z})). \tag{7}$$

The overall loss function for the VAE is the sum of these two terms:

$$\mathcal{L}_{\text{VAE}} = \mathcal{L}_{\text{recon}} + \mathcal{L}_{\text{KL}}. \tag{8}$$

VAEs have been successfully applied in various domains, including image generation, anomaly detection, and natural language processing, due to their ability to capture the underlying data distribution and generate new, plausible samples [11].

2.3 Generative Adversarial Networks (GANs)

Generative Adversarial Networks (GANs) are a class of generative models introduced by Ian Goodfellow et al. in 2014 [5]. GANs consist of two neural networks, a generator G and a discriminator D, which are trained simultaneously through adversarial training. The generator's goal is to produce data that is indistinguishable from real data, while the discriminator's goal is to differentiate between real data and data generated by the generator. The generator G maps a random noise vector $\mathbf{z}$ from a prior distribution $p_{\mathbf{z}}(\mathbf{z})$ to the data space, generating a sample $G(\mathbf{z})$. The discriminator D receives either a real data sample $\mathbf{x}$ or a generated sample $G(\mathbf{z})$ and outputs a probability $D(\mathbf{x})$ or $D(G(\mathbf{z}))$, respectively, representing the likelihood that the sample is real.

The training of GANs involves a minimax game with the following value function $V(G, D)$:

$$\min_G \max_D V(G, D) = \mathbb{E}_{\mathbf{x} \sim p_{\text{data}}(\mathbf{x})}[\log D(\mathbf{x})] + \mathbb{E}_{\mathbf{z} \sim p_{\mathbf{z}}(\mathbf{z})}[\log(1 - D(G(\mathbf{z})))]. \tag{9}$$

Here, $p_{\text{data}}(\mathbf{x})$ represents the distribution of the real data, and $p_{\mathbf{z}}(\mathbf{z})$ represents the prior distribution over the noise vector $\mathbf{z}$.

The generator G aims to minimize the objective function by generating samples $G(\mathbf{z})$ that maximize the probability of the discriminator making a mistake. Conversely, the discriminator D aims to maximize the objective function by correctly distinguishing between real and generated samples. In practice, the training involves iteratively updating the generator and the discriminator using gradient-based optimization methods. The discriminator is trained to maximize the probability of assigning the correct label to both real and generated samples, while the generator is trained to minimize the probability that the discriminator correctly identifies generated samples as fake.

GANs have shown remarkable success in generating high-quality synthetic data across various domains, including image generation, text-to-image synthesis, and style transfer. The adversarial training process helps GANs to produce realistic and diverse outputs, making them one of the most powerful generative models in machine learning [5].

3 The Proposed VAE-GNN Model

This section introduces the VAE-GNN model, which aims to tackle the limitations of current intrusion detection methods. By integrating advanced GNN architectures with VAE-GAN frameworks, VAE-GNN promises a more robust detection of intrusion activities. The proposed VAE-GNN framework in this paper consists of four main components: the data preprocessing module, the data generation module, the feature selection module, and the attack detection module. Each of these modules is designed to address specific challenges in the context of network intrusion detection and to improve the overall performance of the detection system, as shown in Fig. 1.

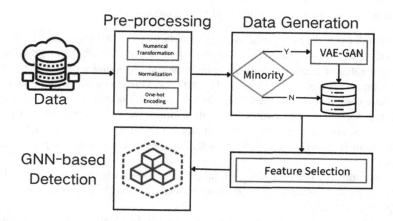

Fig. 1. An overview of the VAE-GNN model. There are four modules in VAE-GNN: a data preprocessing module, a data generation module, a feature selection module, and a GNN-based detection module.

3.1 Data Preprocessing

To enhance the quality of the data, the dataset undergoes a comprehensive preprocessing phase before being fed into the model. This preprocessing step is crucial as it improves the model's convergence speed and ensures the accuracy and reliability of the detection results. The preprocessing includes the following key steps:

Numerical Transformation: The dataset contains several non-numeric features, such as 'protocol type', 'flag', and 'service', which need to be converted into numeric features. This transformation is achieved using label encoding. Label encoding converts categorical variables into numeric format by assigning a unique integer to each category. Additionally, the labels of the data are also transformed into numeric values to facilitate further processing.

Normalization: The dataset features can have widely varying value ranges, which can disproportionately influence the learning process. To mitigate this, a min-max normalization function is applied to each feature. Normalization scales the feature values to the range [0,1], reducing discrepancies between different feature dimensions and ensuring that each feature contributes equally to the learning process. This step is vital for maintaining the effectiveness and accuracy of the intrusion detection model.

3.2 VAE-GAN Based Data Generation

The data generation module leverages a Variational Autoencoder Generative Adversarial Network (VAE-GAN) to generate new samples of minority class attacks. This approach addresses the issue of dataset imbalance, which is a common challenge in intrusion detection systems. The VAE-GAN model is composed of three primary components: an encoder, a generator (which also serves as the decoder in the VAE), and a discriminator. The specific steps involved in this module are as follows:

(*i*) The encoder processes the original data to obtain latent variables. These latent variables capture the essential features of the input data in a lower-dimensional space.
(*ii*) The generator or decoder trains these latent variables, along with noise, to produce new samples labeled with minority class attacks. This step aims to generate realistic synthetic samples that augment the minority class in the dataset.
(*iii*) With the decoder and generator fixed, all samples, including both original and generated ones, are input into the discriminator for training. The discriminator's role is to distinguish between real and synthetic samples, thereby refining the generator's output.
(*iv*) The discriminator is then fixed, and the error it produces is used to train the decoder and generator. This adversarial training process iteratively improves the quality of the generated samples.
(*v*) These steps are repeated until the discriminator's loss value stabilizes around 0.5, indicating that the newly generated data closely resembles the original data. At this point, the model has effectively balanced the dataset, reducing bias towards the majority class.

3.3 Feature Selection

The feature selection module is designed to enhance the representativeness and relevance of the features used for intrusion detection. This module employs a combination of traditional and heuristic methods to select the most informative features.

We consider five traditional feature selection methods to screen features from the dataset.

– *Chi-square Test*: This statistical test measures the association between categorical variables, helping to identify features that significantly contribute to the class labels.
– *Random Forest*: An ensemble learning method that ranks features based on their importance in classification tasks.
– *Mutual Information*: This method measures the mutual dependence between two variables, identifying features that provide the most information about the class labels.

- *Recursive Feature Elimination (RFE)*: A backward selection process that recursively removes the least important features and builds the model until the optimal feature subset is obtained.
- *Genetic Algorithms*: These algorithms mimic the process of natural selection to find optimal feature subsets through iterative evolution.

We also leverage a heuristic feature selection approach, integrating the statistical importance of standard deviation and median mean difference to select features that exhibit high distinctiveness and significant inter-class differences. By considering the overall correlation between features, this method ensures that the selected features collectively enhance the model's detection performance.

3.4 Graph Attention Network for Detection

The Graph Attention Network (GAT). The Graph Attention Network (GAT) [24] is a type of neural network architecture designed specifically for processing graph-structured data. GAT represents a significant advancement in the field of graph neural networks (GNNs) by incorporating attention mechanisms to address the challenge of learning node representations in large and complex graphs.

The key innovation of GAT is its use of self-attention layers to dynamically assign importance to the nodes' neighbors, allowing for more flexible and powerful aggregation of neighbor features. Unlike previous GNN models that treated all neighbors equally or relied on static weighting schemes, GAT can learn to prioritize certain nodes over others based on the structure and features of the graph, leading to more expressive node embeddings.

The operation of a GAT layer can be described as follows. First, compute the attention coefficients for each pair of connected nodes:

$$e_{ij} = \text{LeakyReLU}\left(\mathbf{a}^T[\mathbf{W}h_i \parallel \mathbf{W}h_j]\right), \qquad (10)$$

where h_i is the feature vector of node i, $\mathbf{W}$ is a weight matrix applied to every node, $\mathbf{a}$ is a weight vector for the attention mechanism, $\parallel$ denotes concatenation, and e_{ij} represents the unnormalized attention coefficient between nodes i and j. Then, apply the softmax function to normalize the attention coefficients for each node:

$$\alpha_{ij} = \frac{\exp(e_{ij})}{\sum_{k \in \mathcal{N}(i)} \exp(e_{ik})}, \qquad (11)$$

where α_{ij} is the normalized attention coefficient and $\mathcal{N}(i)$ denotes the neighborhood of node i. Then, we compute the updated node features as a weighted sum of neighbor features, scaled by the attention coefficients:

$$h'_i = \sigma\left(\sum_{j \in \mathcal{N}(i)} \alpha_{ij} \mathbf{W} h_j\right), \qquad (12)$$

where h'_i is the updated feature vector of node i and σ is a non-linear activation function.

GAT's ability to model different levels of importance for different nodes in each node's neighborhood allows it to capture the heterogeneity of real-world graph structures more effectively. This leads to improved performance on a variety of tasks, including node classification, link prediction, and graph classification.

Residual Graph Attention Network (ResGAT). In this work, we employ the Residual Graph Attention Network (ResGAT), which is designed to enhance the feature learning capabilities of Graph Neural Networks (GNNs) by integrating attention mechanisms and residual connections. This architecture builds upon the Graph Attention Network (GAT), a significant development in graph-based deep learning that introduces attention mechanisms to weigh the importance of neighboring nodes during feature aggregation.

The primary innovation of ResGAT lies in its incorporation of residual connections, inspired by the success of residual networks (ResNets) in deep learning for image processing tasks. These residual connections mitigate the vanishing gradient problem, enabling deeper network architectures by facilitating more effective gradient flow through the network. This is particularly beneficial for GNNs, where the complexity of graph structures can make effective feature learning challenging.

The core operation of a Residual Graph Attention Network can be formulated as:

$$h'_i = \sigma \left(\sum_{j \in \mathcal{N}(i) \cup \{i\}} \alpha_{ij} \mathbf{W} h_j \right) + h_i, \tag{13}$$

where h_i is the feature vector of node i before the update, h'_i is the updated feature vector of node i, σ denotes a non-linear activation function such as ReLU, α_{ij} represents the attention coefficient between nodes i and j, computed by the attention mechanism of the GAT, $\mathbf{W}$ is a trainable weight matrix, $\mathcal{N}(i)$ denotes the set of neighbors of node i, The term h_i at the end of the equation represents the residual connection, adding the original feature vector to the updated features.

This architecture effectively captures both local graph topology and node feature information, with the attention mechanism allowing for adaptive importance weighting of neighbors' features, and the residual connections ensuring that deep networks can be trained effectively without the performance degradation often associated with increased depth. The adoption of ResGAT addresses key challenges in GNN architectures, such as over-smoothing and the difficulty of learning deep representations. This makes Res-GAT a powerful tool for a wide range of applications in graph-based machine learning, particularly for network traffic analysis.

3.5 Network Traffic Graph Construction

For the network intrusion detection scheme constructed using the Graph Neural Network (GNN) in this paper, network traffic is first represented as a graph. The general approach is to represent network traffic by using IP addresses as nodes and the network flows between them as edges to construct a network traffic graph. In this graph, S represents the set of source nodes, and D represents the set of destination nodes. However, network traffic features are associated with the network flows, i.e., the edges in the network traffic graph. Intrusion detection aims to determine whether these network flows are indicative of network attacks by classifying the edges. Graph Attention Networks (GATs), however, are typically suited for node classification tasks and cannot be directly applied to edge classification. For edge classification, the network flows are treated as nodes, and the original nodes are treated as edges, resulting in a new network traffic graph. This method ingeniously converts the edge classification problem into a node classification problem. We denote the newly constructed graph as $G'(V', E')$, where V' represents the set of edges from the original graph. Inspired by the E-ResGAT model in literature [7], which uses residual learning to enhance the performance of graph neural networks, our model also integrates the transformation of original node features at each layer. This approach allows the model to learn more deeply and effectively.

4 Experimental Settings

In this section, we discuss the experimental setup, including the benchmark datasets, baselines for comparison, and evaluation metrics.

4.1 Dataset Description and Evaluation Metrics

NSL-KDD Dataset. The NSL-KDD dataset is a refined version of the KDD Cup 1999 dataset, which has been widely used for evaluating the performance of intrusion detection systems. The original KDD Cup 1999 dataset contains a large number of redundant records, which can lead to biased evaluation results and inefficient training processes. To address these issues, Tavallaee et al. introduced the NSL-KDD dataset, which removes redundant records and provides a more balanced and representative dataset for network intrusion detection research [21]. The NSL-KDD dataset includes both normal and malicious network traffic data. It consists of 41 features for each connection record. The NSL-KDD dataset includes four main attack categories: DoS (Denial of Service), Probe, R2L (Remote to Local) and U2R (User to Root).

UNSW-NB15 Dataset. The UNSW-NB15 [17] dataset is a collaborative research initiative aimed at providing a modern and comprehensive dataset for network intrusion detection research. Developed by the Australian Centre for Cyber Security (ACCS) at the University of New South Wales, this dataset addresses the limitations of older datasets by incorporating contemporary attack types that are not present in datasets like KDDâĂŹ99. The UNSW-NB15 dataset includes a wide array of attack scenarios, such as fuzzers, analysis, backdoors, DoS, exploits, generic attacks, reconnaissance, shellcode, and worms, making it a robust resource for evaluating the effectiveness of intrusion detection systems. Its relevance and realism stem from the inclusion of modern attack techniques and traffic features, making it a critical tool for researchers and cybersecurity professionals seeking to enhance IDS capabilities in the face of evolving cyber threats.

Evaluation Metrics. The experiments evaluate the overall performance of the model using *Recall*, *Precision*, *F1-score*, and *Accuracy*, which are widely used in classification tasks.

4.2 Baselines

We use ROS, ADASYN, ENN, WGAN and CGAN as the baselines to compared with our proposed method. Random Over-Sampling (ROS) and Adaptive Synthetic Sampling (ADASYN) are techniques used to address class imbalance in datasets by either duplicating minority class examples or generating synthetic examples, respectively [8, 15]. Edited Nearest Neighbors (ENN) is a data cleaning method that removes noisy or misclassified examples to improve data quality [2]. Wasserstein Generative Adversarial Networks (WGAN) and Conditional Generative Adversarial Networks (CGAN) are advanced GAN variants designed to enhance training stability and allow for conditional data generation, respectively [3].

5 Experimental Results

5.1 Intrusion Detection Performance

The intrusion detection problem can be regarded as a binary classification problem, i.e., determining whether a sample belong to the normal class or the attack class. We show the results of our method and baseline methods in Fig. 2. As we can see, the VAE-GNN method achieves excellent performance on both datasets compared to baselines, particularly in balancing the conflicting metrics of precision and recall. For the NSL-KDD dataset, our method obtained optimal results across all four metrics. Specifically, the accuracy, precision, recall, and F1-score all reached over 99%, outperforming other methods. Notably, the recall improved by up to 25% compared to other class balancing methods. For the UNSW-NB15 dataset, our method achieved near-optimal performance across various performance metrics. In particular, the recall improved by up to 15%, demonstrating excellent detection performance.

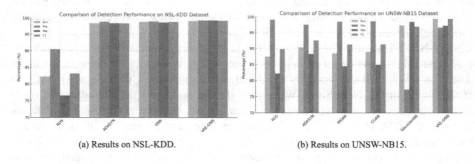

(a) Results on NSL-KDD. (b) Results on UNSW-NB15.

Fig. 2. Binary classification results comparing VAE-GNN with other methods.

The results indicate that VAE-GNN effectively handles the challenge of imbalanced datasets, achieving high precision and recall simultaneously. This is crucial for intrusion detection systems, as it ensures that minority class attacks are detected without compromising the accuracy of normal traffic classification. The substantial improvements in recall, particularly, highlight the robustness of our method in identifying attacks that might otherwise be missed by other approaches.

These findings underscore the efficacy of the VAE-GNN model in enhancing the detection capabilities of network intrusion detection systems, offering a balanced approach that addresses the limitations of traditional methods. By integrating data generation and advanced feature selection, VAE-GNN provides a comprehensive solution that significantly enhances both precision and recall, leading to overall improved performance in detecting network intrusions.

5.2 Attack Attribution Performance

In intrusion detection, simply identifying whether a sample belongs to the normal class or the attack class is insufficient, as we do not know the specific type of attack to which abnormal traffic belongs, i.e., the attack attribution. Therefore, it is necessary to conduct multi-class classification experiments to further differentiate abnormal traffic into specific attack categories. To ensure a more objective and reasonable evaluation, our method is compared with other multi-class balancing methods, with the results shown in Fig. 3.

For the NSL-KDD dataset, the VAE-GNN method achieves accuracy, precision, recall, and F1-score all greater than 99%, attaining the best results across all five metrics. Compared to other methods, our method improved accuracy by up to 22%, precision by nearly 7%, recall by up to 33%, and F1-score by up to 23%. For the UNSW-NB15 dataset, our method achieves the best results for three metrics, excluding the recall. The lower recall can be attributed to the fact that some samples in the original UNSW-NB15 dataset have identical features between the DoS and Exploits classes, as well as between the Normal and Fuzzers classes, which makes it challenging for the model to distinguish them accurately.

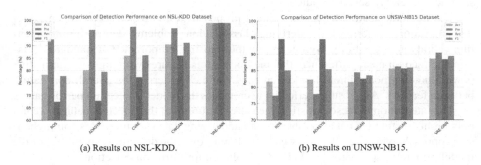

(a) Results on NSL-KDD.   (b) Results on UNSW-NB15.

Fig. 3. Multi-class classification results comparing VAE-GNN with other methods.

The results demonstrate the robustness of the VAE-GNN method in handling multi-class classification tasks, providing a comprehensive solution that significantly enhances the detection and differentiation of various types of network intrusions. By integrating data generation with advanced feature selection, VAE-GNN not only improves the overall performance metrics but also ensures a balanced and effective approach to intrusion detection. The findings underscore the efficacy of the VAE-GNN model in enhancing the detection capabilities of network intrusion detection systems, particularly in multi-class scenarios where the differentiation of attack types is crucial for effective cybersecurity measures. The substantial improvements in accuracy, precision, recall, and F1-score highlight the model's ability to accurately classify different types of attacks, thereby providing a more granular and detailed analysis of network security threats.

5.3 Feature Selection Evaluation

As shown in Fig. 4, compared to existing typical feature selection methods, the heuristic method demonstrates superior detection performance. On the NSL-KDD dataset, the accuracy is improved by 4%. On the UNSW-NB15 dataset, the accuracy is improved by 5%. Precision and recall reflect the model's relevance and sensitivity to the intrusion detection problem. On both datasets, our method achieved the best results for these two metrics, with improvements of up to 5%. Additionally, our method achieved F1-scores of 99.12% and 89.33% on the NSL-KDD and UNSW-NB15 datasets, respectively, indicating a significant enhancement over other methods.

The primary reasons for these improvements are as follows. Chi-square Test is based on the assumption of independence, but the intrusion dataset contains complex non-linear relationships. Therefore, the features selected by the chi-square test may only be locally relevant and fail to consider the overall correlation between features. Mutual Information and Recursive Feature Elimination (RFE), both methods also overlook the correlations between features, which can lead to suboptimal feature selection. Random Forest, as an ensemble model, it may overfit when handling high-dimensional data, resulting in selected features that do not generalize well. Genetic algorithms can easily get trapped in local optima and struggle with multi-objective problems, leading to less satisfactory feature selection outcomes.

The heuristic feature selection method proposed in this paper serves as an effective and rational medium for underlying classification problems, demonstrating scientific validity and feasibility. By integrating the statistical importance of standard deviation and median mean difference, this method selects features with high discriminative power and significant inter-class differences while considering the overall correlation between features. As a result, it effectively enhances the model's detection performance.

The integration of these statistical measures ensures that the selected features are not only highly distinctive but also robust to the underlying data structure. This comprehensive approach to feature selection significantly improves the detection capabilities of the model, providing a balanced and effective solution for intrusion detection.

6 Conclusion

In this paper, we proposed a novel intrusion detection framework leveraging the VAE-GNN model, which integrates Variational Autoencoder (VAE) and Graph Neural Network (GNN) technologies. This framework addresses several key challenges in intrusion detection, including data imbalance, feature representation, and the complexity of

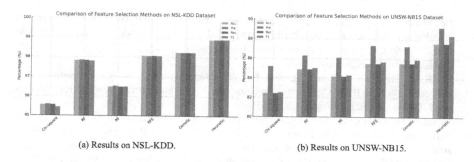

Fig. 4. Comparison of detection performance between the heuristic feature selection method and existing methods.

graph structures. Through comprehensive experiments on the NSL-KDD and UNSW-NB15 datasets, our method demonstrated significant improvements in accuracy, precision, recall, and F1-score compared to existing methods. The results indicate that the VAE-GNN model effectively balances the detection of minority class attacks and normal traffic, substantially enhancing the overall performance of the intrusion detection system. The heuristic feature selection method further contributes to the model's effectiveness by ensuring the selected features are highly discriminative and representative of the underlying data structure.

References

1. Ahmad, I., Basheri, M., Iqbal, M.J., Rahim, A.: Performance comparison of support vector machine, random forest, and extreme learning machine for intrusion detection. IEEE Access **6**, 33789–33795 (2018)
2. Alejo, R., Sotoca, J.M., Valdovinos, R.M., Toribio, P.: Edited nearest neighbor rule for improving neural networks classifications. In: Zhang, L., Lu, B., Kwok, J.T. (eds.) Advances in Neural Networks - ISNN 2010, 7th International Symposium on Neural Networks, ISNN 2010, Shanghai, China, June 6-9, 2010, Proceedings, Part I. Lecture Notes in Computer Science, vol. 6063, pp. 303–310. Springer (2010). https://doi.org/10.1007/978-3-642-13278-0_39
3. Arjovsky, M., Chintala, S., Bottou, L.: Wasserstein generative adversarial networks. In: Proceedings of the 34th International Conference on Machine Learning - Volume 70. p. 214-223. ICML'17, JMLR.org (2017)
4. Chaudhary, A., Mittal, H., Arora, A.: Anomaly detection using graph neural networks. In: Proceedings of the International Conference on Machine Learning, Big Data, Cloud and Parallel Computing (COMITCon). pp. 346–350. IEEE, Faridabad, India (2019)
5. Goodfellow, I., Pouget-Abadie, J., Mirza, M., Xu, B., Warde-Farley, D., Ozair, S., Courville, A., Bengio, Y.: Generative adversarial nets. In: Advances in neural information processing systems. pp. 2672–2680 (2014)
6. Hamilton, W.L., Ying, Z., Leskovec, J.: Inductive representation learning on large graphs. In: Proceedings of the 31st Annual Conference on Neural Information Processing Systems. pp. 1025–1035. Curran Associates Inc., Long Beach, USA (2017)
7. Hamilton, W., Ying, Z., Leskovec, J.: Inductive representation learning on large graphs. Advances in neural information processing systems **30** (2017)

8. He, H., Bai, Y., Garcia, E.A., Li, S.: Adasyn: Adaptive synthetic sampling approach for imbalanced learning. In: 2008 IEEE International Joint Conference on Neural Networks (IEEE World Congress on Computational Intelligence). pp. 1322–1328 (2008). https://doi.org/10.1109/IJCNN.2008.4633969
9. He, K.M., Zhang, X.Y., Ren, S.Q., Sun, J.: Deep residual learning for image recognition. In: Proceedings of the IEEE Conference on Computer Vision and Pattern Recognition (CVPR). pp. 770–778. IEEE, Las Vegas, USA (2016)
10. Hossain, D., Ochiai, H., Fall, D., Kadobayashi, Y.: Lstm-based network attack detection: Performance comparison by hyper-parameter values tuning. In: Proceedings of the 7th IEEE International Conference on Cyber Security and Cloud Computing (CSCloud)/the 6th IEEE International Conference on Edge Computing and Scalable Cloud (EdgeCom). pp. 62–69. IEEE, New York, USA (2020)
11. Kingma, D.P., Welling, M.: Auto-encoding variational bayes. arXiv preprint arXiv:1312.6114 (2013)
12. Kipf, T.N., Welling, M.: Semi-supervised classification with graph convolutional networks. arXiv preprint arXiv:1609.02907 (2017)
13. Li, Y., Zhang, B.: An intrusion detection model based on multi-scale cnn. In: Proceedings of the 3rd Information Technology, Networking, Electronic and Automation Control Conference (ITNEC). pp. 214–218. IEEE, Chengdu, China (2019)
14. Lin, W.H., Lin, H.C., Wang, P., Wu, B.H., Tsai, J.Y.: Using convolutional neural networks to network intrusion detection for cyber threats. In: Proceedings of the IEEE International Conference on Applied System Invention (ICASI). pp. 1107–1110. IEEE, Chiba, Japan (2018)
15. Lunardon, N., Menardi, G., Torelli, N.: Rose: a package for binary imbalanced learning. R J. **6**, 79 (2014), https://api.semanticscholar.org/CorpusID:31227713
16. Mabu, S., Gotoh, S., Obayashi, M., Kuremoto, T.: A random-forests-based classifier using class association rules and its application to an intrusion detection system. Artificial Life and Robotics **21**(3), 371–377 (2016). https://doi.org/10.1007/s10015-016-0281-x
17. Moustafa, N., Slay, J.: Unsw-nb15: a comprehensive data set for network intrusion detection systems (unsw-nb15 network data set). In: 2015 Military Communications and Information Systems Conference (MilCIS) (2015)
18. Simonyan, K., Zisserman, A.: Very deep convolutional networks for large-scale image recognition. In: Proceedings of the 3rd International Conference on Learning Representations (ICLR). pp. 1–14. ICLR, San Diego, USA (2015)
19. Studer, L., Wallau, J., Ingold, R., Fischer, A.: Effects of graph pooling layers on classification with graph neural networks. In: Proceedings of the 7th Swiss Conference on Data Science (SDS). pp. 57–58. IEEE, Luzern, Switzerland (2020)
20. Szegedy, C., Liu, W., Jia, Y.Q., Sermanet, P., Reed, S., et al., D.A.: Going deeper with convolutions. In: Proceedings of the IEEE Conference on Computer Vision and Pattern Recognition (CVPR). pp. 1–9. IEEE, Boston, USA (2015)
21. Tavallaee, M., Bagheri, E., Lu, W., Ghorbani, A.A.: A detailed analysis of the kdd cup 99 data set. In: 2009 IEEE Symposium on Computational Intelligence for Security and Defense Applications. pp. 1–6 (2009)
22. Thakkar, A., Lohiya, R.: Fusion of statistical importance for feature selection in deep neural network-based intrusion detection system. Information Fusion **90**, 353–363 (2023)
23. Tsai, C.F., Hsu, Y.F., Lin, C.Y., Lin, W.Y.: Intrusion detection by machine learning: A review. Expert Syst. Appl. **36**, 11994–12000 (2009)
24. Veličković, P., Cucurull, G., Casanova, A., Romero, A., Lio, P., Bengio, Y.: Graph attention networks. arXiv preprint arXiv:1710.10903 (2017)
25. Wu, Z., Pan, S., Chen, F., Long, G., Zhang, C., Yu, P.S.: A comprehensive survey on graph neural networks. IEEE Transactions on Neural Networks and Learning Systems (2020)

26. Yang, S.: Research on network behavior anomaly analysis based on bidirectional lstm. In: Proceedings of the 3rd Information Technology, Networking, Electronic and Automation Control Conference (ITNEC). pp. 798–802. IEEE, Chengdu, China (2019)
27. Yang, Y., Zheng, K., Wu, B., Yang, Y., Wang, X.: Network intrusion detection based on supervised adversarial variational auto-encoder with regularization. IEEE Access **8**, 42169–42184 (2020)

Semantic-Integrated Online Audit Log Reduction for Efficient Forensic Analysis

Wenhao Liao[1,2], Jia Sun[2,3], Haiyan Wang[2], Zhaoquan Gu[2,3(✉)], and Jianye Yang[2,4]

[1] Shenzhen Institute for Advanced Study, University of Electronic Science and Technology of China, Shenzhen, China
whliao@std.uestc.edu.cn
[2] Pengcheng Laboratory, Shenzhen, China
{sunj01,wanghy01}@pcl.ac.cn, jyyang@gzhu.edu.cn
[3] School of Computer Science and Technology, Harbin Institute of Technology (Shenzhen), Shenzhen, China
guzhaoquan@hit.edu.cn
[4] School of Cyberspace Security, Guangzhou University, Guangzhou, China

Abstract. Audit logs are crucial for revealing and tracking sophisticated cyber threats due to their abundant system-level information. However, the immense scale of logs burdens storage resources and limits their lifecycle to days, which is insufficient for tracking multi-step attacks over months or years. Although log reduction techniques that cater to cold storage can mitigate this issue, many of these are restricted to offline batch processing in data centers. This incurs significant storage and transmission costs at endpoints. Moreover, many log reduction techniques fail to yield a suitable pattern for forensic analysis, which aims to identify signs of malicious activities by scrutinizing past events. In this paper, we present SOPR, an online audit log reduction technique designed to preserve traceability. SOPR enables real-time execution of the entire process, allowing reduction to be performed on raw log data streams. Specifically, our approach can effectively reduce events that lack causal dependence and involve repeated dependency relationships. To achieve this objective, we design a dual-cache architecture that simultaneously models semantically similar files and utilizes a versioned graph to preserve causality between log events. The synergy of these two components enhances the effectiveness of SOPR in log reduction. Our experiments on the DARPA TC datasets show that SOPR can achieve comparable event reduction factor in an online fashion to state-of-the-art offline approaches. Moreover, the runtime overhead and forensic analysis validity meet the deployment requirements for real-world environments.

Keywords: Log Reduction · Provenance Graph · Forensic Analysis

1 Introduction

Advanced Persistent Threats (APTs) are sophisticated cyber attacks meticulously orchestrated by professional groups targeting specific objectives. To defend against such long-term and multi-step attacks, Endpoint Detection and Response (EDR) systems [6,12] utilize system-level activities such as audit logs to continuously monitor system execution status, facilitating timely response to security incidents. Recently, provenance graphs have attracted many researchers because they convert audit log entries into dependency graphs, which provide a clearer illustration of the chronology and causality of execution history on the host [18,25]. Once suspicious behavior is detected, forensic analysis will reconstruct the attack footprint, aiding security analysts in investigating the threat [24].

Unfortunately, the vast quantity and long-term storage requirements of audit logs bring excessive space and processing overhead. The audit data collected from audit frameworks can accumulate at rates reaching GB per day. An enterprise cluster system can easily consume several tens of PB annually, imposing a heavy burden on security budgets [23]. Additionally, the overwhelming quantity of audit logs makes it challenging to detect and trace potential malicious activities on hosts, akin to *needles in the haystack* [7]. The threat detection model's effectiveness and efficiency may be undermined by a considerable amount of redundant logs that do not pertain to valuable system activities [12]. Reducing storage costs and improving the performance of causality analysis by shrinking system events is considered a key approach to addressing the aforementioned issues. Nevertheless, current approaches typically involve offline or partially offline processes, resulting in several significant drawbacks. First, post-hoc compression methods hinder timely response to attacks as the reduced provenance graph is the foundation for efficient forensic analysis. Additionally, many offline algorithms require centralized processing in dedicated data servers, potentially exacerbating the financial cost of network transmission and storage at the endpoints [11].

To this end, we propose SOPR, a semantic-integrated online provenance graph reduction scheme. SOPR enables full-process online processing capabilities for streaming data, including log parsing, provenance graph-based reduction, and data storage. Specifically, SOPR designs a cache-aside pattern based on the Least Recently Used (LRU) cache for parsing audit logs into event triplets in real time, minimizing numerous costly database queries. To effectively and efficiently reduce the volume of log entries, we adopt a dual-cache architecture. We observed a notable frequency of temporary files with similar path names in the audit logs. These files typically stem from massive temporary files generated by processes or substantial libraries and resources loaded by programs during their runtime stages. SOPR adopts a Process-Centric Cache (PCCache) integrated with Hierarchical Path Frequency Tree (HPFTree) to aggregate these temporary files, effectively diminishing the frequency of associated events. Concurrently, we employ a graph versioning technique to capture temporal dependencies and causal relationships between nodes, ensuring coherent traversal flows within the graph. Ultimately, SOPR delivers a streamlined log event sequence that preserves provenance information, facilitating efficient storage and subsequent analysis.

320   W. Liao et al.

To summarize, we make the following contributions:

- We propose SOPR, a full-process online log reduction scheme for forensic analysis. In SOPR, the cache-aside pattern enables efficient preprocessing of raw log streams; while a dual-cache architecture achieves effective log reduction with minimal runtime overhead.
- A novel redundancy elimination approach is introduced for the first time in this study, utilizing the Hierarchical Path Frequency Tree to identify and merge file nodes with similar semantic patterns. This mechanism has a good synergy with the causality-preserving graph in dual-cache architecture, leading to an improved reduction ratio.
- Extensive evaluations demonstrate the effectiveness of our proposed approach. SOPR achives better online reduction performance to state-of-the-art offline approaches. Additionally, it features low runtime overhead and demonstrates good validity for forensic analysis.

2 Background and Motivation

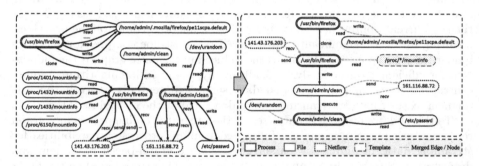

Fig. 1. An example of a provenance graph (left) and the reduced graph after applying SOPR (right). For the sake of brevity, graph versioning has been omitted.

In this section, we report preliminary notions related to audit log and forensic analysis, with the aim of making the paper self-contained. Furthermore, main challenges and the core idea of SOPR will be briefly illustrated.

Audit Log. Kernel-level monitoring frameworks capture system calls to generate system audit logs [12]. Several commonly used provenance data models for logs include Darpa CDM [16] and SPADE JSON [9]. These data models save entities and relations separately in separate entries containing different attributes. Unique identifiers are assigned to both the subject and object entities, while relation entries link specific subject and object entries based on their identifiers to generate complete events. The significant costs associated with log storage and transmission underscore the need for effective log reduction methods [3].

Provenance Graph. A provenance graph is a directed acyclic graph (DAG) in which nodes represent entities (e.g., processes, files, and sockets) and edges represent relations between them (e.g., read and send). For example, Fig. 1 (left) shows a Firefox backdoor attack from the DARPA Engagement3 THEIA dataset [15]. A host accessed a malicious server via `firefox` and was subsequently compromised by a Drakon implant `clean`. Then `clean` connected out to the malicious operator console at `161.116.88.72` and exfiltrated sensitive data.

Forensic Analysis. Once an attack point is identified, forensic analysis is conducted to determine the infiltration points of the attack and its system-wide impact [10]. For example, as shown in Fig. 1 (left), if EDR identifies `clean` as a malicious node, the forward trace can reveal the network port connected by `firefox` as the entry point of the attack, while the backward tracing can uncover that `clean` accessed the critical `passwd` file.

Main Challenges and Core Idea. Current related studys are primarily focused on offline log reduction, however, there is a demand for online algorithms in real-world enterprise environments. Maintaining the dependencies among large-scale log data requires global information, yet real-time reduction of streaming data demands high throughput under limited computational resources, posing a highly challenging task. SOPR successfully achieves an ingenious balance between them. The core idea of SOPR is leveraging the interaction of the dual-cache architecture to aggregate more redundant entities and events. This approach extends SOPR's capacity to preserve more critical dependencies while minimizing resource consumption. As illustrated in Fig. 1, SOPR merges numerous redundent entities and events, while preserving the overall dependency information.

3 Related Work

Previous research primarily focused on minimizing log size while preserving relevant forensic evidence, which can be categorized into four groups thus far [12].

Graph Compression. Such studies concentrate on developing a queryable compression format for investigation beyond cold storage. Elise [5] achieves lossless log compression by employing a Deep Neural Network encoder to determine the most effective character encoding. SEAL [8] achieves a lossless provenance graph compression approach for causality analysis. SEAL enables efficient searching of historic events search. These lossless compression can not address the needle in a haystack problem.

Semantic Pruning. Some log entries are only related to specific processes that are irrelevant to comprehensive forensic analysis. Semantic pruning techniques filter out these routine process behaviors to reduce log size. LogGC [17] performs graph computing method identify and trim *unreachable objects* and *dead end events* [19], which neither influence nor are influenced by other unrelated nodes. NodeMerge [21] attempts to reduce dozens of globally read-only files accessed during application initialization, as these files generate a large volume of redundant events.

Information Flow Preservation. Another significant redundancy manifests in thousands of duplicate system calls that frequently occur within a brief timeframe. Causality Preserving Reduction (CPR) [23] identifies *interleaved flows* which determine whether other information has flowed into the source node between two system calls. An interleaved flow should be preserved as it may be important for forensic analysis. Hossain et al. propose a more relaxed standard called Dependency Preserved Reduction (DPR) [10]. DPR focuses solely on the reachability of nodes rather than all detailed information flows. DPR systems can selectively drop events to achieve correct traversal in every entity's ancestors (S-DPR) or both ancestors and successors (F-DPR). Zhu et al. [24] develop an online DPR and detect suspicious semantics in log events to further reduce events unrelated to attacks.

Causality Approximation. Inspired by lossy compression, some efforts aim to balance space efficiency and forensic analysis accuracy through bounded log approximation. Process-centric Causality Approximation Reduction (PCAR) [23] extends CPR by further excluding redundant events, including interleaved flows, when the number of system calls for a process exceeds a specified threshold. LogApprox [20] utilizes regular expression learning over file I/O events to approximate typical system behaviors while retaining atypical behaviors losslessly.

Building on previous work, some research focuses on real-time performance and the integration of hybrid methods. FAuST [11] proposes a streaming redundant log reduction system that integrates various log reduction techniques. Chang et al. introduce a hybrid compression scheme [2] that utilizes a path tree to extract the fixed redundant structures in system logs and performs information flow analysis to reduce redundant events.

4 SoprDesign

4.1 Approach Overview

SOPR is an online log processing framework that parses logs into provenance graphs, performs reduction approaches, and preserves reduced graphs. Figure. 2 depicts SOPR's architecture consisting of three major components: ① Online Log Parsing (Sect. 4.2). The parser employs a cache-aside pattern to enable the online parsing of raw audit logs into event triplets. ② File Template Generation (Sect. 4.3). This module primarily comprises the Process-Centric Cache

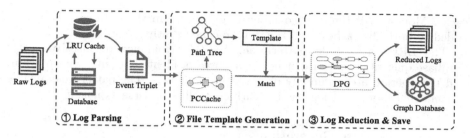

Fig. 2. Overview of SOPR's online log parsing and log reduction architecture. Module ① is the parser, and the combination of modules ② and ③ forms the reducer.

(PCCache) and Hierarchical Path Frequency Tree (HPFTree), which are utilized to extract templates for matching and merging redundant temporary files. ③ Log Reduction and Preservation (Sect. 4.4, Sect. 4.5). This module does not exist independently but requires collaboration with preceding modules. When preceding modules meet the extraction conditions and output partial event triplets, it employs graph versioning techniques to minimize redundant information flows. Subsequently, when these event triplets meet the preservation criteria, this module can store them in multiple file formats.

4.2 Online Log Parsing

To enable online parsing without the need for secondary offline traversal of all entity entries, SOPR implements a Least Recently Used (LRU) cache and SQLite for storing indexes of the raw logs in real time. SOPR first parses raw audit logs by converting them into sets of (subject, relation, object) triplets. These triplets serve as the basic unit of system activities and are subsequently utilized in the construction of provenance graphs. Some essential attributes of audit logs are preserved in the triplets, such as file paths, process image paths, remote IP/Port, and associated event types. Previous studies [4,13] had to use a dual offline traversal method to generate triplets, due to the separability of entities and relations (as elaborated in Sect. 2). The initial step involves an offline traversal of all entities within the datasets to create an identifier-entities mapping. Subsequently, a second online traversal of all relations is conducted to produce final triplets by correlating the identifiers of the subject and object with the entries in the identifier-entities mapping. Traversing all datasets offline is not feasible in a real-world scenario where each log entry is processed in chronological order.

To achieve online parsing, we introduce a *cache-aside pattern* to establish online indexing for mapping identifiers to entities. As illustrated in Fig. 2, the pattern comprises a Least Recently Used (LRU) cache [14] in memory, accompanied by a SQLite database on disk. The cache stores indexes that have been accessed recently, and the least recently accessed index has the highest priority to be evicted. When the queried index is not present in the cache, the system proceeds to query the database and then adds the corresponding index to the cache. Moreover, if the queried index is not found in the database, it will be added to both the cache and the database concurrently.

The LRU cache-aside pattern enables the avoidance of massive costly database queries because frequent occurrences of identical events within a limited time period are commonly observed in audit logs. The pattern can enable efficient real-time audit log parsing with controlled runtime expenses.

4.3 File Template Generation

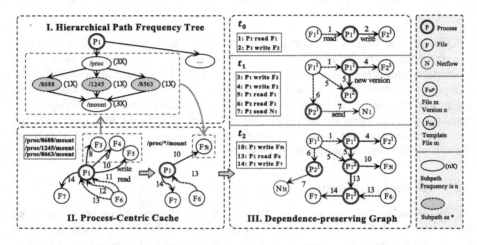

Fig. 3. Example of SOPR's online log reduction steps.

This module primarily uses templates of file path names to match and merge redundant temporary files, which helps alleviate the problem of *dependence explosion* [1]. Dependence explosion stemming from high in-degree and out-degree nodes, as shown in Fig. 1, can significantly hinder the comprehension of graph relationships. Most of these high-degree nodes are caused by *process-centric nodes* that do not interact with any other nodes except the central process (e.g., massive temporary files written by processes, substantial libraries, and resources loaded during the initial stage of processes) [17,21]. Due to their deficiency in structural information and the abundance of redundant semantics, these nodes contribute minimally to subsequent graph representation and forensic analysis, while also increasing computational and storage costs.

In response to this problem, We propose the Process-Centric Cache (PCCache) with Hierarchical Path Frequency Tree (HPFTree) to generate template nodes to address the dependence explosion caused by process-centric nodes. The PCCache stores process-centric nodes within a limited node window W_p. When extraction conditions (introduced in Sect. 4.5) for a particular process are fulfilled, PCCache will extract the entire subgraph of the process, and build an HPFTree to generate templates. HPFTree is constructed based on the hierarchical subdirectories of each file's path and used to identify file nodes that frequently co-occur within paths. Nodes that surpass a certain threshold n in frequency are deemed redundant and converted into template nodes through automatic template creation. Subsequently, PCCache merges all identical edges between the process and the corresponding process-centric nodes, which also include template nodes. Extracted templates associated with specific processes will be utilized to match new incoming nodes.

An example is as illustrated on the left side of Fig. 3, PCCache constructs an HPFTree for P_1, and ergodically calculates the frequency of all nodes in the HPFTree. Subsequently, HPFTree traverses all paths from root P_1 and marks the minimum frequency nodes in a path as template directories, denoted by * (e.g., /proc/8688/mount → /proc/*/mount). The path /proc/*/mount is designated as template F_{5t} for P_1 due to its frequency exceeding the threshold n. Other templates are obtained through the same process. Then, identical edges connecting the same nodes were merged. As shown in Fig. 3 (II), template node F_{5t} and multiple read edges connecting F_6 and P_1 are merged.

4.4 Denpendence-Preserving Reduction

SOPR uses Dependence-Preserving Graph (DPG) with node window W_d to preserve essential information flows and minimize redundancy by eliminating certain repetitive events. DPG utilizes the graph versioning approach proposed in [10] to transfer chronological and causal dependencies among nodes into the directional traversal flow within the graph. Similar to the CPR algorithm [23] mentioned in Sect. 3, DPG tests and retains interleaved flows between nodes, while discarding non-interactive flows. For special circumstances such as thread events or other events that interact with the process itself, DPG directly merges the nodes. Furthermore, DPG treats each remote IP and port combination as a distinct source within W_d, following the method adopted in [1,10].

For instance, as demonstrated in Fig. 3 (III) from $t_0 \rightarrow t_1$, a sequence of chronological events is added to DPG. Considering event 3 (e3) and e4 as repeated events between $P_1^1 \rightarrow F_2^1$ and updating the event timestamp to the most recent value. Since there are no other interleaved flows between the two nodes, and the dependencies are not hindered by merging new events. For e5, DPG does not merge the event but instead creates a new version node P_1^1 of P_1^2, and establishes a connection between two versioned nodes: $F_1^1 \rightarrow P_1^2$. This is because e4 is an interleaved flow between e1 and e5. Directly merging and updating e1 to e5 would lead to an incorrect dependency of e5 → e4. New nodes and events are created for e6 and e7 since the destination nodes were not previously present in the graph.

4.5 Dual-Cache Synergy for Reduction

SOPR introduces a novel online dual-cache architecture, highlighting the synergy between PCCache and DPG, which leads to superior reduction performance. Once a process meets the extraction conditions, PCCache extracts the corresponding process-centric subgraph to generate templates. Subsequently, PCCache merges all identical edges and integrates them into DPG. It is noteworthy that only file nodes enter into PCCache for process-centric node reduction, whereas netflow and process nodes are directly integrated into DPG for redundant dependency reduction.

To ensure the precise and prompt data transfering from PCCache to the DPG, SOPR adheres to three extraction conditions: (1) Upon reaching the maximum node window W_p, PCCache extracts the oldest updated process-centric connected subgraph; (2) When DPG reaches its node window W_d and pops a process node P, the corresponding connected subgraph is extracted from PCCache; (3) If a file node in PCCache interacts with two process nodes simultaneously (i.e., $P_a - F - P_b$), the entire connected subgraph of $P_a - F - P_b$ is extracted. Specifically, F does not need to match the templates of P_a or P_b; instead, all dependencies of F are preserved directly because F is considered to influence critical information flow, necessitating the retention of all information. For instance, at t_2 as depicted in Fig. 3, when a specific extraction condition is met, PCCache consolidates process-centric nodes with their associated events first. Subsequently, it converts them into a series of chronological events, integrating them into the DPG. Furthermore, SOPR can output the reduced logs for downstream tasks, such as saving them as chronological event triplets or storing them in graph databases like Neo4j.

It is important to emphasize that the files aggregated in PCCache exclusively interacted with a single process within a PCCache node window, thereby guaranteeing that all merged nodes are process-centric. This approach retains the critical information flow while minimizing the loss of causality relationships and semantic information for forensic analysis.

5 Evaluation

For better evaluating the efficiency and effectiveness of SOPR. Our experiments focus on addressing the following research questions: **Q1.** Can SOPR effectively reduce audit logs? How does its performance compare to state-of-the-art methods? (Sect. 5.3); **Q2.** How efficient is SOPR? What is the overhead of SOPR when running in real-time? (Sect. 5.4); **Q3.** How do hyperparameters and submodules affect SOPR' log reduction rate and run-time performance? (Sect. 5.5); **Q4.** Can SOPR efficiently conduct intra-graph traversal and sub-graph extraction for forensic analysis tasks? (Sect. 5.6); **Q5.** Can SOPR accurately retain the critical components of the provenance graph after log reduction? (Sect. 5.6).

5.1 Evaluation Setup

SOPR's core function is implemented using 2,400 lines of code in Python 3.11. Various libraries are utilized for different components of SOPR, including SQLite for data parsing, Networkx for provenance graph reduction, and Neo4j for provenance graph preservation. We deploy SOPR on a server with 64 cores (Intel(R) Xeon(R) Gold 5218 CPU @ 2.30GHz) and 125 GB memory. We re-implemented several offline algorithms, including LogGC [17], CPR [23], and F-DPR [10] based on their descriptions in the original papers. Unless specified stated, we empirically selecte the following hyperparameters for all experiments except those in Sect. 5.5, which specifically examine the effect of hyperparameters on SOPR' performance: DPG node window $|W_d| = 5000$, PCCache node window $|W_p| = 5000$ and template threshold $|n| = 5$.

5.2 Dataset

Table 1. The volume of raw logs and parsed datasets.

	Raw Log Size (GB)	Raw Log #Event (Million)	Parsed Log #Event (Million)
E3theia	10.7	113.29	44.40
E3trace	75.0	1049.18	295.58
E3cadets	4.4	44.40	17.86
E5theia	147.0	200	43.44
E5trace	132.0	200	2.96
E5cadets	168.0	200	22.03
ALL	537.1	1806.87	421.06

We use the dataset released by the DARPA Transparent Computing (TC) Program that was collected during Engagements #3 and #5 [15,22]. The DARPA dataset is the most recent and extensive open-source dataset in the field of audit log security [11]. It has been widely utilized in related research in recent years [12]. We selected E3cadets, E3theia, E3trace, E5cadets, E5theia, and E5trace datasets to evaluate SOPR, all of which were collected from systems running Linux and Unix. The selected audit logs are considered sufficient for experiments due to their substantial volume and temporal continuity. The total size and number of events in raw logs are 537.1 GB and over 1.8 Billion, respectively. Detailed statistical information for each dataset is provided in Table 1. Notablely, we retain almost all event types between process, netflow, and file, rather than manually selecting some types, like [11,20]. And SOPR also allows multiple edges with different types between two nodes. This ensures that SOPR is more suitable for deployment in real production settings.

5.3 Reduction Performance

To evaluate the efficacy of SOPR in reduction performance, we analyze two distinct statistics utilized in previous studies [11,24]: the log event reduction factor $\times.\text{Red} = |E_p|/|E_r|$; the log event reduction percentage $\%.\text{Red} = 1 - (|E_r|/|E_p|)$. Here, $|E_p|$ represents the quantity of events in the parsed log, while $|E_r|$ denotes the quantity of events in the reduced log. The primary objective of this study

Table 2. The reduction performance for each technique on separate datasets. − indicates the experiment failed, primarily due to out-of-memory or stack overflow.

	E3thiea		E3trace		E3cadets		E5thiea		E5trace		E5cadets	
	%.Red.	×.Red.	%.Red.	×.Red.	%.Red.	×.Red.	%.Red.	×.Red.	%.Red.	×.Red.	%.Red.	×.Red.
LogGC [17]	3.35%	1.03×	−	−	−	−	0.16%	1.00×	12.37%	1.14×	−	−
CPR [23]	45.10%	1.82×	99.34%	152.83×	81.17%	5.31×	35.89%	1.55×	59.27%	2.45×	70.78%	3.42×
F-DPR [10]	**88.00%**	**8.33×**	99.67%	304.52×	81.35%	5.36×	**69.32%**	**3.26×**	63.92%	2.77×	77.78%	4.50×
SOPR	78.26%	4.58×	**99.81%**	**533.04×**	**86.91%**	**7.64×**	54.14%	2.18×	**83.18%**	**5.94×**	**90.75%**	**10.82×**

is to reduce the volume of log entries, represented as events, rather than implementing additional operations at the storage layer to enhance compression ratios. To ensure clarity and effectiveness, we opt to assess the efficacy based on the quantity of events.

In Table 2, we compare the reduction performance using DARPA datasets with three state-of-the-art methods: LogGC [17], CPR [23] and F-DPR [10], which are briefly introduced in Sect. 3. The results show that SOPR achieved a higher compression rate than F-DPR in four out of six datasets, outperforming LogGC and CPR in all datasets. It is worth noting that SOPR operates online, while the comparison algorithms function offline, requiring the entire graph to be loaded into memory for reduction. This process resulted in significant bottlenecks in terms of real-time performance and memory consumption.

One particularly unique dataset is the E3-Trace dataset, where various compression algorithms demonstrate significantly higher compression rates compared to other datasets. Upon inspection, it was observed that the parsed E3-Trace dataset contains a substantial number of network connection events, as well as numerous events involving interactions with the process itself, such as EVENT_UNIT. These events are highly redundant *dead-end* information [17], easily mergeable by all algorithms. Therefore, we consider it a special dataset, and subsequent ablation experiments can validate this conclusion. In summary, the SOPR achieves superior real-time compression performance, even when compared with state-of-the-art offline log reduction methods.

5.4 Runtime Overhead

To evaluate the real-time performance of SOPR during runtime operation, a series of experiments were conducted on both the parser and reducer modules. As shown in Fig. 4 (b), the throughput of the parser is significantly higher than the reducer in the majority of datasets, with the exception of E5trace and E5cadets where it is slightly slower than the reducer. Given that both the parser and reducer can be executed concurrently through multi-processing and multi-threading, any potential impact on the reducer is expected to be minimal. We further verified the runtime overhead and parsing efficiency of the online log parser are nearly constant. The average throughput across all datasets is 63,387 events per sec, while the average memory overhead is 190.5 MB. These values are considered entirely acceptable for real-time operations. Therefore, for

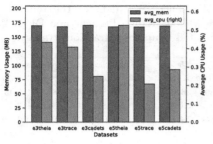 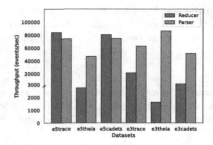

(a) CPU and memory overhead of reducer (Online).   (b) Throughput of parser and reducer (Offline).

Fig. 4. Runtime overhead for parser and reducer.

the remaining experiments, we primarily focus on discussing and comparing the reduction algorithms. Furthermore, we simulate a real-time log stream referring to the timestamp of events to verify the average CPU utilization and memory overhead, as depicted in Fig. 4 (a). When running in real-time, the memory usage is relatively constant for each dataset. Overall CPU utilization is low when the parser streams data into the reducer due to the low output throughput of the parsed data stream, often causing the reducer to be in a blocked state awaiting data. The CPU usage increases only when pushing the data in PCCache to DPG. Experiments demonstrate that SOPR exhibits strong real-time capabilities with an acceptable level of resource consumption for typical hosts.

5.5 Ablation Study

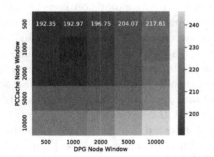 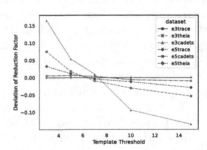

(a) Memory overhead with different node window.   (b) Effect of template thresholds on reduction factor.

Fig. 5. Hyper parameter analysis. The deviation of the reduction factor in (b) is calculated as (Reduction factor - Average reduction factor).

Experiments in this section primarily analyze the impact of internal parameters and modules on SOPR. To validate the proposed PCCache and DPG dual

Table 3. Ablation study of reduction performance. WT denotes the removal of HPFTree-based template extraction strategy while retaining Pccache and DPG; WP represents the removal of Pccache and its affiliated template extraction strategy.

	E3thiea		E3trace		E3cadets		E5thiea		E5trace		E5cadets	
	%.Red.	×.Red.	%.Red.	×.Red.	%.Red.	×.Red.	%.Red.	×.Red.	%.Red.	×.Red.	%.Red.	×.Red.
SOPR(WT)	74.70%	3.95×	99.81%	533.04×	84.14%	6.30×	36.89%	1.58×	83.18%	5.94×	86.54%	7.43×
SOPR(WP)	40.96%	1.69×	99.81%	533.04×	78.99%	4.76×	8.08%	1.08×	83.18%	5.94×	86.52%	7.42×
SOPR	78.26%	4.58×	99.81%	533.04×	86.91%	7.64×	54.14%	2.18×	83.18%	5.94×	90.75%	10.82×

caching architecture in the model, as well as the effectiveness of our HPFTree-based template extraction strategy, we conducted ablation experiments on these two components separately. The ablation study result can be found in Table 3, it demonstrates that both components contribute to the overall performance. It can be observed that overall, the impact of PCCache on the reduction rate is greater than that of the template, as PCCache implicitly includes the processing of the template. Combining the two in the model can achieve the best reduction performance. The ablation of the E3trace and E5trace datasets showed no variation, confirming the presence of a significant number of similar netflow events and process events in the datasets, which are not influenced by PCCache and thus remain unaffected after ablation.

The hyper-parameters analysis primarily evaluates the performance of three parameters: DPG node window $|W_d|$, PCCache node window $|W_p|$ and template threshold $|n|$. As shown in Fig. 5 (a), we evaluate memory overhead on the E3theia dataset with varying node window sizes, including $|W_d|$ and $|W_p|$. As expected, larger windows resulted in larger graphs in memory and higher memory consumption. It is noteworthy that increasing the window size from 500 to 10,000 only resulted in an approximate 60 MB increase in memory usage. This observation suggests that our algorithm's window expansion is not highly memory-sensitive. Therefore, SOPR could be applicable for broader applications that require expanding the window size to accommodate more nodes in the graph. Figure. 5 (b) illustrates the impact of adjusting the template threshold $|n|$ on the reduction factor across different datasets. A subtle trend can be observed, indicating that a higher $|n|$ setting results in a lower reduction factor. This aligns with expectations, as a larger $|n|$ makes it more challenging to match nodes for reduction. It is reasonable to infer that the deviation of the two trace datasets remains unchanged as they are not affected by the template setting. Experimental results demonstrate the effectiveness of our module settings, indicating that SOPR is robust to variations in the three hyperparameters, making it suitable for a variety of scenarios.

5.6 Forensic Analysis Validity

To validate SOPR's suitability for forensic analysis tasks, two experiments were conducted on the reduced graph. These experiments aimed to assess the efficiency of random walks on the reduced graph and the preservation of critical

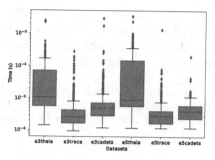 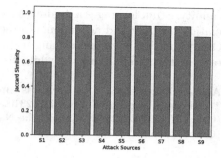

(a) 3-hop traversal time consumption.   (b) Forensic validity on attack graphs.

Fig. 6. The result of forensic analysis validity experiments. The S on the x-axis of Figure (b) denotes distinct attack source nodes.

traceback paths, such as the node paths formed during malicious attacks. First, We randomly selected source nodes from various datasets and executed traversal to obtain its three-hop neighbors. 1,000 experiments were conducted on each dataset, and the time taken for each walk was recorded. The results as shown in Fig. 6 (a), the efficiency of traversing the reduced graph is high, with an average traversal time of no more than 10^{-5} seconds and the maximum quartile time around 10^{-4} seconds. It is noted that the traversal cost of theia datasets is approximately 10 times higher than others, primarily due to the higher outdegree of certain nodes, resulting in increased complexity of the reduced graph.

To evaluate the extent to which critical traceability paths are preserved, we conducted an experiment via the Firefox Backdoor attack in E3theia based on ground truth reports provided by DARPA. We identified attack events in the corresponding logs as our ground truth labels. Subsequently, in the reduced graph, we randomly selected malicious nodes and performed forward and backward traversal to detect a set of malicious nodes they reached within the attack scope. The obtained set was then compared to the malicious points in the logs using the Jaccard similarity coefficient to assess the preservation of critical traceability paths. As illustrated in Fig. 6 (b), the vast majority of forensic analyses exhibit a matching degree exceeding 90%. Some malicious outliers are primarily due to the online nature of SOPR, resulting in certain nodes becoming discontinuous after being evicted from memory. Increasing the node window can mitigate this issue, while also need to take into account the runtime overhead. Overall, SOPR's reduced graph demonstrates a high efficiency in graph traversal, meeting the requirements for forensic analysis in real-world environments.

6 Conclusion

SOPR is an online log reduction scheme for endpoints, distinguished by its low runtime overhead, full process real-time processing nature, and forensic analysis validity. SOPR leverages a dual-cache architecture that integrates semantic information merging and dependency-preserving reduction to perform log reduction. Our evaluation in various settings demonstrates that SOPR can effectively reduce raw log in an online manner while preserving causality for forensic analysis.

Acknowledgments. This work was supported in part by the Shenzhen Science and Technology Program (No. KJZD20231023094701003), the Major Key Project of PCL (Grant No. PCL2023A07-4), and the National Natural Science Foundation of China (Grant No. 62372137).

References

1. Altinisik, E., Deniz, F., Sencar, H.T.: ProvG-Searcher: A graph representation learning approach for efficient provenance graph search. In: CCS. pp. 2247–2261 (2023)
2. Chang, B., Zhou, F., Wang, Z., Wen, Y., Zhang, B.: A distributed storage system for system logs based on hybrid compression scheme. In: ISPA/BDCloud/SocialCom/SustainCom. pp. 724–735 (2023)
3. Chen, T., et al.: System-level data management for endpoint advanced persistent threat detection: Issues, challenges and trends. Comput. Secur. **135**, 103485 (2023)
4. Cheng, Z., et al.: KAIROS: practical intrusion detection and investigation using whole-system provenance. In: SP (2024)
5. Ding, H., Yan, S., Zhai, J., Ma, S.: ELISE: a storage efficient logging system powered by redundancy reduction and representation learning. In: USENIX Security. pp. 3023–3040 (2021)
6. Dong, F., et al.: Are we there yet? an industrial viewpoint on provenance-based endpoint detection and response tools. In: CCS. pp. 2396–2410 (2023)
7. Dong, F., et al.: DISTDET: A cost-effective distributed cyber threat detection system. In: USENIX Security. pp. 6575–6592 (2023)
8. Fei, P., Li, Z., Wang, Z., Yu, X., Li, D., Jee, K.: SEAL: storage-efficient causality analysis on enterprise logs with query-friendly compression. In: USENIX Security. pp. 2987–3004 (2021)
9. Gehani, A., Tariq, D.: SPADE: support for provenance auditing in distributed environments. In: Middleware. pp. 101–120 (2012)
10. Hossain, N., et al.: Dependence-preserving data compaction for scalable forensic analysis. In: USENIX Security. pp. 1723–1740 (2018)
11. Inam, M.A., et al.: FAuST: striking a bargain between forensic auditing's security and throughput. In: ACSAC. pp. 813–826 (2022)
12. Inam, M.A., et al.: SoK: history is a vast early warning system: Auditing the provenance of system intrusions. In: SP. pp. 2620–2638 (2023)
13. Jia, Z., Xiong, Y., Nan, Y., Zhang, Y., Zhao, J., Wen, M.: MAGIC: detecting advanced persistent threats via masked graph representation learning. In: USENIX Security (2024)
14. Jiang, B., Nain, P., Towsley, D.F.: On the convergence of the TTL approximation for an LRU cache under independent stationary request processes. TOMPECS **3**, 20:1–20:31 (2018)
15. Keromytis, A.D.: Transparent computing engagement 3 data release (2018)
16. Khoury, J., Upthegrove, T., Caro, A., Benyo, B., Kong, D.: An event-based data model for granular information flow tracking. In: TaPP (2020)

17. Lee, K.H., Zhang, X., Xu, D.: LogGC: garbage collecting audit log. In: CCS. pp. 1005–1016 (2013)
18. Li, Z., Chen, Q.A., Yang, R., Chen, Y., Ruan, W.: Threat detection and investigation with system-level provenance graphs: A survey. Comput. Secur. **106**, 102282 (2020)
19. Ma, S., Zhang, X., Xu, D.: ProTracer: Towards practical provenance tracing by alternating between logging and tainting. In: NDSS (2016)
20. Michael, N., Mink, J., Liu, J., Gaur, S., Hassan, W.U., Bates, A.: On the forensic validity of approximated audit logs. In: ACSAC. pp. 189–202 (2020)
21. Tang, Y., et al.: NodeMerge: Template based efficient data reduction for big-data causality analysis. In: CCS. pp. 1324–1337 (2018)
22. Torrey, J.: Transparent computing engagement 5 data release (2020)
23. Xu, Z., et al.: High fidelity data reduction for big data security dependency analyses. In: CCS. pp. 504–516 (2016)
24. Zhu, T., et al.: General, efficient, and real-time data compaction strategy for apt forensic analysis. IEEE Trans. Inf. Forensics Secur. **16**, 3312–3325 (2021)
25. Zipperle, M., Gottwalt, F., Chang, E., Dillon, T.S.: Provenance-based intrusion detection systems: A survey. ACM Comput. Surv. **55**, 135:1–135:36 (2022)

Extended Abstracts

Truth Discovery for Spatio-Temporal Data from Multiple Sources

He Zhang[1,3](✉), Shuang Wang[1], Yufei Wang[2], Tianxing Wu[1], Lu Jixiang[3], and Long Chen[1]

[1] Southeast University, Nanjing, China
[2] Huawei Noah Ark Lab, Montreal, Canada
[3] NARI Research Institute, NARI Technology Co., Ltd., Nanjing, Jiangsu, China

1 Introduction

Truth discovery in big data aims to find the most reliable information among conflicting sources, with crowd sensing data being crucial due to its real-time and geospatial nature [3]. Existing methods have limitations, such as ignoring spatial-temporal dynamics or source correlations [1]. Our paper introduces an unsupervised, iterative approach using probabilistic graphical models [4] and a mass-spring system for truth discovery that considers source quality and spatio-temporal relationships. We employ the EM framework [3] to evaluate source reliability and a geo-spatial fusion model to integrate data from multiple sources, accounting for their geographical correlations and deriving final truth values based on source weights.

2 Methods

In a spatial-temporal context, sources S report objects O like temperature and humidity at times m and locations l. Key elements include source quality (reliability and performance), location popularity (affecting data collection), and claims influenced by location, source, and truth label (z_j^m). The objective is to aggregate claims (c) to ascertain the truth t_j^m for each object o_j at time m across geo-distributed sources.

In this paper, we introduce an EM-based mass-spring model to address the challenge of aggregating geographic data with assessments of source reliability, as shown in Fig. 1. Our framework uses a Probability Graphical Model to represent claims, locations, and sources. Source reliability is determined by coverage (V_i) and freshness (E_i), where coverage measures a source's data extent and freshness reflects the update frequency, crucial for real-time data reliability. The true positive (α_i^m) and false positive (β_i^m) probabilities capture the reliability of participants' reporting behaviors. Reliable sources have high α_i^m and low β_i^m values, indicating accurate reporting of true events. If a source reports a claim $c_{i,j}^{m,n}$, it is considered present ($h_{i,j}^{m,n} = 1$). In the absence of prior data, default values for β_i^m and α_i^m are set, and initial values for V_i and E_i can be uniform.

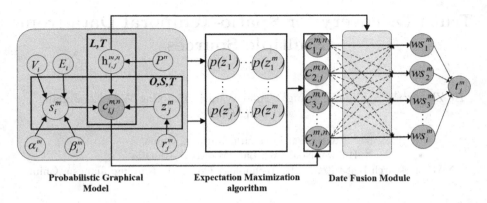

Fig. 1. Overview of the EM based mass spring: A probabilistic graphical model is constructed to measure the claim reliability where the probabilities of latent variables are computed by Expectation Maximization algorithm. To infer the latent truth from multiple values, data fusion is adopted by mass spring model.

The prior probability of a claim being true, r_j^m, is determined using a "major voting" approach based on the proportion of reporting sources. The EM algorithm is employed to iteratively estimate latent variables, assuming users report events at specific locations and times. The EM framework estimates the latent variable z_j^m, representing the event's true state. To improve upon previous truth discovery algorithms that focused solely on the most reliable source's claim values, we propose a data fusion stage with weighted claim values based on source credibility. We also incorporate geographic data from crowdsensing and introduce a mass-spring model for enhanced fusion. The model simulates geospatial interactions through node-spring dynamics, where nodes represent entities with mass and position, and springs signify connections. This framework allows for the adjustment of entity weights with geographic information, effectively integrating and analyzing systemic geographical attributes.

3 Results

In our experiments, we tested our algorithm against seven state-of-the-art and one practical method SG using 2021 weather data from three cities in Shandong, China. The datasets, with claims from unreliable sources, were meant to enhance renewable energy prediction. Our algorithm outperformed others, showing notable reductions in error metrics like RMSE, MAE, MSLE, and RMSLE, especially in the Tai'an dataset. The results, as shown in Table 1, indicate our method's effectiveness in handling real-world spatio-temporal data challenges.

Table 1. Performance Comparison on RMSE, MAE, MSLE, RMSLE.

Algorithms	Tai An				Zi Bo				Dong Ying			
	RMSE	f MAE	MSLE	RMSLE	RMSE	MAE	MSLE	RMSLE	RMSLE	MAE	MSLE	RMSLE
random	0.077	0.057	0.0033	0.058	0.097	0.069	0.0050	0.071	0.064	0.045	0.0023	0.048
Voting	0.052	0.037	0.0015	0.039	0.061	0.041	0.0020	0.044	0.049	0.033	0.0013	0.036
TruthFinder	0.048	0.033	0.0013	0.036	0.063	0.042	0.0021	0.046	0.053	0.037	0.0016	0.040
TwoEstimates	0.063	0.045	0.0022	0.048	0.079	0.055	0.0033	0.058	0.060	0.042	0.0020	0.045
CSS	0.047	0.033	0.0012	0.035	**0.060**	**0.040**	**0.0019**	**0.044**	**0.045**	**0.031**	**0.0011**	**0.033**
EMMutiF	**0.046**	**0.032**	**0.0012**	**0.035**	0.061	0.041	0.0020	0.045	0.048	0.033	0.0013	0.036
SRTD	0.065	0.047	0.0024	0.049	0.083	0.058	0.0036	0.060	0.065	0.046	0.0024	0.049
SG	0.065	0.047	0.0024	0.035	0.083	0.058	0.0036	0.045	0.065	0.046	0.0024	0.036
EM_mass_spring	**0.042**	**0.029**	**0.0010**	**0.032**	**0.053**	**0.035**	**0.0015**	**0.039**	**0.043**	**0.029**	**0.0010**	**0.032**

4 Conclusion

In this paper, we address important challenges in truth discovery for spatio-temporal events by incorporating fine-grained quality assessment and geospatial modeling. Our proposed framework demonstrates promising results in inferring accurate truth values from sources in spatio-temporal events. However, the mass-spring system used for data fusion considers source reliability but may not fully capture the complex spatial relationships and uncertainties associated with geographical information which can be studied in future.

References

1. Garcia-Ulloa, D.A., Xiong, L., Sunderam, V.: Truth discovery for spatio-temporal events from crowdsourced data. Proc. VLDB Endow. **10**(11), 1562–1573 (2017)
2. Wang, D., Abdelzaher, T., Kaplan, L., Aggarwal, C.C.: Recursive fact-finding: a streaming approach to truth estimation in crowdsourcing applications. In: 2013 IEEE 33rd International Conference on Distributed Computing Systems, pp. 530–539 (2013)
3. Wang, S., et al.: A survey on truth discovery: concepts, methods, applications, and opportunities. IEEE Trans. Big Data 1–20 (2024)
4. Zhao, B., Rubinstein, B.I.P., Gemmell, J., Han, J.: A Bayesian approach to discovering truth from conflicting sources for data integration. Proc. VLDB Endow. **5**(6), 550–561 (2012)

An Efficient Wind Power Prediction Based on Improved Feature Crossover Mechanism, N-BEATSx and LightGBM

Kaibo Zhang[1]($\boxtimes$)[iD], Feng Ye[1], Lina Wang[2], Nadia Nedjah[3], Xuejie Zhang[1], and Shulei Yu[4]

[1] School of Computer and Software Engineering, Hohai University, Nanjing 210098, Jiangsu, China
[2] Nanjing University of Information Science and Technology, Nanjing 210044, Jiangsu, China
[3] Universidade do Estado do Rio de Janeiro (UERJ), Rio de Janeiro, RJ 20550-013, Brazil
[4] Nanjing Zhiyichao Network Technology Co., Ltd., Nanjing 210000, Jiangsu, China

1 Introduction

Wind power generation is easily affected by weather factors such as wind speed and temperature. Moreover, the variation law shows complexity, among which the volatility is an important aspect. The traditional model structure is simple. However, it is difficult for a single model to accurately analyze the data with large volatility changes. Although the common combined models can improve the accuracy, they cannot excavate the influence of different factors on the power. To mitigate these problems, a wind power prediction method based on improved feature crossover mechanism, N-BEATSx and LightGBM is proposed.

2 Method

The model proposed in this study is generally divided into four modules (see Fig. 1). Firstly, an improved feature crossover mechanism is introduced to increase feature dimension. Then, some new features generated after feature crossover are added to the data set, and N-BEATSx [1] and LightGBM [2] are combined by error reciprocal method [3] to form the FC-N-BEATSx-LightGBM model.

Among them, for the improved feature crossing mechanism, the Pearson correlation coefficient formula [4] is first used for analysis in this study (see formula (1)). The comprehensive correlation coefficient is then used to show the deep correlation between the predicted value and the features, thereby screening out the features that need to be crossed (see formula (2)). In this way, we can obtain new features that can express nonlinear correlation and give full play to the hidden function of weak correlation features.

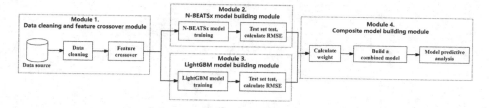

Fig. 1. The wind power prediction framework using the combined model

$$r_{x,y} = \frac{\sum\limits_{i=1}^{n}(x_i - \bar{x})(y_i - \bar{y})}{\sqrt{\sum\limits_{i=1}^{n}(x_i - \bar{x})^2 (y_i - \bar{y})^2}} \quad (1)$$

$$r_{(x_1 \rightarrow x_2)y} = (1 - r_{x_2 y})(r_{x_1 x_2} r_{x_1 y}) \quad (2)$$

3 Results

The data set selected in this study is from Longyuan Power Group, and the effect of the proposed model is tested by forecasting power data for the next 24 h. The model is compared with different single models (see Fig. 2) and different combined models (see Fig. 3), and the results are shown in Table 1. In addition, the model is applied to ten different wind farm data sets, and the results are shown in Table 2.

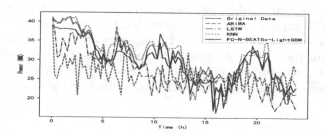

Fig. 2. Comparison of different separate models

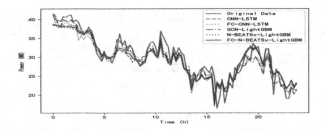

Fig. 3. Comparison of different combined models

Table 1. Evaluation of prediction results of each model

Model	ARIMA	LSTM	KNN	CNN-LSTM	GCN-LightGBM	N-BEATSx-LightGBM	FC-N-BEATSx-LightGBM
RMSE	7.001	5.509	2.513	2.227	2.064	1.813	1.769
MAE	5.772	4.432	2.011	1.756	1.624	1.312	1.260
FA	81.346%	85.035%	92.559%	93.614%	94.147%	94.982%	95.448%

Table 2. Evaluation of prediction results of FC-N-BEATSx-LightGBM model in different wind farms

Serial number	01	02	03	04	05	06	07	08	09	10
RMSE	1.133	3.271	2.489	1.242	3.899	1.769	1.331	2.234	3.337	2.233
MAE	0.951	2.301	2.002	0.991	3.015	1.260	0.996	1.595	2.645	1.791
FA	91.372%	92.617%	95.068%	91.260%	95.898%	95.448%	92.538%	92.282%	95.318%	95.120%

4 Conclusion

This study proposed a combined prediction method based on improved feature crossover mechanism, N-BEATSx and LightGBM. The FC-N-BEATSx- LightGBM model is constructed and applied to the power prediction of wind power generation. The experimental results show that the FA of this prediction model is higher than that of other models in the experiment, and its RMSE and MAE are lower than them. The generalization test proves that the model has high accuracy and generalization. Therefore, the research fully demonstrates that the model proposed in this paper has significant advantages in forecasting performance.

References

1. Oreshkin, N.-B., Carpov, D., Chapados, N.: N-BEATS: Neural basis expansion analysis for interpretable time series forecasting. CoRR **abs/1905.10437** (2019)
2. Andres, B., Gorka, U., Perez, J.M.: Smart optimization of a friction-drilling process based on boosting ensembles. J. Manuf. Syst. **48**, 108–121 (2018)
3. Phillip, S., Christine, D., Brian, H.: Effects of mixing weights and predictor distributions on regression mixture models. Struct. Equ. Model. **29**(1), 70–85 (2022)
4. López, C.-S., Trejo, H.-M., Román, C.-M.: Analysis of the steelmaking process via data mining and Pearson correlation. Materials **17**(11), 1–15 (2024)

A Novel Hash Hypercube-Based Attribute Reduction Approach For Neighborhood Rough Sets

Qin Si[1()], Haoyan Qiu[2], Yuanyuan Dong[3], Xiang Ding[3], Haibo Li[4], and Xiaojun Xie[1]

[1] College of Artificial Intelligence, Nanjing Agricultural University, 1 Weigang Road, Nanjing 210095, China
[2] School of Finance, Tianjin University of Finance and Economics, 25 Zhujiang Ave, Tianjin 300222, China
[3] College of Sciences, Nanjing Agricultural University, 1 Weigang Road, Nanjing 210095, China
[4] School of Informatics, Xiamen University, 4221 Xiang'an South Road, Xiamen 361102, China

1 Introduction

The existing attribute reduction algorithms for neighborhood rough sets have low efficiency due to the high time complexity of computing positive regions. To address the limitations, this paper proposes a novel attribute reduction approach, which employs an idea of enhanced hash, i.e., each sample is mapped into a hypercube. A fast hypercube-based computing positive regions algorithm (HH-POS) is designed with a lower average time complexity. On the foundations of HH-POS, a hypercube-based attribute reduction approach (HHARA) for neighborhood rough sets is raised. Experiments reveal that our algorithm attains state-of-the-art performance on some large-scale datasets.

2 Methods

Definition 1. *In a NDS $= <U, C \cup D, \theta>$, U represents the set of samples, C and D are the sets of conditional attributes and decision attributes, respectively, and θ is the neighborhood radius. K attributes are used to partition U into $\lceil \frac{1}{\theta} \rceil^K$ hypercubes. If the attribute values of sample x_i are $\{x_{i1}, x_{i2}, \cdots, x_{iK}\}$, the corresponding hash hypercube mapping is defined as:*

$$HCube(x_i) = H_{\lceil \frac{x_{i1}}{\theta} \rceil, \lceil \frac{x_{i2}}{\theta} \rceil, \cdots, \lceil \frac{x_{iK}}{\theta} \rceil} = \bigcap_{t=1}^{K} \left\{ x_j | x_j \in U \text{ and } \lceil \frac{x_{it}}{\theta} \rceil = \lceil \frac{x_{jt}}{\theta} \rceil \right\} \quad (1)$$

In Fig. 1, two samples are partitioned by using the hash hypercube model and each of them has two conditional attributes c_1, c_2.

Theorem 1. In a $NDS = <U, C \cup D, \theta>$, if $\forall x_i, x_j \in U, \exists t \in [1, K]$, st.

$$\left|\lceil \frac{x_{it}}{\theta} \rceil - \lceil \frac{x_{jt}}{\theta} \rceil\right| \geq 2 \tag{2}$$

then $x_i \notin \theta(x_j)$.

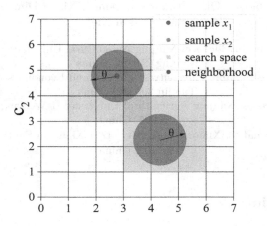

Fig. 1. An example of hash hypercube model

Utilizing Theorem 1, the neighborhood search space for each sample can be significantly reduced. Therefore, we have designed a fast hypercube-based computing positive regions algorithm (HH-POS). Naturally, a hypercube-based attribute reduction approach (HHARA) for neighborhood rough sets is raised.

3 Results

We explored the efficiency of HHARA under different neighborhood radius. FHARA [2] and F-FAR [1] are used as comparative algorithms, where F-FAR is the most effective algorithm we are aware of. The results are shown in Fig. 2, with the horizontal axis representing θ and the vertical axis representing the running times (in seconds). From Fig. 2, it can be seen that the running times of HHARA were less than that of FHARA and F-FAR under any large-scale dataset and θ.

4 Conclusion

This paper introduces a novel method, HH-POS, leveraging the hash hypercube to efficiently compute the positive region, with an average time complexity of $O(N \log NM)$. HHARA has been proven to be efficient across a range of datasets through comparative experiments, achieving state-of-the-art performance particularly on some large-scale datasets.

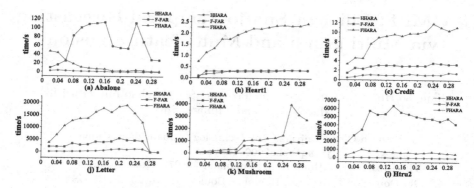

Fig. 2. Running times of three algorithms on various datasets under different neighborhood radius

References

1. Peng, X., Wang, P., Xia, S., Wang, C., Pu, C., Qian, J.: FNC: a fast neighborhood calculation framework. Knowl.-Based Syst. **252**, 109394 (2022)
2. Yong, L., Wenliang, H., Yunliang, J., Zhiyong, Z.: Quick attribute reduct algorithm for neighborhood rough set model. Inf. Sci. **271**, 65–81 (2014)

MGMFF: Efficient Spatio-Temporal Forecasting via Multi-graph and Multi-feature Fusion

Jianqiao Hu[1](✉), Qian Tao[1], Songbo Wang[2], Chenghao Liu[1], Lusi Li[3], Hao Yang[1], and Xiuhang Shi[1]

[1] South China University of Technology, Guangzhou, China
jianqiaohu2000@163.com
[2] The University of Hong Kong, Pokfulam 999077, Hong Kong
[3] Old Dominion University, Hampton Boulevard, Norfolk 5115, USA

1 Introduction

Spatio-temporal sequence prediction refers to the task of forecasting how a sequence of events or patterns evolves over both space and time. Examples of such tasks include predicting traffic conditions at various locations over time in intelligent transportation systems [1, 3], forecasting city-wide parking availability within a city [4], or understanding changes in air pollutant concentration across spatial regions and time intervals [2]. The goal is to capture and leverage the complex interactions between spatial and temporal dimensions for accurate predictions.

MGMFF is characterized by the use of a Multi-static-dynamic Graph fused adjacency learning Module (MGM) to fuse dynamic temporal dependencies and static node feature relationships. For the node-to-node correspondence between nodes, we add a Temporal Feature extraction Module (TFM). The main contributions of MGMFF can be summarized as follows:

- We propose a novel framework to extract information from various aspects for fusing dynamic and static adjacency information in both spatial and temporal dimensions.
- We present a TFM module, which can preserve the temporal sequence information between node and node without GNN architecture.
- We propose a parameter perturbation-based contrastive learning method, which can enhance the robustness of the prediction method.
- To handle the specific scenario of spatio-temporal sequence prediction, we propose a node-level loss function to regularize the prediction results.

2 Method

We propose the overall structure of the multi-graph and multi-feature fusion (MGMFF) framework as Fig. 1 which contains two main submodules. The first submodule is the spatial location extraction module based on the graph convolutional neural network. The second submodule is a dependency extraction module based on relational memory units at the temporal level. In the first structure, we refer to the graph learning module in MTGNN as the MGM module to fuse static and dynamic graphs to build an adjacent matrix. In the second structure, we construct the TFM module and the relationship memory extraction module. To extract the basic physical feature relationships among the nodes in the graph as the dynamic dependency relationships at the graph level.

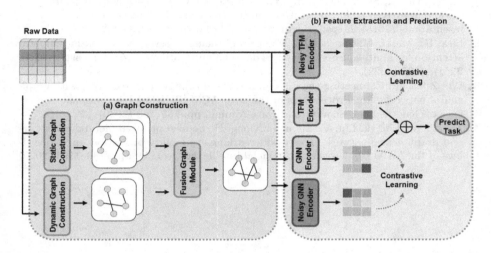

Fig. 1. Architecture of the MGMFF Model with Static and Dynamic Graphs

3 Experiments and Conclusion

From the comparison of the experimental results in Table 1, it can be concluded that the prediction accuracy of MGMFF in these four datasets outperforms the comparison models, especially in RMSE and corr. Our proposed method has a

Table 1. Forecasting results on different datasets

ECG				nyc-taxi			
Model	MAE ↓	RMSE↓	corr↑	Model	MAE↓	RMSE↓	corr↑
MTGNN	0.340	0.564	0.442	MTGNN	5.492	10.141	0.840
STGNN	**0.327**	0.572	0.446	STGNN	3.793	16.541	0.105
AutoFormer	0.351	0.590	0.406	AutoFormer	**3.729**	12.757	0.174
GMAN	0.340	0.604	0.403	GMAN	15.976	29.672	0.048
GTS	0.328	0.568	0.455	GTS	5.069	20.172	0.075
DCRNN	0.348	0.594	–	DCRNN	9.913	115.170	–
LSTMNN	0.645	0.996	0.082	LSTMNN	18.789	33.674	0.588
ours (MGMFF)	0.329	**0.562**	**0.462**	ours (MGMFF)	5.349	**9.798**	**0.842**

large improvement in CORR results, indicating that the model extracts deeper spatio-temporal correlations and ultimately improves the predictive ability of the model.

References

1. Li, Y., Yu, R., Shahabi, C., Liu, Y.: Diffusion convolutional recurrent neural network: data-driven traffic forecasting. In: International Conference on Learning Representations (2018). https://openreview.net/forum?id=SJiHXGWAZ
2. Lira, H., Martí, L., Sanchez-Pi, N.: A graph neural network with spatio-temporal attention for multi-sources time series data: an application to frost forecast. Sensors **22**(4), 1486 (2022)
3. Wu, Z., et al.: Connecting the dots: multivariate time series forecasting with graph neural networks. In: Proceedings of the 26th ACM SIGKDD International Conference on Knowledge Discovery and Data Mining, pp. 753–763 (2020)
4. Zhang, W., Liu, H., Liu, Y., Zhou, J., Xiong, H.: Semi-supervised hierarchical recurrent graph neural network for city-wide parking availability prediction. In: Proceedings of the AAAI Conference on Artificial Intelligence, vol. 34, pp. 1186–1193 (2020)

A Few-Shot Relation Extraction Approach for Threat Intelligence Field

Haiyan Wang[1], Junchi Bao[2], Liyi Zeng[1], Wenying Feng[1], Weihong Han[1], and Zhaoquan Gu[1(✉)]

[1] Peng Cheng Laboratory, Shenzhen 518000, China
wanghy01@pcl.ac.cn, zengly@pcl.ac.cn, fengwy@pcl.ac.cn, hanwh@pcl.ac.cn, guzhq@pcl.ac.cn
[2] H3C Technologies Co., Ltd., Hangzhou 310000, China
3421681587@qq.com

1 Introduction

As the scale of Internet applications continues to expand, the risk of cyber attacks is growing globally. Security incidents such as data breaches, ransomware attacks, message tampering, and identity theft are becoming more frequent, posing serious threats to the security of assets in human society. Cyber Threat Intelligence (CTI) analysis has become particularly important. Security analysts can leverage big data to collect, analyze, and process information about cyber threats from various sources to understand the nature, origins, targets, and methods of these threats, and to provide early warning, response, and mitigation measures against them. Therefore, extracting structured information from unstructured data in cyber threat intelligence analysis is crucial. Cyber threat intelligence relation extraction technology analyzes the contextual information of cyber threat intelligence texts to identify semantic relationships between cybersecurity entities, which is a key step in analyzing cybersecurity incidents. However, relation extraction tasks in the threat intelligence domain face challenges such as limited data and high annotation costs. To address these issues and challenges, we utilize a small amount of threat intelligence training data to complete the relation extraction task and implement a corresponding prototype system.

2 Methods

This paper proposes a Few-shot Relation Extraction Based on Threat Intelligence Field (FRTI-Matt, as shown in Fig. 1). The model is built upon a foundation of pre-trained language models and prototype networks to address the challenges of limited data and high annotation costs. It rapidly learns from meta-training tasks, enabling swift generalization to new classification tasks.

The model's core component is the pre-training language module, which employs BERT [3]. BERT, with its Transformer encoder architecture, uses self-attention mechanisms to calculate attention weights for each word, capturing semantic relationships within the context. This process is detailed by a series of formulas, including the attention mechanism formula and the feedforward neural network formula.

The Prototype Networks Module enhances the model with multi-level attention mechanisms and a hybrid prototype network [1, 2]. This approach ensures that sentence instances more relevant to the query instance are given greater attention weights during prototype generation. The model also incorporates relationship text features during training, which is crucial for improving performance in threat intelligence scenarios where data is scarce.

We further refines the model through self-training and contrastive learning techniques. The self-training module expands the annotated dataset by assigning pseudo-labels to unlabeled data, enhancing the model's relation extraction capabilities and reducing the risk of overfitting. The contrastive learning module introduces a novel relationship-prototype contrastive learning method, treating the relationship prototype as a positive sample and other class prototypes as negative samples within the embedding space.

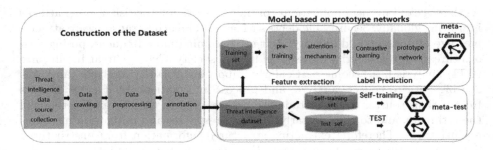

Fig. 1. FRTI-Matt model.

3 Results

The experimental results of the FRTI-Matt model demonstrated significant performance improvements in the threat intelligence domain. The model initially achieved an accuracy of 68.95% for the 5-way 1-shot task and 70.03% for the 5-way 5-shot task, outperforming existing few-shot relation extraction models by over 4.85%. Further enhancements through self-training and contrastive learning led to additional accuracy gains of 5.11% and 5.41% in the respective few-shot scenarios. The refined model showcased substantial accuracy improvements, highlighting the effectiveness of the integrated approach in addressing data scarcity and annotation challenges. The model's ability to acquire more domain-specific knowledge and its enhanced classification capabilities for similar pattern sentences were particularly noteworthy.

4 Conclusion

We combined relation extraction models with few-shot learning techniques to propose a few-shot relation extraction algorithm tailored for the threat intelligence field. This algorithm enables the model to quickly capture common knowledge among meta-tasks, enhancing the model's generalization ability in new tasks with limited samples of relation types. By employing multi-level attention mechanisms, it improves upon the drawbacks of the original baseline prototype network, which did not directly consider query instance features and relationship text features when computing class prototypes using all instances in the support set. By replacing the ordinary prototype network with a hybrid prototype network, it more fully utilizes effective features of the data for training in scenarios with limited training corpora. Additionally, through further model optimization, self-training and contrastive learning methods are employed to improve the encoding layer and prototype network layer of the threat intelligence relation extraction model. This allows the model to acquire more domain-specific features during the encoding phase and adjust prototype representations in the metric space based on relationship features, leading to more accurate predictions for challenging classification tasks. In the future, we will investigate how to automatically perform named entity recognition in raw threat intelligence datasets and establish joint relation extraction models.

References

1. Chen, Y., et al.: Revisiting self-training for few-shot learning of language model, pp. 9125–9135 (2021)
2. Gao, H., Huang, J., Tao, Y., Hussain, W., Huang, Y.: The joint method of triple attention and novel loss function for entity relation extraction in small data-driven computational social systems. IEEE Trans. Comput. Soc. Syst. **9**(6), 1725–1735 (2022)
3. Kenton, J.D.M.-W.C., Toutanova, L.K.: BERT: Pre-training of deep bidirectional transformers for language understanding, pp. 4171–4186 (2019)

A Subgroup Framework of Interpatient Arrhythmia Classification Using Deep Learning Network

Xia Yu[✉], Zi Yang, Ning Shen, Youhe Huang, and Hongru Li

College of Information Science and Engineering, Northeastern University, Shenyang, China
yuxia@ise.neu.edu.cn

1 Introduction

This paper proposes a subgroup framework based on a multi-feature deep learning network for interpatient arrhythmia classification that will respond to the actual requirements for aided diagnosis. In particular, it balances and compromises both generalization and specificity in pursuing the classification model. Careful analysis of the individual biological characteristics (such as BMI, age, gender, medical information, etc.) can exhibit correlations between individual differences and ECG records [1, 2]. Therefore, we choose the more commonly utilized and readily available gender and age categories to discuss the framework of subgroup division. To the best of our knowledge, it is the first time that a subgroup approach has been adopted to improve the accuracy of arrhythmia classification. Following this, CNN and LSTM are exploited under the sub-population framework to extract ECG signals' morphological and rhythmical features, respectively. Additionally, the network-trained features are combined with hand-crafted features (RR intervals) to effectively reduce the advent effects of individual specificity and improve the accuracy of arrhythmia classification. The overview of the proposed approach, including the subgroup division and the subgroup-based classification flowchart, is shown in Fig. 1.

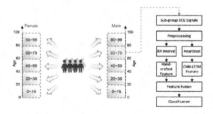

Fig. 1. Overview of the subgroup division and the classification flowchart.

2 Methods

2.1 Materials

This work adopts the MIT-BIH arrhythmia database [3], which is publicly accessible via the web. The dataset contains 48 two-channel ambulatory ECG recordings from 47 distinct patients. These recordings are truncated from 24-hour records into approximately 30 minutes (each), sampled at 360 Hz. It can be challenging and time-consuming to distinguish heartbeats on ECG signals due to corruption from noises such as power line interference and muscular artifacts [4, 5]. Hence, the ECG signals are subjected to discrete wavelet transformation for noise reduction (for the precise approach, refer to Shi et al. [6]).

2.2 Classification Model Architecture

Firstly, we divide the population by individual characteristics to highlight personal differences and erect more detailed subgroup classification models. Then, we use three convolutional layers, three pooling layers, one LSTM layer, and three fully connected layers. Single-beat ECG signals with 200 samples are fed into the CNN and LSTM to extract morphological and temporal features, respectively. Those network-trained features are concatenated with the RR intervals for more comprehensive features, with a length of 15. Finally, the fused features are converted into fully connected layers to classify N, S, and V class arrhythmias as shown in Fig. 2.

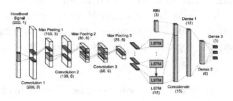

Fig. 2. The proposed CNN-LSTM arrhythmia classification model structure.

3 Results

Figure 3 shows the accuracy and loss plots on the train set and validation set, and there is no obvious overfitting. The confusion matrices for the validation set and test set are depicted in Fig. 4, revealing the performance of the multi-class model. It can be seen that the model operates better on the validation set, accurately predicting almost all heartbeats.

The efficiency of the proposed network and other methods in classifying three different classes of ECG (Class N, Class S and Class V) has been represented in Fig. 4. As shown in Fig. 4, our method demonstrates acceptable performance for all the classes in terms of Accuracy and Specificity. Compared with the higher Precision values for

(a) accuracy (b) loss

Fig. 3. The proposed CNN-LSTM arrhythmia classification model structure.

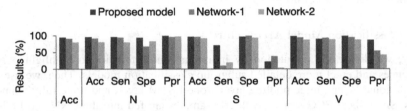

Fig. 4. Comparison of classification results between the proposed model and two compared networks.

Class N and Class V, a lower index for Class S indicates the higher risk of false positives in this class type. The low-performance metrics in F1-score and MCC also mean the proposed model cannot perform well in this imbalanced class. Overall, the system works excellent for Class N and Class V and a relatively average version for Class S. The classification results between the proposed subgroup model and other methods under the interpatient paradigm are summarized in Table 1.

Table 1. Table captions should be placed above the tables.

Methods	Acc (%)	N (%)				S (%)				V (%)			
		Acc	Sen	Spe	Ppr	Acc	Sen	Spe	Ppr	Acc	Sen	Spe	Ppr
Proposed model	94.74	95.42	95.64	93.56	99.21	96.35	70.75	96.68	21.76	97.71	89.39	98.56	86.47
Network-1	90.94	91.42	94.18	67.26	96.18	96.04	9.69	99.38	37.87	94.42	92.79	94.53	54.27
Network-2	79.98	80.33	80.00	83.09	97.55	92.13	19.73	93.07	3.58	87.51	88.17	87.45	41.94

4 Conclusion

In this paper, a subgroup framework is designed for automatic arrhythmia classification to distinguish it from the traditional interpatient paradigm. First, we divide subgroups according to age and gender. Then, aiming at sub-populations, the corresponding interpatient model is established, where CNN-LSTM and RR intervals are used for network-trained features and hand-crafted features, respectively. This learning model can take advantage of the characteristics of subgroup patients with better accuracy than traditional interpatient models. The proposed framework can not only realize an automatic arrhythmia classification. Still, it can also balance the contradiction between the limited accuracy of general models and the inferior practicality of individual models.

References

1. Sakai, T., Yagishita, A., Morise, M., et al.: Impact of exercise capacity on the long-term incidence of atrial arrhythmias in heart failure. Sci. Rep. **11**, 18705 (2021). https://doi.org/10.1038/s41598-021-98172-9
2. Lucci, V.E.M., et al.: Markers of susceptibility to cardiac arrhythmia in experimental spinal cord injury and the impact of sympathetic stimulation and exercise training. Auton. Neurosci-Basic **235** (2021)
3. Moody, G.A., Mark, R.G.: The impact of the MIT-BIH arrhythmia database. IEEE Eng. Med. Biol. **20**(3), 45–50 (2001)
4. Acharya, U.R., et al.: Automated detection of arrhythmias using different intervals of tachycardia ECG segments with convolutional neural network. Inf. Sci. **405**, 81–90 (2017)
5. Acharya, U.R., et al.: A deep convolutional neural network model to classify heartbeats. Comput. Biol. Med. **1**(89), 389–396 (2017)
6. Shi, H.T., et al.: A hierarchical method based on weighted extreme gradient boosting in ECG heartbeat classification. Comput. Methods Programs Biomed. **171**, 1–10 (2019)

Text and Image Multimodal Dataset for Fine-Grained E-Commerce Product Classification

Ajibola Obayemi[1](✉)[iD] and Khuong Nguyen[2][iD]

[1] University of Brighton, Brighton, UK
e.a.obayemi@brighton.ac.uk
[2] Royal Holloway, University of London, Egham, UK
Khuong.Nguyen@rhul.ac.uk

1 Introduction

Fine-grained classification is a challenging task that aims to reduce the misclassification errors in the visual classification of similar image samples. Multimodal learning can improve results by combining text and image data. This approach can help minimise misclassification errors caused by intra-class variability and inter-class similarity. For example, consumer products such as grocery products and printer cartridges may have several product variations that are similar enough to make it hard to classify them using SOTA Computer Vision and NLP models [1–3].

In recent research works, multimodal learning has been shown to effectively improve the results of fine-grained classification tasks when compared to the unimodal alternatives [4–6]. A rich and well-curated dataset has significant potential to push forward the research in this area [3–7]. However, such datasets are scarce and difficult to curate, due to the laborious effort in data collection and the challenge of consistency and accuracy of the labels across different modalities.

The main contributions of our paper are: (1) We created a new multimodal dataset with 17,000 image-text pairs. (2) We propose a generalised pipeline for collecting text and image multimodal datasets to simplify the data collection process and encourage more researchers to curate such datasets. (3) We provide the baseline results using a CNN-based unimodal architecture (ResNet-152) and a text and image multimodal architecture (CLIP & MultiModal BiTransformers) to quantitatively demonstrate how the fusion of text and image modalities work to improve the results.

2 Methods

2.1 Data Collection

Over two months, 17,000 images were captured using a Canon DSLR camera, and a photo lightbox (for consistent lighting and image quality). Using the EOS digital SDK software, we automated several aspects of the image capture process, such as lens focus and shutter control. The captured images were saved locally, and their corresponding metadata was stored in a Microsoft SQL Server database. Additionally, to improve the quality and consistency of the dataset, we pre-processed the images by applying centre cropping to remove unnecessary background elements and focus on the object of interest. To create our text & image multimodal dataset, we extracted the text from the product labels using an OCR pipeline (based on a modified version of CRAFT & CRNN + CTC). We generated image and text pairs by combining the images and extracted texts. The dataset is publicly available at https://github.com/multimodal-research/TAIMD-17k.

2.2 Unimodal and Multimodal Architecture

For the unimodal experiment, a pre-trained ResNet-152 model was fine-tuned for image classification. The model was trained on the dataset with early stopping to prevent overfitting, and training was conducted on two NVIDIA RTX 4090 GPUs.

For the multimodal experiment, we used Multimodal BiTransformers (MMBT) for vision-language modelling. CLIP was used for image encoding and BERT for text encoding, and MMBT was trained on the dataset using Hugging Face and PyTorch. Training was also conducted on two NVIDIA RTX 4090 GPUs.

3 Results

We evaluate the results from our experiments using the F1-score, precision and recall metrics and compare the results from the image based unimodal architecture with the results from the image and text based multimodal architecture. We observed that the multimodal model performed better than the unimodal model on our dataset, achieving a 75% precision, 71% recall and 67% F1-score respectively on our dataset (see Table 1).

Table 1. The performance of CNN-based Image Classification and Multimodal Image & Text Classification. Multimodal architecture outperformed unimodal architecture in all metrics.

Method	Precision	Recall	F1-Score
Unimodal (ResNet-152)	0.70	0.61	0.59
Multimodal (MMBT + CLIP)	**0.75**	**0.71**	**0.67**

4 Conclusion

This paper introduces a novel multimodal dataset for fine-grained e-commerce product classification. It comprises 17,000 images categorised into 31 distinct product classes. We further present a generalised pipeline for collecting and annotating text-image multimodal datasets. This pipeline utilises OCR to extract and annotate the data, generating a large collection of image-text pairs.

Furthermore, we compare the unimodal and multimodal architectures applied to our dataset, demonstrating a significant improvement in performance with the multimodal approach. Utilising CLIP and MMBT based architecture, we achieve up to a 10% increase in precision, recall, and F1-score compared to the unimodal architecture. This dataset is publicly available, aiming to contribute to research and understanding of multimodal learning for fine-grained product classification.

The results of our experiments demonstrate that there is potential for improvement in fine-grained product classification using multimodal learning. In the future, We plan to explore co-learning which is a relatively new area of multimodal learning to further improve fine-grained product classification.

References

1. Baz, I., Yoruk, E., Cetin, M.: Context-aware hybrid classification system for fine-grained retail product recognition. In: 2016 IEEE 12th Image, Video, and Multidimensional Signal Processing Workshop (IVMSP). IEEE (2016)
2. Xuan, Q., et al.: Evolving convolutional neural network and its application in fine-grained visual categorization. IEEE Access **6**, 31110–31116 (2018)
3. Zahavy, T., et al.: Is a picture worth a thousand words? A deep multi-modal architecture for product classification in e-commerce. In: Proceedings of the AAAI Conference on Artificial Intelligence, vol. 32, issue number 1 (2018)
4. Fu, J., et al.: CMA-CLIP: cross-modality attention clip for text-image classification. In: 2022 IEEE International Conference on Image Processing (ICIP). IEEE (2022)
5. Jiang, X., et al.: Delving into multimodal prompting for fine-grained visual classification. In: Proceedings of the AAAI Conference on Artificial Intelligence, vol. 38, issue number 3 (2024)
6. Lu, Z., et al.: StreamSketch: exploring multi-modal interactions in creative live streams. In: Proceedings of the ACM on Human-Computer Interaction, vol. 5, issue number CSCW1, pp. 1–26 (2021)
7. Kim, E., McCoy, K.F.: Multimodal deep learning using images and text for information graphic classification. In: Proceedings of the 20th International ACM SIGACCESS Conference on Computers and Accessibility (2018)

Enhanced Network Traffic Prediction with Transformer-Based Models and Clustering

Peiqi Jin[1], Kun Zheng[1], Yixin Che[1], Chenming Qiu[2], Ling Jin[2], Yunkai Wang[3], Tian Hu[2], Shouguo Du[2], and Yiming Tang[1,2(✉)]

[1] Shanghai Lixin University of Accounting and Finance, Shanghai, China
[2] Shanghai Municipal Big Data Center, Shanghai, China
tang.yiming@lixin.edu.cn
[3] Shanghai Data Group Co., Ltd., Shanghai, China

1 Introduction

Time series forecasting is widely applied in fields such as sensor networks, transportation, and finance. We obtained historical network traffic data from cloud service logs, including source IPs and destination IPs, with the aim of using time series models to predict and improve the network capacity of the destination IPs. Initially, we employed the LSTM model to predict four metrics for 430 destination IPs:client traffic, representing the traffic generated by the address as a client; server traffic, representing the traffic generated by the address as a server in response; request counts, indicating the total number of requests within an hour; and the number of user IPs, representing the number of users accessing the IP within an hour. However, the LSTM model demonstrated slow training speeds and high memory demands. To enhance computational efficiency and prediction accuracy, we introduced attention-based models such as Autoformer, Informer, Reformer, FEDformer, TDformer, iTransformer, TimeMixer, and PatchTST. Given the significant differences in data characteristics across various IPs, we used the K-shape clustering model to preprocess the network traffic data, dividing them into 10 groups. The features within each group exhibited high similarity as shown in Fig. 1, and clustering helped reduce noise and improve model robustness. Finally, we compared the mean squared errors (MSE) of nine prediction models, selected the most suitable prediction model for the clustered data, and verified the effectiveness of the clustering method in improving prediction accuracy.

2 Methods and Results

We trained and tested various models on the data of each cluster separately and evaluated the performance of the models on each cluster. Specifically, we used nine transformer models, including Transformer, Autoformer, Informer, Reformer, FEDformer, TDformer, iTransformer, TimeMixer, and PatchTST. The prediction results of all models and the weighted average metrics are listed

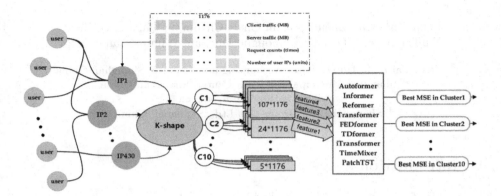

Fig. 1. The overall architecture of the network traffic time series data with clustering and prediction.

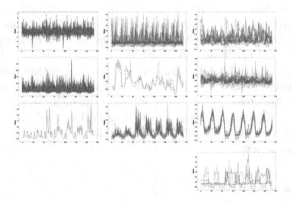

Fig. 2. Plot of the average client traffic for each cluster over seven days. Each line represents a target IPs average weekly traffic.

Table 1. Prediction accuracy (MSE, MAE), training loss (Loss) and training time (Time) for the server traffic, the client traffic, the numbers of requests and client IPs of each target IPs using different models after clustering.

Metrics		Transformer	Autoformer	Informer	Reformer	FEDformer	TDformer	iTransformer	TimeMixer	PatchTST	Optimal
Server	MSE	1.663	1.417	1.718	1.713	1.284	1.226	1.283	1.225	1.174	**1.133**
	MAE	0.706	0.527	0.585	0.586	0.503	0.411	0.393	0.421	0.382	**0.376**
	Loss	0.680	0.728	0.723	0.815	0.778	0.747	0.794	0.815	0.804	**0.635**
	Time	**11.398**	15.458	11.707	14.319	24.846	17.818	20.650	28.781	93.083	
Client	MSE	1.411	1.094	1.427	1.405	1.001	0.936	0.968	0.948	0.893	**0.882**
	MAE	0.581	0.504	0.589	0.578	0.478	0.383	0.377	0.397	0.360	**0.357**
	Loss	0.744	0.731	0.726	0.808	0.777	0.730	0.794	0.815	1.609	**0.685**
	Time	**11.886**	16.065	12.181	17.819	27.500	28.279	22.578	25.284	90.640	
Request	MSE	1.715	1.271	1.690	1.678	1.223	1.081	1.050	1.069	1.018	**1.001**
	MAE	0.750	0.625	0.734	0.735	0.604	0.525	0.516	0.529	0.499	**0.495**
	Loss	0.679	0.708	0.625	0.757	0.745	0.689	0.759	0.781	0.771	**0.609**
	Time	**11.773**	16.917	14.592	15.616	26.117	23.733	14.733	29.819	87.780	
Unique	MSE	1.102	0.829	1.080	1.105	0.815	0.798	0.754	0.798	0.737	**0.724**
	MAE	0.672	0.571	0.844	0.660	0.560	0.507	0.503	0.518	0.492	**0.489**
	Loss	0.702	0.693	0.669	0.759	0.722	0.675	0.757	0.778	0.771	**0.642**
	Time	**15.766**	23.024	15.808	15.791	25.239	26.567	23.536	28.941	92.913	

in Table 1. By selecting the most suitable method based on the characteristics of each cluster, we obtained the best prediction results.

The results in Table 1 highlight the effectiveness of the clustering approach. Specifically, the PatchTST model achieved the suboptimal performance in terms of MSE and MAE for periodic data, with an MSE of 1.174 and MAE of 0.382. For non-periodic data, the TDformer and iTransformer models showed comparable performance. Weighted average metrics confirm that clustering enhances prediction accuracy, with the "Optimal" metrics demonstrating the lowest MSE and MAE, underscoring the benefits of using cluster-specific models.

3 Conclusion

This study examines how time series clustering impacts multivariate time series forecasting accuracy. By applying K-Shape clustering to categorize 430 IPs into 10 clusters and using nine forecasting models for each, we found that clustering significantly improves prediction accuracy over non-clustered models. Future research will aim to integrate clustering with forecasting to further enhance both clustering effectiveness and prediction accuracy.

BSCL: A Model for Solving Math Word Problems Based on Contrastive Learning

Tiancheng Zhang[✉], Yuyang Wang, Yijia Zhang, Minghe Yu, Fangling Leng, and Ge Yu

School of Computer Science and Engineering, Northeastern University, Shenyang 110819, China
tczhang@mail.neu.edu.cn, 2272043@stu.neu.edu.cn, neu_yjzhang@163.com, yuminghe@mail.neu.edu.cn, lengfangling@cse.neu.edu.cn, yuge@mail.neu.edu.cn

1 Introduction

Solving Math Word Problems (MWPs) in Natural Language Processing (NLP) involves converting natural language into mathematical expressions. MWPs are classified as Arithmetic Word Problems (AWPs) with single unknowns or Equation Set Problems (ESPs) with multiple unknowns.

Traditional methods, including rule-based and machine learning approaches, use predefined templates, limiting their adaptability. Deep learning models, such as SAU-Solver [1], improve performance but confuse MWPs types, impacting accuracy. They also fail to handle subtle distinctions between similar MWPs.

We propose BSCL (BERT-SAU with Contrastive Learning), a contrastive learning model that improves MWP understanding by recognizing problem pattern differences, providing accurate solutions for both AWPs and ESPs (Table 1).

2 Methods

The BSCL framework consists of four modules: Encoder Module, Decoder Module, Supervision Module, and Contrastive Learning Module.

Encoder Module uses pre-trained language models to encode MWPs in natural language form into vector representations. Decoder Module employs the tree-structured decoder with semantic-aligned regularization proposed by SAU-Solver to generate a Universal Expression Tree (UET) representing equations for AWPs and ESPs. Supervision Module uses a multi-head attention mechanism to verify mathematical consistency and distinguish MWPs with similar problem descriptions but different solving objectives by generating similar yet incorrect expressions. Contrastive Learning Module applies contrastive learning to differentiate between AWPs and ESPs, helping the model learn the similarities and differences between different problem types (Fig. 1).

Table 1. Examples of MWP

	Example 1
AWP	In order to create a good campus learning environment, Zhaohua Middle School has increased the area of the campus for planting flowers and trees by 69% in two years. What is the average annual growth rate of the area of the campus for planting flowers and trees in these two years?
	SAU-Solver: $y*(1+x)^2 = y*(1.0+0.69)$ (x) Ours: $1.0*(1+x)^2 = 1.0*(1.0+0.69)$ (✓)

	Example 2
ESP	The sum of a two-digit number is 7. When this two-digit number is subtracted by 27, the digits in its tens and units places are exchanged. What are the values of the tens and units digits of this two-digit number?
	SAU-Solver: $10*x + x + 2.0 = 10*x + x$ (x) Ours: $\begin{cases} x+y=7 \\ 10*y+x-27=10*x+y \end{cases}$ (✓)

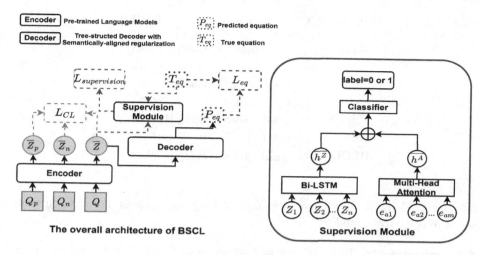

Fig. 1. The overall architecture of BSCL, where Q represents a MWP.

3 Results

We have conducted extensive experiments on Chinese dataset HMWP and English dataset ALG514. The performance of BSCL outperforms all baselines and dialogue models fine-tuned using the P-tuning v2 [2] (Tables 2 and 3).

Table 2. BSCL vs. baselines

Model	HMWP		ALG514	
	Acc_{eq}	Acc_{ans}	Acc_{eq}	Acc_{ans}
Stack-Decoder	-	0.274	-	0.289
GroupATT	0.252	0.332	0.331	0.376
DNS	0.240	0.327	0.370	0.418
GTS	0.316	0.420	0.204	0.547
SAU-Solver	0.323	0.429	0.208	0.549
RoBERRTa2UET	0.370	0.500	0.212	0.619
BSCL(Our)	0.375	0.512	0.228	0.640

4 Conclusion

In this paper, we propose a universal solver for various MWPs-BSCL. It improves performance using consistency verification and contrastive learning to distinguish MWP types. Future work should focus on enhancing model interpretability and

Table 3. BSCL vs. Dialogue models on HMWP

Model	ChatGLM-6B		ChatGLM2-6B	
	Acc_{eq}	Acc_{ans}	Acc_{eq}	Acc_{ans}
P2E&A	0.300	0.257	0.366	0.326
E2A	-	0.483	-	0.487
P2E-E2A	0.341	0.368	0.370	0.404
P2E	0.341	0.441	0.370	0.483
BSCL(Our)	<u>0.375</u>	<u>0.512</u>	<u>0.375</u>	<u>0.512</u>

integrating domain knowledge with natural language understanding to further boost MWP-solving ability.

Acknowledgments. This work was partially supported by National Natural Science Foundation under Grant (Nos. 62272093, 62372097, 62137001).

References

1. Qin, J., Lin, L., Liang, X., Zhang, R., Lin, L.: Semantically-aligned universal tree-structured solver for math word problems. arXiv preprint arXiv:2010.06823 (2020)
2. Liu, X., Ji, K., Fu, Y., et al.: P-tuning v2: prompt tuning can be comparable to fine-tuning universally across scales and tasks. arXiv preprint arXiv:2110.07602 (2021)

Predicting Student Success in Learning Management Systems: A Case Study of the Madrasati Platform

Abdullrahman Alabdali[1,3(✉)], Mohammad A. Alshehri[2], Matthew Stephenson[1], and Paulo Santos[1]

[1] Flinders University, College of Science and Engineering, Adelaide, South Australia
[2] University of Jeddah, Jeddah, Saudi Arabia
[3] Ministry of Education, Riyadh, Saudi Arabia

1 Introduction

Student performance evaluation is critical, especially during their formative educational years. In recent years, Artificial Intelligence (AI), through it machine learning (ML)algorithms, This paper focuses on evaluating the performance of eighth-grade students in Madrasatia, a Saudi Arabian Learning Management System (LMS). We analysed their interaction metrics and performance in mathematics and science questions from the Trend in International Mathematics and Science Study (TIMSS. The aim of the present study is to determine the minimal amount of data required for effective student performance prediction.

Recent research highlights the growing use of ML algorithms to predict student success by analysing students data at an early stage and enhancing student outcomes [6]. Studies have applied methods such as Artificial Neural Network (ANN), Decision Tree(DT), and Random Forest(RF)across various datasets [2, 3]. Integrated Virtual Learning Environment(VLE) data with demographic factors to further enhance prediction accuracy [4]. Overall, adopting ML techniques like regression and boosting models shows improvement in predicting student performance at an early stage of their learning journeys [1, 5, 7].

2 Methods

The data was collected from 150,003 students on the Madrasati platform in Jazan City. The dataset includes interaction metrics from several educational activities (such as resolved quizzes, assignments, etc.) from primary, secondary, and high school students. A second dataset including exam results from 12391 eighth grade students who answered 15 mathematics and 15 science questions was also included. After preprocessing to remove duplicate and missing data, the final dataset consisted 5928 students. The performance prediction was based on both interaction metrics and exam results. In terms of model training, four ML algorithms were used: Linear Regression (LR), Decision Tree (DTR), Random Forest (RFR) and Gradient Boosting (GBR). The data was divided using K-fold cross-validation (k = 10) to ensure reliable evaluation. Models were trained across sixteen training phases, the first phase included only student interaction metrics, while every other phase added two new exam question results (one mathematics and one science). This increase was cumulative, with the final phase including 15 mathematics question results and 15 science question results along with all student interaction metrics.

3 Results

Our results indicate that student interaction metrics alone are not sufficient to achieve high prediction accuracy. However, including a small number of exam questions helps us to achieve an acceptable level of prediction accuracy shown in Fig. 1. Based on our findings, pre-exams or recurring assessments into courses can help predict student performance and identify those needing support early in the learning process. Using only two pre exam questions, we can reasonably predict students final exam scores.

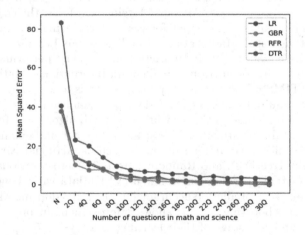

Fig. 1. The figure indicates the Mean Squared Error and Number of Questions with LR, DTR, RFR, GBR.

4 Conclusion

This study demonstrates that while traditional LMS interaction metrics alone provide limited predictive power, incorporating quiz results significantly improves model accuracy. This is critical for identifying at-risk students at an early stage to allows for timely interventions, which can improve educational outcomes. Future research will explore the reasons behind the low correlation between interaction metrics and final grades and investigate other variables that may improve the model predictability.

References

1. Alhazmi, E., Sheneamer, A.: Early predicting of students performance in higher education. IEEE Access **11**, 27579–27589 (2023)
2. Ghorbani, R., Ghousi, R.: Comparing different resampling methods in predicting students' performance using machine learning techniques. IEEE Access **8**, 67899–67911 (2020). https://doi.org/10.1007/s42001-024-00281-8
3. Mengash, H.A.: Using data mining techniques to predict student performance to support decision making in university admission systems. IEEE Access **8**, 55462–55470 (2020)
4. Merchant, A., Shenoy, N., Bharali, A., Kumar, M.A.: Predicting students' academic performance in virtual learning environment using machine learning. In: 2022 Second International Conference on Power, Control and Computing Technologies (ICPC2T), pp. 1–6. IEEE (2022)
5. Montaha, S., et al.: TimeDistributed-CNN-LSTM: a hybrid approach combining CNN and LSTM to classify brain tumor on 3D MRI scans performing ablation study. IEEE Access **10**, 60039–60059 (2022)
6. Pande, S.M.: Machine learning models for student performance prediction. In: 2023 International Conference on Innovative Data Communication Technologies and Application (ICIDCA), pp. 27–32. IEEE (2023)
7. Song, H., Kim, M., Park, D., Shin, Y., Lee, J.G.: Learning from noisy labels with deep neural networks: a survey. IEEE Trans. Neural Netw. Learn. Syst. (2022)

IEP: An Intelligent Event Prediction Model Based on Momentum

Lize Zheng[✉] and Yanxi Li

School of Statistics, Renmin University of China, Beijing, China
2022201469@ruc.edu.cn, liyanxi5895@ruc.edu.cn

1 Introduction

The prediction of sports events significantly impacts culture and economy [1, 2], but the randomness of the outcomes presents challenges. Sudden positive performances can boost the confidence of players, creating a momentum that affects the dynamics of the game [3]. Therefore, understanding this psychological inertia is crucial to capture and analyze the unpredictable nature of sporting events. Existing methods, including the Elo Rating System [4, 5], primarily focus on historical performance and lack real-time updates and sensitivity to current psychological and physiological factors. To address these limitations, we proposed an Intelligent Event Prediction (IEP) model that incorporates Momentum to better capture the relationships between various indicators, aiding decision-making in different industry applications.

2 Methods

Our IEP model consists of three main components: an Automatic Evaluation System (AES), a Momentum Capture Algorithm (MCA), and an Inflection Point Detection (IPD) model.

First, AES uses factor analysis to reduce the dimensionality of high-dimensional data. Then it uses the Analytic Hierarchy Process (AHP) to determine the judgment matrix, followed by the Order Preference by Similarity to Ideal Solution (TOPSIS) technique combined with the weights obtained from AHP to calculate the Comprehensive Performance Score (CPS).

Second, MCA defines Momentum as CPS minus the overall average CPS, because we have noticed that the players' CPS generally hover around a certain "baseline" but exhibit significant fluctuations due to changing conditions at times. MCA also provides rich methods for visualizing the match flow. Through empirical verification on the real-world dataset, the Spearman Correlation Coefficient between Momentum and performance and the P-value prove the high validity of Momentum.

Third, the IPD model detects the inflection points of the competition trend through the BinSeg algorithm. After processing the data using down-sampling techniques, this paper uses the random forest algorithm to analyze the importance of factors on match breakpoints and subsequently predicts whether the next point will be scored and the match outcomes.

3 Results

Our dataset was sourced from IBM's point-by-point records of the gentlemen's singles matches at the 2023 Wimbledon Championships.

AES reduced the 11 variables (λ_i, i = 1,2,3...11) in the dataset to 5 dimensions, and the cumulative rates of contribution to the communal variance of these 5 factors were 75.82%.

FAC_1 was positively correlated with serving aggressiveness (λ_3), direction (λ_8), and depth (λ_9), and was named "Offensive Aggressiveness". FAC_2, linked to the number of strokes (λ_1) and the total running distance (λ_{11}), was termed "Stamina Consumption". FAC_3, showing a positive correlation with serve speed (λ_2) and untouchable shots (λ_{10}), and a negative correlation with the second serve rate (λ_6), was called "Hitting Proficiency". FAC_4 correlated with the number of sets (λ_4) and games won (λ_5), named "Scoring Advantage". FAC_5, highly correlated with break success rate (λ_7), was named "Break Capability".

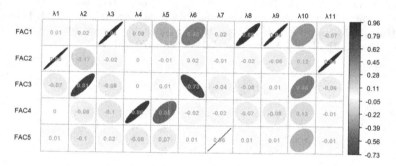

Fig. 1. Rotated factor load matrix. Blue represents a positive correlation; the flatter the ellipse, the stronger the positive correlation. Similarly, red indicates a negative correlation.

According to the IPD model, the prediction model analysis revealed that the feature importance scores of the five factors were 0.217, 0.205, 0.187, 0.194, and 0.198. The model achieved an accuracy of 0.867, with a recall of 0.867 and a precision of 0.862.

4 Conclusion

Our proposed IEP model effectively reduces the complexity of sports events to capture Momentum while retaining most of the information, introducing a methodology for real-time prediction of event trends with good accuracy. We demonstrated the model's robustness and generalization ability by introducing additional variable groups, data replacement, and Gaussian perturbations. In the future, this model requires further training, validation, and refinement on more datasets to extend its applicability beyond tennis to other sports events.

Acknowledgments. We would like to thank our advisor, Dr. Zhou Feng, for his guidance throughout the research process. We are grateful for the support of the Qiushi Academic Foundation and the School of Statistics, Renmin University of China.

References

1. Luckner, S., Schröder, J., Slamka, C.: On the forecast accuracy of sports prediction markets. In: Gimpel, H., Jennings, N.R., Kersten, G.E., Ockenfels, A., Weinhardt, C. (eds.) Negotiation, Auctions, and Market Engineering. LNBIP, vol. 2, pp. 227–234. Springer, Heidelberg (2008). https://doi.org/10.1007/978-3-540-77554-6_17
2. Wen, Y.F., Hung, K.Y., Hwang, Y.T., Lin, Y.S.F.: Sports lottery game prediction system development and evaluation on social networks. Internet Res. **26**(3), 758–788 (2016)
3. Mamassis, G., Doganis, G.: The effects of a mental training program on juniors pre-competitive anxiety, self-confidence, and tennis performance. J. Appl. Sport Psychol. **16**(2), 118–137 (2004)
4. Hubacek, O., Sourek, G., Zelezny, F.: Forty years of score-based soccer match outcome prediction: an experimental review. IMA J. Manage. Math. **33**(1), 1–18 (2022)
5. Hvattum, L.M., Arntzen, H.: Using ELO ratings for match result prediction in association football. Int. J. Forecast. **26**(3), 460–470 (2010)

Identifying Sources in Complex Dynamic Networks Through Label Propagation and ODE Integration

Fuyuan Ma[1], Yuhan Wang[2], Junhe Zhang[1], and Ying Wang[1(✉)]

[1] The College of Computer Science and Technology, Jilin University, Changchun 130000, China
[2] The School of Artificial Intelligence, Jilin University, Changchun 130000, China

1 Introduction

Social networks have facilitated faster and more convenient communication among individuals as typical examples of complex dynamic networks. However, they have also intensified the spread of rumors and malicious information, posing significant threats to society. Therefore, to completely eradicate malicious information and safeguard social stability, the accurate tracing of the origins and dissemination channels for such malicious information is crucial. Current techniques for source tracing without known propagation models rely on information propagation characteristics and label propagation, primarily focusing on static networks and neglecting real-world network dynamics. Moreover, the utilization of solely the infection status of nodes as node label results in an inadequate amount of label information, thereby compromising the accuracy of source tracing. To address the aforementioned limitations, this paper introduces an ODESI model for source tracing, which combines label propagation techniques with ordinary differential equations. The model first assigns integer labels to nodes to represent their infection status. These integer labels are then expanded into vector labels using a label expansion algorithm, enriching the label information. Lastly, the propagation of vector labels in dynamic networks is facilitated through ordinary differential equations and message-passing mechanisms, determining the information source based on the propagation results. This paper presents experiments validating the ODESI model's performance, demonstrating its superiority and effectiveness.

2 Methods

In this paper, we propose a novel source identification model called ODESI, which utilizes label propagation algorithm and ordinary differential equation(ODE) to achieve source identification in dynamic networks. The goal of ODESI is to accurately identify the source of information by modeling the characteristics of information dissemination. Initially, ODESI utilizes a label propagation algorithm to convert integer labels into vector labels, where each dimension of the vector corresponds to distinct information, thus enriching the label information. Subsequently, the propagation of vector labels in dynamic networks is modeled by employing graph neural networks-based ODE and message passing mechanisms. Finally, the predicted set of information sources is determined based on the propagation results of vector labels. The framework of ODESI is illustrated in Fig. 1.

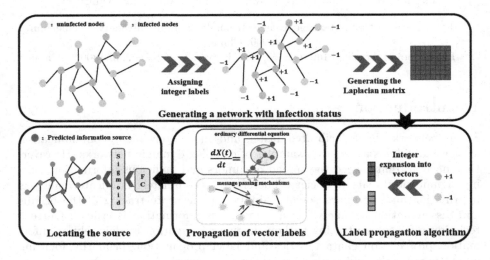

Fig. 1. The framework of ODESI

3 Results

We conducted experiments on the email social networks of two small communities, Enron and CollegeMsg. Tables 1 and 2 report the F1 scores for source tracing by different models under varying proportions of rumor sources, as well as the minimum error hop count, Err Hops, from the predicted sources to the true sources for our model.

Experimental results show that ODESI outperforms all baseline methods on both datasets. And our model achieved a minimum error hop count of 2 or less on both datasets when the proportion of rumor sources was set at 5% and 10%. ODESI benefits from enriching label information and leveraging ordinary differential equations for vector label propagation in dynamic networks. The use of the label propagation algorithm enhances node representations with features crucial for information dissemination, thereby improving accuracy in identifying information sources. Additionally, modeling vector label propagation captures both network structure and temporal evolution patterns.

Table 1. The results of the comparative experiments on the Enron dataset

	K=5%	K=10%	K=15%
LPSI	0.278	0.323	0.301
GCNSI	0.133	0.150	0.277
GATSI	0.340	0.203	0.353
ODESI	**0.730**	**0.667**	**0.736**
Err Hops	1.47	1.88	2.21

Table 2. The results of the comparative experiments on the CollegeMsg dataset

	K=5%	K=10%	K=15%
LPSI	0.227	0.221	0.223
GCNSI	0.375	0.349	0.393
GATSI	0.396	0.372	0.382
ODESI	**0.475**	**0.512**	**0.457**
Err Hops	1.73	1.96	2.46

4 Conclusion

This paper introduces a source identification model called ODESI, which leverages label propagation and ordinary differential equations. This model employs a label propagation algorithm to transform the initial integer labels of nodes into vector labels, which are then utilized as the initial node representations. In addition, the model employs ordinary differential equation and message passing mechanisms to achieve label propagation in complex dynamic networks. Lastly, this paper conducts multiple experiments for ODESI. The experimental results demonstrate that the model exhibits superior source identification accuracy compared to other baseline methods, thereby substantiating the efficacy of the proposed model.

Acknowledgments. This work was supported by the National Science and Technology Major Project under Grant No. 2021ZD0112500, the National Natural Science Foundation of China under Grant No. 62272191, the Science and Technology Development Program of Jilin Province No. 20220201153GX, the International Science and Technology Cooperation Program of Jilin Province under Grant No. 20240402067GH.

Unsupervised Domain Adaptation for Entity Blocking

Yaoshu Wang[1](✉) and Mengyi Yan[2]

[1] Shenzhen Institute of Computing Sciences, Shenzhen, China
yaoshuw@sics.ac.cn
[2] Beihang University, Beijing, China
yanmy@act.buaa.edu.cn

1 Introduction

Entity resolution (ER), also known as record linkage, focuses on retrieving all potentially matching tuple pairs within relational tables. It plays a crucial role in data cleaning, information integration, and data processing pipelines for training machine learning models. When dealing with a large volume of tuples, ER typically employs entity blocking as a filtering step to filter unmatched tuple pairs, thereby avoiding the computationally expensive Cartesian product operation. Therefore, entity blocking, as a key step in ER, need to achieve high speed for efficient retrieval of candidate tuple pairs and high recall to prevent missing any matched pairs.

In this paper, we design an unsupervised domain adaptation approach for the task of entity blocking, denoted by UDAEB. We tackles the challenge by designing a mechanism for learning representations across different domains, complemented by the integration of LLMs to enhance the process. To elaborate, this mechanism is designed to fine-tune the representation model's parameters. It does so by mapping the embeddings of pairs of tuples into a space where their similarities can be measured and then aligning these similarity vectors across the source and target domains. This alignment is achieved by using an adversarial loss function. The experimental results show UDAEB outperforms the existing methods in terms of effectiveness.

2 Method

2.1 Representation Learning with Enrichment by Large Language Models

Given a tuple t of basic attributes $\bar{A}$ in the relational table, we first leverage the powerfulness of LLMs to enrich t with additional attributes $\bar{B}$, such that $\bar{B}$ and enriched values of t are generated by LLMs via prompt optimization. We then adopt the SentenceBert model [4] $\mathcal{M}$ to map each tuple t to the high-dimensional embedding vector. In detail, we serialize t to a sequence as $\mathsf{serial}(t) =$

$[COL]A_1[VAL]t[A_1]\ldots[COL]A_m[VAL]t[A_m]$, where $A_1,\ldots,A_m \in \bar{A} \cup \bar{B}$. Then we fed serial($t$) into $\mathcal{M}$ and return its embedding, s.t., $\mathbf{emb}_t = \mathcal{M}(\text{serial}(t))$.

To fine-tune $\mathcal{M}$ in the source domain, we use the contrastive learning loss [3]. Given the source labeled training data D^S, we process it into a set CL^S of triplets, s.t. $\mathsf{CL}^S = \{(a, \mathcal{P}_a, \mathcal{N}_a) | \forall a, (a, b_1, \text{True}) \in D^S, (a, b_2, \text{False}) \in D^S\}$, where $\mathcal{P}_a$ and $\mathcal{N}_a$ represent the sets of tuples that match (positive) and do not match (negative) with a, respectively. Specifically, for each tuple a in the left table, we scan the right-side tuples in the right table to identify those that either match or mismatch with a, forming pairs (a, b_1, True) and (a, b_2, False). After aggregating all such tuples a, we obtain the set CL^S. we then fine-tune $\mathcal{M}$ using contrastive learning as $\mathcal{L}_{\mathsf{CL}} = \sum_{b \in \mathcal{P}_a} -\log \frac{\exp(\langle \mathbf{emb}_a, \mathbf{emb}_b \rangle/\tau)}{\sum_{b' \in \mathcal{P}_a \cup \mathcal{N}_a} \exp(\langle \mathbf{emb}_a, \mathbf{emb}_{b'} \rangle)/\tau}$, where τ is the temperature parameter.

2.2 Cross-Domain Representation Learning

We further propose a similarity neural network to aggregate embeddings of two tuples into their similarity representation. More specifically, given (a^S, b^S, y^S) of the source training data D^S and (a^T, b^T) of the target training data D^T, we first transform each tuple into an embedding, i.e., $\mathbf{emb}_{a^S}$, $\mathbf{emb}_{b^S}$, $\mathbf{emb}_{a^T}$, and $\mathbf{emb}_{b^T}$. Next, we use a feature extractor $\mathcal{F}$, implemented as a MLP-based fully-connected layer NN with the activation function ReLU, to generate the similarity representation. For $(a, b) \in \{(a^S, b^S), (a^T, b^T)\}$, we have $\mathbf{emb}_{(a,b)} = \mathcal{F}(\mathbf{emb}_a \oplus \mathbf{emb}_b) = \text{ReLU}(\text{NN}(\mathbf{emb}_a \oplus \mathbf{emb}_b))$, where $\oplus$ is the concatenation operation by row. After generating the similarity embeddings, we also adopt a binary classifier $\mathcal{G}(\cdot)$ to transfer the similarity representation to a binary decision. Finally, we have $\mathcal{G}(\mathcal{F}(\mathbf{emb}_{a^S} \oplus \mathbf{emb}_{b^S}))$, and $\mathcal{G}(\mathcal{F}(\mathbf{emb}_{a^T} \oplus \mathbf{emb}_{b^T}))$ as the logits of (a^S, b^S) and (a^T, b^T), respectively. Following CGANs [1], we adopt the domain adversarial neural network approach as a two-player game using the adversarial loss function, so we could align the distribution in the similarity representation space of the source and target domains.

3 Results

Table 1 presents the effectiveness results on Walmart-Amazon (WA), Amazon-Google (AG) and Abt-Buy (AB) [2], where UDAEB consistently outperforms the other methods. Notably, its average PC and PQ are 6% and 19% higher than those of the baselines, with maximum improvements of 7% and 22%, respectively. These results highlight the effectiveness of the entity blocking approach, which leverages features such as similarity alignment between the source and target domains and enhancement through LLMs. UniBlocker ranks second in performance in most cases, benefiting from its pre-training that is adaptable to various domains.

Table 1. The Performance of Effectiveness with metrics PC, PQ and K.

Datasets	DeepBlocker [5] PC / PQ	K	Sudowoodo [6] PC / PQ	K	STransformer [4] PC / PQ	K	UniBlocker [7] PC / PQ	K	UDAEB PC / PQ	K
WA-AG	85.69 / 3.67	20	90.06 / 9.80	8	91.60 / 15.69	5	90.38 / 17.24	5	**92.46 / 19.80**	4
AB-AG	85.69 / 3.67	20	90.83 / 11.29	7	91.52 / 13.06	6	90.38 / 17.24	5	**90.83 / 25.92**	3

4 Conclusion

This paper presents **UDAEB**, an unsupervised domain adaptation framework for entity blocking. UDAEB includes an enrichment step that enhances tuples using the capabilities of LLMs and a cross-domain representation learning step that aligns feature representations in the similarity space between the source and target domains. We conduct extensive experiments on benchmark datasets, demonstrating the superior performance of **UDAEB** over state-of-the-art methods using standard evaluation metrics.

Acknowledgments. This work was supported by Longhua Science and Technology Innovation Bureau 10162A20220720B12AB12.

References

1. Long, M., Cao, Z., Wang, J., Jordan, M.I.: Conditional adversarial domain adaptation. In: NeurIPS (2018)
2. Mudgal, S., et al..: Deep learning for entity matching: a design space exploration. In: SIGMOD, pp. 19–34 (2018)
3. Oord, A.V.D., Li, Y., Vinyals, O.: Representation learning with contrastive predictive coding. arXiv preprint arXiv:1807.03748 (2018)
4. Reimers, N., Gurevych, I.: Sentence-BERT: sentence embeddings using Siamese BERT-networks. In: EMNLP, November 2019
5. Thirumuruganathan, S., et al.: Deep learning for blocking in entity matching: a design space exploration. Proc. VLDB Endow. **14**(11), 2459–2472 (2021)
6. Wang, R., Li, Y., Wang, J.: Sudowoodo: contrastive self-supervised learning for multi-purpose data integration and preparation. In: ICDE (2023)
7. Wang, T., et al.: Towards universal dense blocking for entity resolution. CoRR **abs/2404.14831** (2024)

Author Index

A
Al Ahmad, Hussain 193
Alabdali, Abdullrahman 367
Alshehri, Mohammad A. 367

B
Bao, Junchi 349

C
Cao, Jiasheng 80
Chao, Pingfu 19
Che, Yixin 359
Chen, Long 337
Chen, Lvying 174
Chen, Xingguo 34
Chen, Zhenyu 254
Chen, Zulong 49
Chu, Mingze 224

D
Ding, Xiang 343
Ding, Xuefeng 224
Dong, Yuanyuan 343
Du, Shouguo 359

F
Fan, Chunlong 269
Feng, Wenying 349

G
Gao, Xiangyun 19
Geng, Shisong 254
Gu, Zhaoquan 238, 318, 349
Guo, Yunlong 3

H
Han, Weihong 349
He, Tao 80
Hu, Dasha 224
Hu, Jianqiao 346

Hu, Tian 359
Huang, Qiangbo 287
Huang, Subin 19
Huang, Xiaoyang 80
Huang, Xvxin 287
Huang, Youhe 352

J
J. M. Shehada, Dina 193
Jiang, Yu 128
Jiang, Yuming 224
Jin, Ling 359
Jin, Peiqi 359
Jixiang, Lu 337

K
Kong, Chao 19

L
Leng, Fangling 363
Li, Bohan 174
Li, Chenyu 95
Li, Haibo 343
Li, Haiying 34
Li, Hongru 352
Li, Junyu 302
Li, Lusi 346
Li, Miao 269
Li, Qianmu 128, 143, 208
Li, Weimin 3
Li, Yanxi 370
Li, Zhao 49
Lian, Zhichao 208
Liang, Yanming 287
Liao, Peixin 287
Liao, Wenhao 318
Lin, Junfa 110
Lin, Shicong 158
Liu, Chenghao 346
Liu, Fan 208
Liu, Nan 128

Liu, Runfei 224
Liu, Xiao 128
Liu, Xinglin 95
Liu, Xuejun 80
Long, Meixiu 110
Lou, Jiazhen 49
Lv, Fuyu 49
Lv, Jingsong 49

M
M Anzar, S. 193
Ma, Fuyuan 373
Mansoor, Wathiq 193
Meng, Shunmei 34, 128, 208
Mu, Su 158

N
Nedjah, Nadia 340
Nguyen, Khuong 356
Nguyen, Thanh Tam 65

O
Obayemi, Ajibola 356

P
Panthakkan, Alavikunhu 193

Q
Qi, Youhan 95
Qiu, Chenming 359
Qiu, Haoyan 343
Qiu, Shuhan 143
Qu, Liang 65

S
Santos, Paulo 367
Shan, Zhilong 158
Shen, Ning 352
Shen, Yuncheng 224
Shi, Xiuhang 346
Si, Qin 343
Song, Xiangyu 238
Stephenson, Matthew 367
Sun, Jia 318
Sun, Xianlan 19

T
Tan, Runnan 238
Tang, Yiming 359

Tao, Qian 346
Teng, Yiping 269
Tian, Hongyu 3

W
Wang, Haiyan 238, 318, 349
Wang, Haoxi 302
Wang, Huan 269
Wang, Huihui 34
Wang, Jiahai 110
Wang, Lina 340
Wang, Shiqing 269
Wang, Shuang 337
Wang, Songbo 346
Wang, Tao 208
Wang, Yaoshu 376
Wang, Ying 95, 373
Wang, Yufei 337
Wang, Yuhan 373
Wang, Yunkai 359
Wang, Yuyang 363
Wang, Zhenhai 3
Wei, Shanming 143
Wen, Hong 49
Wu, Guowei 254
Wu, Jia 49
Wu, Tianxing 337

X
Xie, Xiaojun 343
Xie, Yushun 238
Xing, Zhuoya 80
Xu, Chen 19
Xu, Zhouying 80

Y
Yan, Mengyi 376
Yang, Chaoqun 65
Yang, Hao 346
Yang, Jianye 318
Yang, Ming 287
Yang, Zi 352
yao, Lin 254
Ye, Feng 340
Yin, Hailian 174
Ying, Zijian 208
Yu, Ge 363
Yu, Minghe 363
Yu, Shulei 340

Author Index

Yu, Sujie 174
Yu, Xia 352
Yuan, Wei 65

Z
Zeng, Liyi 287, 349
Zhang, Denghui 287
Zhang, Junhe 373
Zhang, He 337
Zhang, Jing 49
Zhang, Kaibo 340
Zhang, Tiancheng 363
Zhang, Xuejie 340
Zhang, Yijia 363
Zheng, Kun 359
Zheng, Lize 370
Zhu, Haibei 19
Zhu, Zitong 110
Zhuo, Junnan 174
Zong, Chuanyu 269

Printed in the United States
by Baker & Taylor Publisher Services